Periglacial processes and environments

Periglacial processes and environments

A. L. Washburn
Director, Quaternary Research Center, University of Washington

St. Martin's Press New York

 For information, write:
St. Martin's Press, Inc., 175 Fifth Avenue, New York, N. Y. 10010
Printed in Great Britain
Library of Congress Catalog Card Number: 73-75113
First published in the United States of America in 1973

AFFILIATED PUBLISHERS: Macmillan and Company, Limited, London–
also at Bombay, Calcutta, Madras and Melbourne

Contents

Preface

Much of this book was inspired by Professor Richard Foster Flint of Yale University. As a teacher just prior to World War II, as a fellow officer during that war, as a colleague at Yale many years later and throughout as a friend, Dick Flint encouraged the writer in many ways. An interdisciplinary Pleistocene seminar at Yale in the early 1960s was also a stimulant in providing a fruitful discussion of periglacial processes and environments.

Professor Emeritus Hugh M. Raup of Harvard University and later of The Johns Hopkins University is another close friend who is responsible for the orientation of this book. As colleagues in fieldwork in Northeast Greenland in the period 1956–66, we had the opportunity to study and discuss many periglacial problems in detail. Numerous facets of this book benefited from this collaboration and some are based on the resulting publication in the *Meddelelser om Grønland.*

The writer is also indebted to Dr Duwayne Anderson of the U.S. Army Cold Regions Research Laboratory, to Professor Larry W. Price of Portland State University, and to students at the University of Washington for helpful suggestions.

The publishers simplified the writer's task in many ways, and he deeply appreciates the cooperative attitude they exhibited throughout. He is particularly grateful to Mrs Brenda Hall for preparing the comprehensive index and to Mrs Ruth Hertz of the University of Washington, Quaternary Research Center, who contributed immensely by typing manuscript drafts and proof reading galleys.

Finally the book could not have been written without Tahoe Washburn's wifely understanding and patient acceptance of many uncommunicative evenings and weekends.

Except as otherwise noted, temperatures are in degrees centigrade and the metric system is used throughout. Equivalent measurements in other units are given in parentheses if the original observations were reported in these units.

The periglacial literature is enormous and varied, and the present writer is fully aware of his uneven presentation of the subject and his cursory treatment of many aspects. He hopes readers of this volume will give him the benefit of their criticisms and suggestions so that appropriate improvements can be made in a revised edition that is already planned.

Acknowledgements

The author and publishers gratefully acknowledge permission given by the following to reprint or modify copyright material:

Industrial Press for Figure 3.1; the National Research Council of Canada for Figures 3.2, 3.4, 3.7, 3.15 and 10.1; McGill-Queen's University Press for Figure 3.3; the United States Department of the Interior for Figure 3.6, the United States Cold Regions Research and Engineering Laboratory for Figures 3.8, 3.16, 3.23 and 12.10; the Association of American Geographers for Figure 3.9; the Arctic Institute of North America for Figure 3.17; the American Geophysical Union for Figures 3.18 and 3.22; the National Research Council of Canada for Figures 3.19, 3.20, 4.2 and 4.3; the Japanese Society of Soil Mechanics and Foundation for Engineering for Figure 4.4; the Geological Society of America for Figures 4.5 and 4.14; Geologiska Foreningen for Figure 4.12; the American Association for the Advancement of Science for Figure 4.13; the National Academy of Sciences, National Academy of Engineering, National Research Council for Figures 4.25, 4.43, 4.44 and 4.45; Niedersächsisches Landesamt für Bodenforschung for Figure 4.28; Franz Steiner for Figures 4.33, 4.46, 4.53, 5.20 and 6.2; the Editor, *Geografiska Annaler* for Figures 4.48 and 4.50; the Editor, *Journal of Geomorphology* for Figure 4.61; Dr Fritz Müller for Figures 4.64 and 4.65; the Editor, *Erdkunde* for Figures 9.1, 12.1, 12.2, 12.3, 12.5, 12.6, 12.7 and 12.8; the University of Chicago Press for Figure 10.3; the Editor, *Quaternary Research* for Figures 10.5 and 10.6; Fachbereich Geowissenschaften der Philipps-Universität Geologie for Figure 12.4; the Editor, *Zeitschrift für Geomorphologie* for Figure 12.9.

1 Introduction

I DEFINITION OF PERIGLACIAL

The term periglacial was introduced by Łoziński (1909, 10–18) to designate the climate and the climatically controlled features adjacent to the Pleistocene ice sheets.

Many investigators have extended the term to designate nonglacial processes and features of cold climates regardless of age and of any proximity to glaciers. As a result there have been varying usages (cf. Butzer, 1964, 105; Dylik, 1964*a*; 1964*b*). Although not without criticism because of its lack of precision (Linton, 1969), the term is being widely used in the extended sense, as here, because of its comprehensiveness.

The term has no generally recognized quantitative parameters, although some rough estimate of precipitation and temperature limits have been given. According to Peltier's (1950, 215, Table 1) estimate, the periglacial morphogenetic region is characterized by an average annual temperature ranging from −15° (5°F) to −1° (30°F) and an average annual rainfall (excluding snow) ranging from 127 mm (5 in) to 1397 mm (55 in). Peltier's (1950, 222, Figure 7) diagram of morphogenetic regions, which has been widely reproduced, is reasonably consistent with these figures for the periglacial region except that the lower limit of rainfall is given as 0 mm (0 in). According to Lee Wilson's (1968*a*, 723, Figure 9; 1969, 308, Figure 3) scheme, the precipitation range is from some 50 to 1250 mm and the temperature range from some −12° (10°F) to 2° (35°F) in variable combinations. The diagnostic criterion is a climate characterized by significant frost action and snow-free ground for part of the year. As stated by Tricart (1967, 9), '*les pays froids son ceux où l'action géomorphologique de l'eau est commandée par son existence à l'état solide, permanente ou périodique.*'

Tricart (1967, 29, Figure 4; 30) stressed permafrost (whether or not it is in balance with the present climate) as the primary characteristic of the periglacial domain. However, he recognized a distribution of minor periglacial features, such as earth hummocks, as lying outside the periglacial domain as defined by permafrost; also Tricart (1967, 56–67) defined as periglacial some climates lacking permafrost, and it is clear that he did not consider permafrost to be a necessary condition in a definition of periglacial. Péwé (1969, 2, 4) came closer to regarding permafrost as a necessary criterion. However, such a limitation seems overly restrictive, since many features such as gelifluction, frost creep, and several forms of patterned ground that are related to frost action, are commonly regarded as periglacial but are not necessarily associated with permafrost. Furthermore, it is common to speak of former periglacial environments; yet in the present state of our knowledge there are very few criteria by which a former permafrost condition can be proved. Although a cold-climate process, glaciation is not periglacial by definition. Glaciation and 'periglaciation', where both are present, are complementary aspects of cold environments.

As used in the following, the term periglacial designates cold-climate, primarily terrestrial,[1] nonglacial processes and features regardless of date or proximity to glaciers (Figures 1.1–1.2).

II OBJECTIVES

The objectives of periglacial research are to (1) determine the exact mechanism of periglacial processes, (2) determine the environmental significance of the processes, (3) apply the information to reconstruct Quaternary environments, and (4) use these historical and process approaches as an aid in predicting environmental changes.

III PROCESSES

1 General

Many different processes are responsible for periglacial effects, but for the most part these processes are not peculiar to periglacial environments. Rather they are common to many environments that have a climate sufficiently cold to leave physical evidence of its influence. It is the combination and intensity of these processes that characterize periglacial environments.

Given low enough temperatures both the poleward and upper altitudinal limits of periglacial features, where limits exist, are determined by precipitation blanketing the land with perennial snow and ice. On mountains this limit is the snowline. On the other hand the equatorial and lower altitudinal limits are limited by temperature rather than precipitation.

2 Frost action

By far the most widespread and important periglacial process is frost action. Actually, frost action is a 'catch-all' term for a complex of processes involving freezing and thawing including, especially, frost cracking, frost wedging, frost heaving, and frost sorting.

[1] Marine ice-shove ridges are included as periglacial features by the present writer.

1 Regions of accumulation with underlying syngenetic permafrost containing ground ice 2 Regions of accumulation with underlying seasonal frost only 3 Regions of equilibrium between processes of accumulation and denudation on syngenetically frozen rocky substrate containing ice wedges 4 Flat regions of equilibrium between processes of accumulation and denudation (on syngenetically frozen rocky substrate) 5 Regions of altiplanation, stable with respect to accumulation and denudation (on epigenetically frozen rocky substrate) 6 Regions stable with respect to accumulation and denudation (on seasonally frozen rocky substrate) 7 Regions without frozen-ground processes modelling the landscape 8 Regions of dominant denudation with underlying permafrost 9 Regions of dominant denudation with underlying seasonal frost only 10 Glaciers 11 Polygons with ice veins 12 Polygons with ice veins, associated with [agrémentés] thermokarst forms 13 Peat bogs with flat hummocks [buttes gazonées] 14 Baydjarakhs 15 Alases 16 Polygons with soil veins 17 Reduced (deformed) polygons with soil veins – forms infilled with cover loam 18 Forms similar to hummocks [buttes gazonées] with gaps [enfoncements] resulting from degradation 19 Pseudo kames 20 Hummocks [buttes gazonées] – mounds 21 Altiplanation terraces 22 Nonsorted circles [formes tachetées – médaillons] 23 Sorted polygons [polygons de pierres], circles, and other forms sorted by freezing 24 Solifluction stripes on slopes 25 Stratified Aufeis ['Nalédi'] 26 Seasonal hummocks [buttes gazonées] resulting from soil heaving 27 Perennial hummocks [buttes gazonées] resulting from soil heaving 28 Hummocks [buttes gazonées] due to water migration toward the frozen surface 29 Solifluction forms related to soil flow 30 Present limit of permafrost

1.1 (*Opposite*) Present and upper Holocene periglacial features in USSR (*after Markov, 1961; Popov, 1961; key translated*)

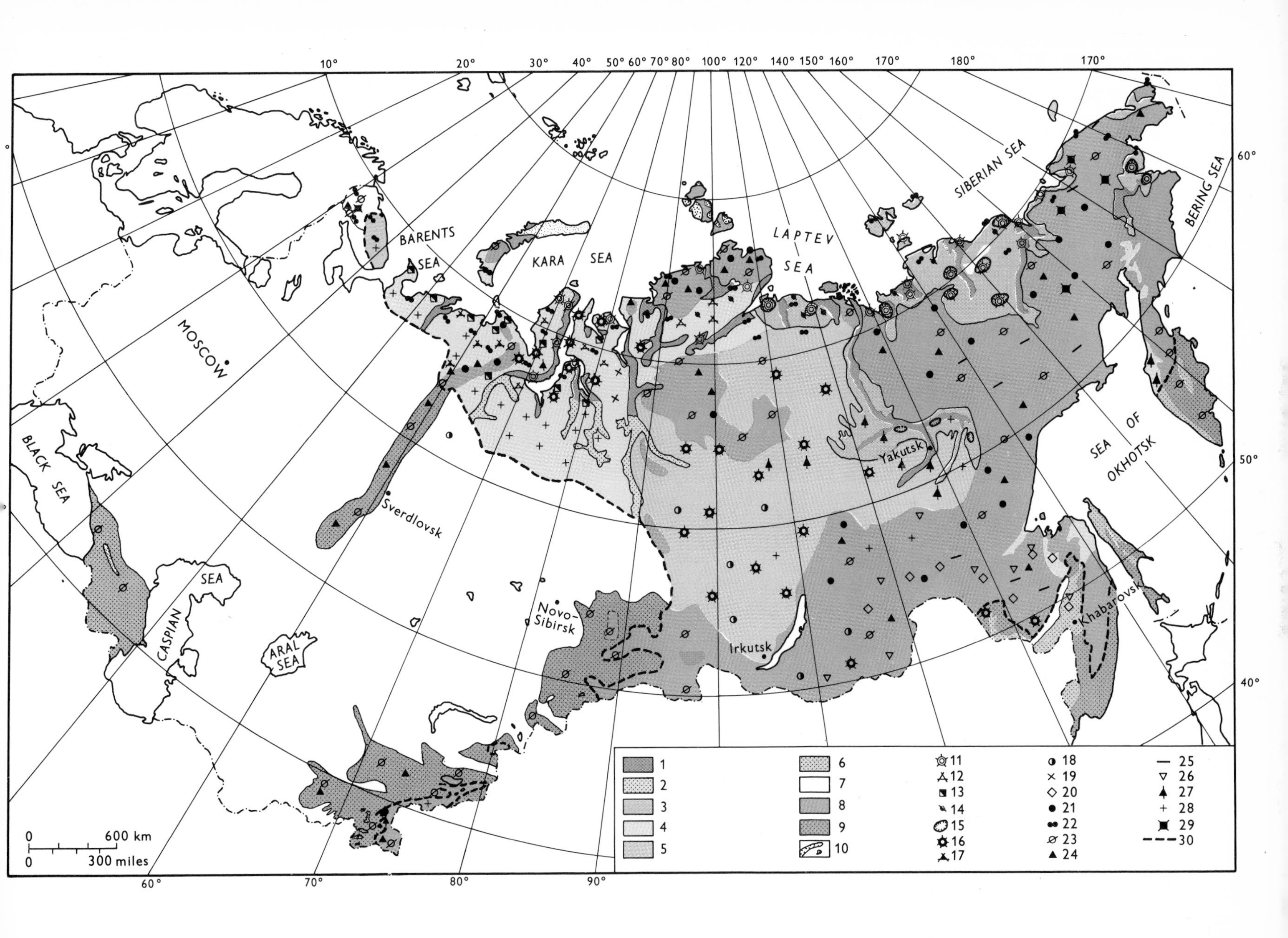

BARENTS SEA
KARA SEA
LAPTEV SEA
SIBERIAN SEA
BERING SEA
SEA OF OKHOTSK
BLACK SEA
CASPIAN SEA
ARAL SEA
MOSCOW
Sverdlovsk
Novo-Sibirsk
Irkutsk
Yakutsk
Khabarovsk
0 600 km
0 300 miles
1
2
3
4
5
6
7
8
9
10
11
12
13
14
15
16
17
18
19
20
21
22
23
24
25
26
27
28
29
30

3 Other processes

In addition to frost action, certain aspects of mass-wasting, nivation, fluvial action, lacustrine action, marine action, and wind may produce characteristic periglacial features.

IV ENVIRONMENTS

1 General

Many factors make up environment but for periglacial environments the overriding controls are regional climate and topography. Local factors may modify the regional climate and in this and other ways strongly influence processes.

Tricart (1967, 44–67; 1969, 19–27) recognized the following periglacial climates:

A *Cold dry climate with severe winters*

Encompasses elements of the D and E climates of Köppen, discussed later. Characteristics include (*a*) very low winter temperatures, (*b*) short summers, (*c*) permafrost, (*d*) low precipitation, (*e*) violent winds. Consequently there is (*a*) intense freezing, (*b*) reduced or even negligible activity of running water, and (*c*) important wind action.

B *Cold humid climates with severe winters*

(i) *Arctic type:* Corresponds to the most humid parts of Köppen's ET climates, excluding those without marked seasons. Characteristics include (*a*) similar mean temperatures to A but with a tendency for smaller annual range, (*b*) permafrost, (*c*) great climatic irregularities that tend to be masked by mean figures, (*d*) greater humidity with annual totals almost always exceeding 300 mm, resulting in appreciable snow cover and some rain. Consequently, as compared with A, (*a*) freezing is less intense and less long, (*b*) wind action is reduced by the snow cover, (*c*) running water is more important.

(ii) *Mountain type:* Corresponds to prairie-alpine zone of temperate zone. Characteristics include (*a*) monthly temperature trends similar to B(i) but with higher means, (*b*) precipitation is much greater than in B(i) and tends to inhibit wind action and deep penetration of frost, (*c*) freeze-thaw cycles are less common in the summit areas than in the valleys. Consequently (*a*) frost action is important but permafrost is commonly lacking, (*b*) running water is an important geologic agent, (*c*) wind action is slight.

C *Cold climates with small annual temperature range*

(i) *High-latitude island type:* Characteristics include (*a*) mean annual temperature near 0° with small annual range (generally on the order of 10°), numerous freeze-thaw cycles, (*b*) instability of weather, (*c*) considerable precipitation, generally exceeding 400 mm, which tends to inhibit wind effects. Consequently (*a*) frost action is characterized by many freeze-thaw cycles of short duration and slight penetration into the ground, (*b*) wind action is slight.

(ii) *Low-latitude mountain type:* Characteristics include (*a*) lack of seasonal temperature variations, (*b*) considerable variations in diurnal temperature, far exceeding the seasonal variations, (*c*) high precipitation except in arid mountains such as Puna de Atacama. Consequently (*a*) there is considerable frost action because of the frequent freeze-thaw cycles, (*b*) only slight frost penetration into the ground, (*c*) absence of permafrost, (*d*) lack of wind action in high-precipitation areas.

As stated by J. L. Davies (1969, 13–14), this classification brings out some marked contrasts. Permafrost is characteristic of A, absent in C. Annual temperature cycles are of large amplitude and extend to considerable depth in A; the opposite is true in C. Freeze-thaw cycles are much fewer in A than in C. Running water is much less important in A than in C. In each case conditions in B tend to be intermediate.

In some respects Tricart's classification may be superior to Köppen's but the latter has the advantage of more precise boundaries between categories.

The Köppen classification of climates (Köppen, 1936; Köppen-Geiger, 1954; cf. Strahler 1969, 224–30) is widely used and is the one on which the following scheme is based for polar, subpolar, and middle-latitude lowland climates and geographic zones. That these climates and zones can be similarly designated arises from the fact that the zones are climatically defined. The addition of highlands is to emphasize the role of altitude in determining climate and topography. Only the most general kind of classification of climates and zones is used here in view of the limited knowledge concerning the distribution and frequency of periglacial processes and features. Gerdel (1969) has cited some of the practical limitations to climatic classifications and has presented a useful general description of the characteristics of cold regions, especially in the Northern Hemisphere.

2 Polar lowlands

In the polar zone, the average temperature of the coldest month is $< -3°$ and of the warmest month $< 10°$. The zone, which is controlled by polar and arctic air masses in the Northern Hemisphere, lies roughly north of lat. 55°N and south of lat. 50°S. It includes an ice cap (Köppen's EF) climate, dominated by arctic and antarctic air masses, in which the average temperature of the warmest month is $< 0°$, and, of more concern to periglacial research, a tundra (Köppen's ET) climate, in which the average temperature of this month is $> 0°$.

The zone is characterized by ice caps, bare rock, and/or vegetation of tundra types – mainly grasses, sedges, small flowering plants, and in places herbaceous shrubs.

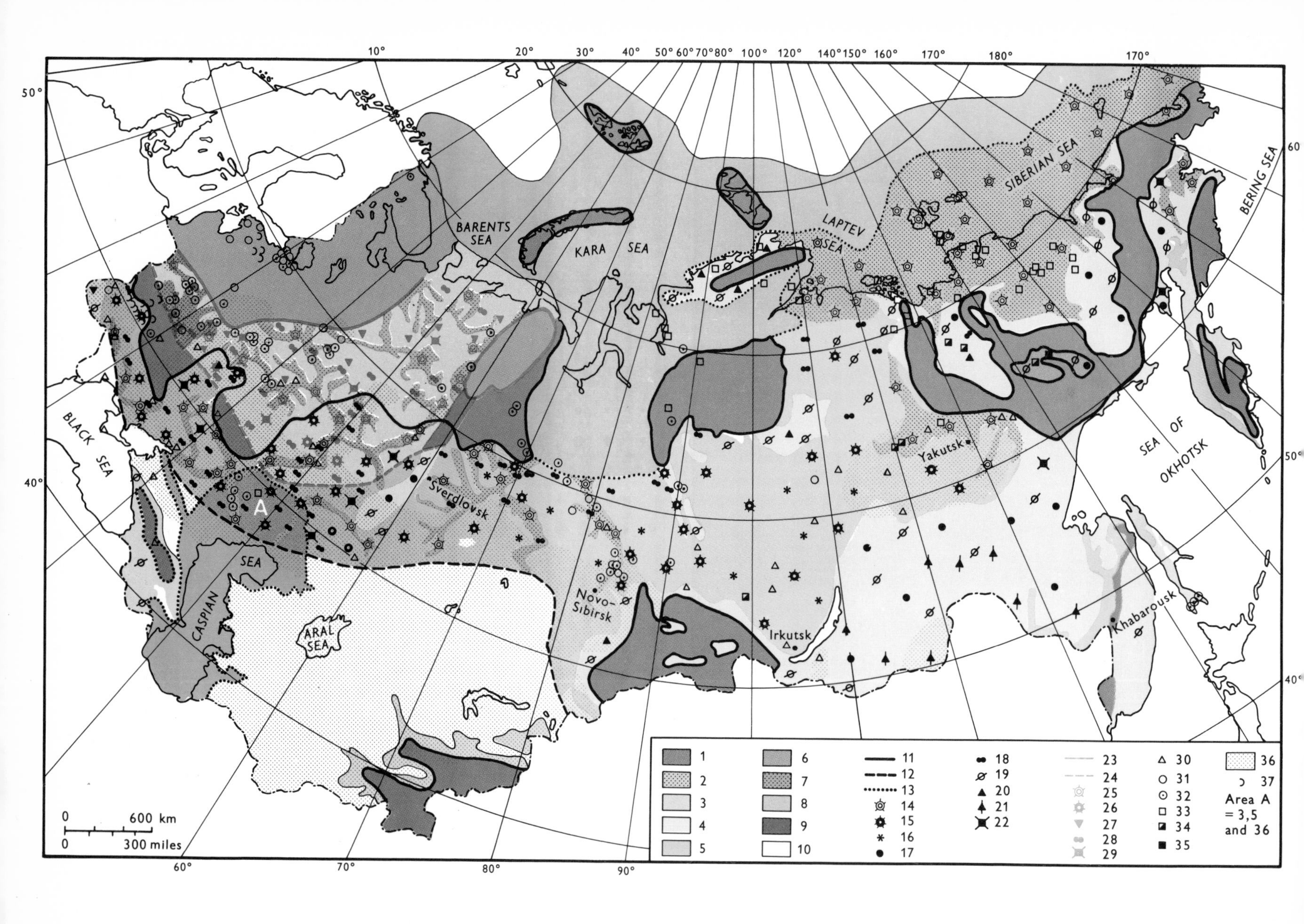
BARENTS SEA
KARA SEA
LAPTEV SEA
SIBERIAN SEA
BERING SEA
SEA OF OKHOTSK
BLACK SEA
CASPIAN SEA
ARAL SEA
Sverdlovsk
Novo-Sibirsk
Irkutsk
Yakutsk
Khabarousk
A
0 600 km
0 300 miles
Area A = 3,5 and 36

1 Regions of maximum glaciation 2 Regions of sediment accumulation corresponding to maximum glaciation (with underlying syngenetic permafrost) 3 Regions of equilibrium relative to processes of sediment accumulation and transportation corresponding to maximum glaciation (with underlying epigenetic permafrost) 4 Regions of dominant transportation processes corresponding to maximum glaciation (with underlying permafrost of destructive origin ['d'origine génétique destructive']) 5 Extension of ocean during period of maximum glaciation 6 Regions of Valdaï Glaciation 7 Regions of accumulation corresponding to Valdaï Glaciation (with underlying syngenetic permafrost) 8 Regions of equilibrium relative to accumulation and transportation corresponding to Valdaï Glaciation (with underlying epigenetic permafrost) 9 Regions of dominant transportation corresponding to Valdaï Glaciation (with underlying permafrost of destructive origin ['d'origine génétique destructive']) 10 Regions without permafrost phenomena 11 Maximum limit of glaciation 12 Southern limit of permafrost during maximum glaciation 13 Shoreline during maximum glaciation 14 Ice-wedge polygons ['Polygones a fissures remplies de glace fossile'] developed during maximum glaciation 15 Soil-wedge polygons ['Polygones a fissures remplies de sol'] corresponding to maximum glaciation 16 Hummocks [buttes gazonées] and concave forms corresponding to maximum glaciation 17 Altiplanation terraces corresponding to maximum glaciation 18 Nonsorted circles ['formes (taches) en médaillon'] corresponding to maximum glaciation 19 Sorted polygons ['Polygones de pierres'], circles, etc., corresponding to maximum glaciation 20 Solifluction stripes ['Bandes de solifluxion sur versant'] corresponding to maximum glaciation 21 Perennial hummocks [buttes gazonées], due to frost heaving, corresponding to maximum glaciation 22 Solifluction forms corresponding to maximum glaciation 23 Limits of Valdaï Glaciation 24 Southern limit of permafrost during Valdaï Glaciation 25 Ice-wedge polygons ['Polygones a fissure remplies de glace fossile'] developed during Valdaï Glaciation 26 Soil-wedge polygons ['Polygones a fissure remplies de sol'] corresponding to Valdaï Glaciation 27 Degrading polygons with soil veins (pebbles in cover loam) corresponding to Valdaï Glaciation 28 Nonsorted circles ['Formes (taches) en médaillon'] corresponding to Valdaï Glaciation 29 Solifluction forms corresponding to Valdaï Glaciation 30 Discovery site of Upper Paleolithic fauna 31 Periglacial fossil macroflora 32 Periglacial fossil pollen and spores 33 Mammoth remains in permafrost ['tjäle'] 34 Rhinoceros remains in permafrost ['tjäle'] 35 Horse (*Equus caballus*) in permafrost ['tjäle'] 36 Regions of loess [d'apparition de loess et de formations loessiques'] 37 Regions with ancient continental dunes

1.2 (*Opposite*) Pleistocene periglacial features in USSR (*after Markov, 1961; Popov, 1961; key translated*)

3 Subpolar lowlands

In the subpolar zone the average temperature of the coldest month is $< -3°$, the temperature of the warmest month is $> 10°$ but there are less than four months above this temperature (Köppen's Dfc, Dfd, Dwc, Dwd climates). The zone is controlled by continental air masses and extends roughly from lat. 50° to 70°N (northern taiga zone) and from lat. 45° to 60°S. Because the climatic gradients vary with longitude there is some latitudinal overlap.

The 10° isotherm for the warmest summer month, which is accepted here as the boundary between polar and subpolar lowlands, tends to coincide with treeline in the Northern Hemisphere. Characteristically, coniferous forest predominates. However, treeline (i.e. the northern limit of scattered trees) can be some $1\frac{1}{2}$ degrees of latitude north of the coniferous forest (cf. Washburn, 1951, 270–1).

4 Middle-latitude lowlands

In the middle-latitude zone the average temperature of the coldest month is $< -3°$ but there are more than four months with average temperature $> 10°$ (Köppen's Dfa, Dfb, Dwa, Dwb climates). This zone is controlled by both polar and tropical air masses and extends roughly from lat. 35° to 60°N.

5 Highlands

A highland climate may differ in important ways from climates primarily controlled by latitude. For instance, highlands are commonly colder than lowlands in the same zone, some highlands have stronger diurnal than seasonal temperature changes, and the orientation of mountain slopes tends to exert a strong climatic influence. Particularly in periglacial environments where topography is a critical factor, it is desirable to make a marked distinction between lowland zones controlled primarily by latitude and highland zones controlled by altitude as well as latitude.

In a general way counterparts to cold lowland zones can be found in highlands, and parallels are frequently cited between arctic and alpine zones and between subarctic and subalpine zones. However, there is considerable difference of opinion regarding zone classifications, many of which are based on biological considerations reflecting climate rather than being based directly on climatic parameters (Löve, 1970). No consistent attempt is made in the following to adopt any particular highland zonation except to recognize that the altitudinal treeline, like the latitudinal treeline, is a critical boundary for certain periglacial processes. Rather, highlands are cited in relation to polar, subpolar, middle-latitude, and low-latitude zones to indicate that certain periglacial processes and features are more common in highlands than in lowlands of the same latitudinal zone, either because of a more rigorous climate in the highlands or because of topography. As an order of magnitude, 1000 m will be considered the minimum altitude difference between a highland and a lowland in the same general region.

V REFERENCES

1 General

Although periglacial research began to emerge as a recognized field following Łoziński's (1909) work, only recently has the subject been comprehensively treated. The Polish and Russian periglacial literature is particularly rich. Unfortunately the present writer has had to forego much of it that has not yet been translated, including Jahn's (1970) comprehensive treatment, which was received as the present volume was being submitted for publication. General references include: Cailleux and Taylor (1954), J. L. Davies (1969), Embleton and King (1968), Jahn (1970), Kaplina (1965), Peltier (1950), L. W. Price (1972), Tricart (1963; 1967; 1969), Troll (1944; 1958), and the Biuletyn Peryglacjalny (1954–). An illustrated English and French glossary of periglacial features has been compiled by Hamelin and Cook (1967).

Regional studies of contemporary periglacial phenomena include Sekyra (1960), Bird (1967, 161–270), Bout (1953), Boyé (1950), Büdel (1960), Kelletat (1969), Jan Lundqvist (1962), Malaurie (1968), Markov and Bodina (1961; 1966), Popov (1961), and many others.

2 Frost action

Frost action, including permafrost, is at the centre of much periglacial research. The literature here is voluminous and only some of the most comprehensive bibliographies and references are cited below.

American Meteorological Society (1953), Arctic Institute of North America (1953–), Beskow (1935; 1947), Highway Research Board (1948; 1952*a*; 1952*b*; 1957; 1959; 1962; 1970), Muller (1947), National Academy Sciences-National Research Council (1966), Popov (1967), Terzaghi (1952), US Army Cold Regions Research and Engineering Laboratory (1951–), J. R. Williams (1965), P. J. Williams (1967).

2 Environmental factors

I Introduction: II Basic factors: *1 Climate: a* Scale, *b* Climatic parameters, *c* Zonal climate, *d* Local climate, *e* Microclimate; *2 Topography; 3 Rock material: a* General, *b* Structure, *c* Mineral composition, *d* Texture, *e* Colour; *4 Time.* **III Dependent factors:** *1 Snow cover; 2 Liquid moisture; 3 Vegetation.*

I INTRODUCTION

The most nearly independent environmental factors influencing the various periglacial processes and the development and distribution of frozen ground are climate, topography, and rock material. Their degree of independence depends on the scale. For instance, zonal climate – a climate determined by the largest-scale factors of latitude, atmospheric circulation systems, and widespread highlands – is an independent factor. However, smaller-scale climatic effects are dependent on superimposed influences, some of which may result in an azonal climate – a climate that is atypical and not truly representative on a zonal basis.

The mutual interaction of climate, topography, and rock material is illustrated by countless examples. Thus topography, through altitude and exposure, modifies zonal climate but climate can modify topography by determining the processes acting on a region – for example, glaciation produces cirques that create variations of exposure and thereby influence the local climate. The nature of a rock influences the effect climate may have on it but climate may determine the kind of rock developed – for example, an evaporite formed in an arid basin.

In contrast to the foregoing factors, snow cover, liquid moisture, and vegetation are always dependent. However, they can be critical controls of periglacial processes, especially frost action, since they can determine whether or not a process is climatically zonal or azonal in its effect.

The following generalized review is to set the stage for later discussion of the environmental implications of periglacial processes.

II BASIC FACTORS

1 Climate

a Scale In discussing the influence of climate, it is useful to recognize three scales of climate: zonal climate, local climate, and microclimate. As noted, zonal climate reflects only large-scale effects such as latitude and widespread highlands; it is the critical element to establish when determining past climates and reconstructing climatic changes. Local climate represents the combined influence of zonal climate and local topography. Microclimate is at a still smaller scale in that it incorporates the additional influence of ground-surface characteristics such as vegetation, moisture, and air-earth or air-water interface effects (Geiger, 1965). As indicated below, local and microclimatic influences can be highly significant and must be evaluated in reconstructing past zonal climates from biologic and geologic evidence.

b Climatic parameters The most important climatic parameters controlling periglacial processes are temperature, precipitation, wind, and their seasonal distribution. The past influence of wind may be readily discernible through erosional and depositional effects but temperature and precipitation may interact to the extent that their relative importance may be difficult to determine. For instance, both sufficiently low temperature and sufficiently high moisture are required for glaciation and for certain frost-action effects, but within limits one parameter may substitute for the other. Thus increased snow accumulation and nivation may result from lower summer temperatures and less melting as well as from increased winter snowfall. On the other hand, increased permafrost may be due as much to lower winter snowfall, and hence less insulation of the ground, as to lower air temperatures.

c Zonal climate Present-day climate can be studied and described by quantitative observations that allow for local and microclimatic effects in describing zonal climate. This is much more difficult to do in reconstructing past zonal climates, since the evidence at any one place may be very strongly influenced and in some cases dominated by purely local and microclimatic effects.

d Local climate The extent to which local topography can influence climate is seen in the differing precipitation regime of windward and lee slopes and in the differing temperature regime of north and south slopes of mountains, especially in high latitudes. For instance, other factors being constant, depth of thawing in the Northern Hemisphere may be 50 to 60 per cent greater on south-facing than on north-facing slopes (Khesthova *et al.*, 1961, 50; 1969, 8). Such differences may strongly affect periglacial processes as discussed later.

e Microclimate Meteorological observations at the common shelter height (1·7–2·0 m in Europe, 4 ft in the United States) do not adequately indicate the climate at the ground surface, yet this is where some critical processes occur. For instance, the number of freeze-thaw cycles at the surface may be very different from those a few feet higher in a shelter. Wind velocities are also significantly different.

2 Topography

The way topography can influence climate and thereby, indirectly, process is outlined above. Topography can also exert a direct effect on process. For instance, the configuration and gradient of a hillside can determine whether the dominant mass-wasting process is landsliding as opposed to frost creep or solifluction. A very low-angle slope favours retention of moisture and development of certain forms of patterned ground.

3 Rock material

a General The term rock material as used here covers structure, mineral composition, texture, and colour, and includes both bedrock and unconsolidated material.

Bedrock is essentially an independent factor in relation to periglacial processes, since it is usually a pre-existing, very much older feature.

Unconsolidated material can be a dependent or independent factor. Since it results from processes acting on bedrock, it is dependent on climate to the extent that the processes are climatically controlled. The character of humus-bearing soils is strongly dependent on climate. However, in many places the unconsolidated material, like bedrock, predates the situation being considered and in this sense is independent. Furthermore there is such a variety of ways in which the unconsolidated material can vary that even though it is a function of one periglacial process it could be considered an independent factor in relation to a different periglacial process.

Rock material is less dependent on topography than vice versa, and in this respect is also essentially an independent factor. Therefore despite intimate interactions between climate, topography, and rock material, it is not circular reasoning to regard rock material as an independent as well as a dependent variable in considering periglacial processes.

b Structure The structure of rock material comprises its gross features: consolidated bedrock or unconsolidated sediments, nature of jointing in bedrock or of fissuring in unconsolidated sediments, or attitude of stratification planes in either case. In many places structure determines the relative importance of processes. For instance, joints and fissures favour ingress of moisture and they localize weathering and subsequent erosion, and the attitude of joints and fissures determines the direction and inclination of the resulting features.

c Mineral composition The mineral composition of rock material strongly influences its reaction to weathering and abrasion and thus to erosion. Bedrock or sediment in a warm humid climate may weather very differently from material of the identical mineral composition in a cold dry climate, and chemically different materials in the same climate may have diverse reactions. In either case the quantitative effect on a given geomorphic process may be very significant. Wind action may produce numerous ventifacts in a limestone region but comparatively few where there are only harder rocks.

d Texture The texture of rock material (i.e. the grain size and arrangement of its mineral constituents or particles) strongly influences its characteristics in any climate, whether the material is bedrock or unconsolidated. In general, fine-grained bedrock is more resistant than coarse grained to weathering in a periglacial environment, other conditions being comparable. On the other hand, fine-grained unconsolidated material may be more subject than coarse to frost action. For instance, as discussed under Freezing process in the chapter on Frost action, it is common engineering practice to regard soils as susceptible to frost heaving if they contain several per cent of particles finer than 0·07 mm (passing the 200 mesh screen, *c.* the upper limit of silt) (Terzaghi, 1952, 14), or (Casagrande criterion) containing more than 3 to 10 per cent particles finer than 0·02 mm (Casagrande, 1932, 169).

e Colour Although colour is a function of mineral composition and texture, it is worth listing separately to emphasize its effect on temperature: namely dark-coloured materials absorb radiant heat and warm their surroundings whereas the opposite is true for light-coloured materials. For instance, thawing of snow has been observed adjacent to dark-coloured objects at air temperatures as low as $-16{\cdot}5^\circ$ (Taylor, 1922, 47) and even -20° (Souchez, 1967, 295).

4 Time

Time is an independent factor that is usually overlooked in periglacial studies. However, in the opinion of P. J. Williams (1961, 346) 'observed variations in density of occurrence of specific fossil frozen ground phenomena are not so much indicative of particular climatic conditions but of length of time during which the features could be formed'. This view is based on the belief that the processes responsible for periglacial phenomena such as solifluction and patterned ground are very slow; for instance, that it would take, say, 12000 years for solifluction to move materials 10–20 m downslope.

On the other hand, observations in Northeast Greenland indicate that present rates of solifluction on a slope of 10–14° could move materials 9–37 m downslope in 1000 years, the amount depending on the moisture conditions, which varied along the contour. These and other observations on solifluction by various investigators (cf. Washburn, 1967, 93–8, 118) and the rapidity with which some forms of patterned ground can develop (as indicated by occurrences on recently emerged shores and by direct observation), show that more information is required before quantitative inferences can be made regarding the extent to which time affects the processes. In any event, as stressed by Williams, the time factor certainly merits consideration in paleoclimatic reconstructions.

III DEPENDENT FACTORS

1 Snow cover

The amount and distribution of snow cover are functions of climate and topography. The effect of the latter is particularly important in the distribution of snowdrifts.

The insulating effect of snow cover on the temperature at the snow ground interface can be estimated in cgs units by applying Lachenbruch's (1959, 28; Plate 1) equation

$$\Delta\tilde{\Theta} = \frac{1}{\pi} A^*[1 - A(X)/A^*]$$

where $\Delta\tilde{\Theta}$ = change in mean annual temperature at the ground surface as the result of the snow, A^* = amplitude of mean annual temperature at bare-ground surface in summer and upper snow surface in winter, $A(X)$ = steady amplitude resulting from wave with amplitude A^* passing through snow of thickness X.

By moderating the ground temperature, snow can protect vegetation from frost action; it can also provide the critical moisture for growth. Thus snow favours vegetation in several ways. However, if lasting too long, snow can also completely inhibit vegetation. Each of these effects can be critical for a given periglacial process.

2 Liquid moisture

The amount of liquid moisture and its seasonal distribution are functions of climate, topography, and rock material. Climate exerts the large-scale control, topography modifies the large-scale climatic influence, and rock material can determine the amount of moisture entering and remaining in the ground. The texture of the rock material may be diagnostic in this respect and either favour or inhibit frost action and vegetation, each of which also influences the other in complex ways.

3 Vegetation

Vegetation, like liquid moisture, is a function of climate, topography, and rock material. Some plant species are much more sensitive to climate than others. Zonal climate sets the broad pattern but local climate can dominate a given situation, and there is an intimate interaction and mutual influence between vegetation and microclimate. For instance, 'unvegetated dry soils can be 1° or 2° warmer than a nearby site underlain by vegetated wet soils' (Ferrians, Kachadoorian, and Greene, 1969, 8). Destruction of the vegetation cover can have far-reaching effects, especially where there is permafrost (Jerry Brown, Rickard, and Vietor, 1969). Aside from its influence on local climate, topography is critical in that the angle of slope affects the continuing contest between the ability of plants to take root and opposing processes such as mass-wasting. Soil characteristics also exert a strong influence on the ability of plants to survive opposing processes.

The especially important ways in which vegetation influences periglacial processes are through its insulating effect and its binding effect on soils. As with the other factors that influence periglacial processes, there is an intimate competitive interaction. For instance, a coherent vegetation cover impedes frost action and erosion but, concurrently, frost action and erosion impede the growth of vegetation. Details of the interaction between vegetation and frost action in soils are complicated but of critical importance in places. The problem is discussed by Benninghoff (1952; 1966), R. J. E. Brown (1966*a*), and Raup (1965; 1969; 1971a) among others.

3 Frozen ground

I INTRODUCTION

Frozen ground is central to a consideration of periglacial processes and environments. In particular, knowledge of its nature sets the stage for discussing the various processes collectively known as frost action, considered in the next chapter.

In many respects it is important to distinguish between seasonally frozen ground and permafrost. The latter is especially significant in periglacial studies and is therefore discussed in some detail.

II SEASONALLY FROZEN GROUND

1 General

'Seasonally frozen ground is ground frozen by low seasonal temperatures and remaining frozen only through the winter' (Muller, 1947, 221). Some investigators (cf. Muller, 1947, 6, 213) include the zone of annual freezing and thawing above permafrost (i.e. active layer), others (cf. Black, 1954, 839) restrict or appear to restrict the term to a non-permafrost environment. Because of the ambiguity and difficulty of trying to differentiate between two zones of annual freezing and thawing by a descriptive term equally applicable to both, the present writer uses the term seasonally frozen ground in the broad sense and specifies, where necessary, whether the reference is to a permafrost or non-permafrost environment. In most instances, use of the term active layer (discussed later) for the seasonally frozen ground above permafrost helps to avoid ambiguity.

2 Depth of seasonal freezing and thawing

Depth of freezing and thawing is controlled by many of the factors reviewed previously, and is therefore subject to considerable local variation. On a larger scale, there is a clear increase in depth of seasonal freezing with increasing latitude in a non-permafrost environment, the range being from a few millimetres to over 1·8 m (72 in) in the United States (Figure 3.1) and up to a depth of 3 m in Canada (Crawford and Johnston, 1971, 237). In a permafrost environment the trend is towards a decrease in depth with increasing latitude, and the permafrost tends to approach the surface of the ground.

The climatic parameters responsible for deep seasonally frozen ground and for permafrost may be very similar in the transition zone, but the resulting products can be quite different in a non-permafrost or a permafrost environment. Construction problems, for instance, can be many times more difficult in a permafrost environment than where there is seasonally frozen ground without permafrost.

It may not always be practicable to obtain direct measurements of the depths to which seasonal freezing and thawing extend, and several indirect approaches have been suggested. Some of these involve a number of variables, and others are relatively simple yet useful.

For example, the depth of winter freezing can be estimated by the equation (Terzaghi, 1952, 41)

$$Z = \sqrt{\frac{2tTk_h}{72n \text{ cal/cm}^3 + T/2(0{\cdot}53 + 0{\cdot}47n) \text{ cal/cm}^3}}$$

where Z = depth of winter freezing (cm), t = freezing period (sec), T = mean temperature of freezing period (°C), k_h = thermal conductivity (cal/cm sec °C), n = porosity. By inserting figures for the thawing period, the equation can also be used to estimate depth of summer thawing. It applies to both the permafrost and non-permafrost environments.

There are several potential sources of error, such as taking all temperatures as representative of the air-ground interface and disregarding the effect of snow or vegetation. Ground-surface temperatures permit closer approximations, and Lachenbruch's (1959, 28, Plate 1) equation, cited under discussion of Dependent factors in the chapter on Environmental factors, can be applied to evaluate the effect of snow cover. In addition to the mean annual temperature, this equation also considers the amplitude of the temperature change from summer to winter, which can be a critical variable in evaluating the effect of a disturbance of the ground surface on depth of thawing. 'The larger the amplitude, the worse the thawing problem' (Lachenbruch, 1970*b*, J5).

3.1 (*Opposite*) Maximum depth of frost penetration in United States (*after Strock and Koral, 1959, 1–102*)

MAXIMUM DEPTH OF FROST PENETRATION (CM; CONVERTED FROM INCHES)

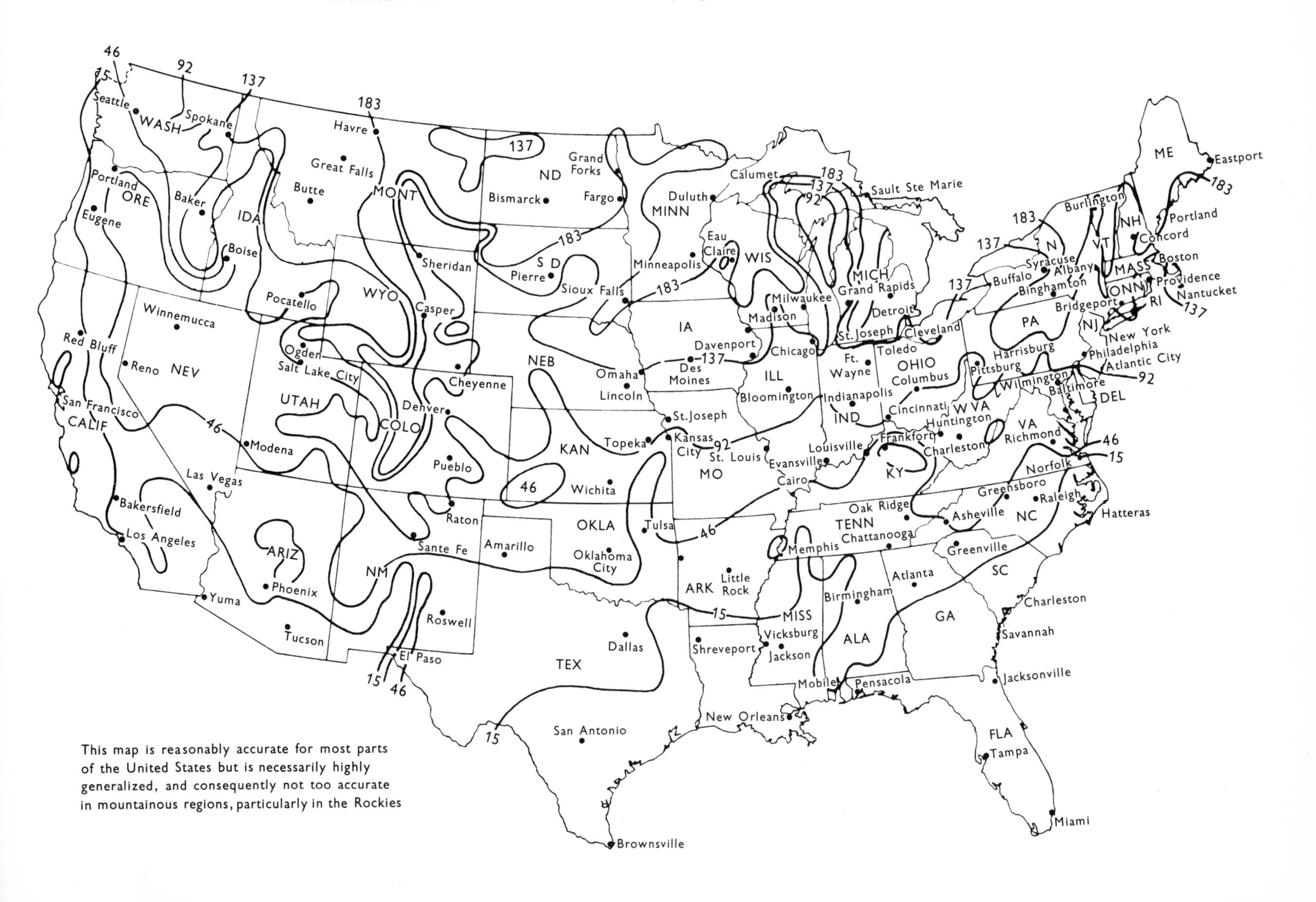

This map is reasonably accurate for most parts of the United States but is necessarily highly generalized, and consequently not too accurate in mountainous regions, particularly in the Rockies

III PERMAFROST

1 General

The term permafrost, also known as pergelisol (Bryan, 1946, 635, 640) and perennially frozen ground, was first defined by Muller (1947, 3, cf. 219):

> Permanently frozen ground or permafrost is defined as a thickness of soil or other superficial deposit, or even of bedrock, at a variable depth beneath the surface of the earth in which a temperature below freezing has existed continually for a long time (from two [years] to tens of thousands of years).[1] Permanently frozen ground is defined exclusively on the basis of temperature, irrespective of texture, degree of induration, water content, or lithologic character.

The term permafrost has been widely adopted but defined as perennially frozen rather than permanently frozen ground, since changes in climate and surface conditions can cause rapid thawing of permafrost. Muller's definition in which texture, induration, water content, and lithology are eliminated as factors (cf. also Muller, 1947, 30) supports the view that 0° is the basic criterion rather than the exact freezing point, which can vary with such factors including pressure. Most workers cite a temperature below 0° in defining permafrost (cf. R. J. E. Brown, 1967a; Ferrians, 1965), but in practice it is doubtful if ground that rises to 0° for part of the year would be excluded if the temperature rose no higher and its mean annual temperature remained below 0°. Amended definitions of permafrost have been suggested. According to Stearns (1966, 1–2)

> The term 'permafrost' is defined as a condition existing below ground surface, irrespective of texture, water content, or geological character, in which:
> *a*. The temperature in the material has remained below 0°C continuously for more than 2 years, and
> *b*. If pore water is present in the material a sufficiently high percentage is frozen to cement the mineral and organic particles.
>
> A temperature definition alone is not considered sufficient, for often a geothermal situation exists in which a frozen, or cemented, state is not obtained even though the temperature of the material is well below 0°C. . . .
>
> The definition of permafrost given above includes both dry-frozen and wet-frozen ground. In the dry-frozen, or dry frost, condition there is very little or no water contained in the pores so that temperature becomes the only criterion. In the wet-frozen condition some cementing ice must be present.

It remains to be seen to what extent Stearns' redefinition is accepted. From the viewpoint of engineering requirements it has the merit of specifying cementation where there is wet-frozen ground. However, the last two paragraphs pinpoint a problem in that temperature alone is not considered sufficient for wet-frozen ground but is the only criterion for dry-frozen ground.

[1] The Institut Merzlotovedeniya in Yakutsk, USSR, one of the world's leading permafrost research agencies, specifies three years or more in accordance with the recommendations of the Commission on Terminology of the Institut Merzlotovedeniya im. V. A. Obrucheva (1956, 10; 1960, 6; Corte, 1969*a*, 130).

Furthermore cementation is a subjective criterion unless physical parameters defining cementation are established. Muller's definition is more precise and for mapping purposes, at least, the easier to apply. In any event, for the time being, it behoves authors to specify which definition they are adopting if there is any possibility of misunderstanding. For the purposes of the present review, Muller's definition will be followed, although in most of the discussion the distinction would not be critical.

Glaciers whose temperature does not reach 0° are permafrost by either definition but they are frequently omitted from discussions of permafrost and treated separately. They will be mentioned here only incidentally.

2 Significance

Civil engineering problems connected with building and highway construction, sewage disposal, water supply, and hot oil pipelines are magnified in permafrost regions (cf. Jerry Brown, Rickard, and Vietor, 1969; R. J. E. Brown, 1970; Corte, 1969*a*; Crawford and Johnston, 1971; Ferrians, Kachadoorian, and Greene, 1969; Lachenbruch, 1970*a*; J. R. Williams, 1970), and much attention has been devoted to permafrost as a result. The problems are both civilian and military. The military interest that began with World War II led to the establishment of the US Army Snow, Ice, and Permafrost Research Establishment (SIPRE), later renamed the US Army Cold Regions Research and Engineering Laboratory (CRREL), which has been responsible for much of the basic research on permafrost in the United States. The US Geological Survey has also been very active in permafrost research. Yet our knowledge is still inadequate as illustrated by the controversy aroused by the trans-Alaska pipeline (Alyeska) project.

Aside from the engineering aspects, there are many intriguing scientific questions related to permafrost – its origin, its climatic implications, its effect on life (Péwé, 1966*c*), and what it can tell us about past, present, and future environments. Much of the record is there to be read, and part of the task of periglacial research is to interpret that record.

3 Distribution

Because permafrost becomes thinner and breaks up into patches as it merges with seasonally frozen ground where permafrost is lacking, permafrost is classified into continuous permafrost and discontinuous permafrost (Figure 3.2).

The distribution of present-day permafrost in the Northern Hemisphere, disregarding high-altitude occurrences, is illustrated in Figures 3.3–3.6. Stearns (1966, 9–10), following Black (1954, 839–42), concluded that permafrost underlies some 26 per cent of the land surface (including glaciers) of the world, based on figures compiled from various sources (Table 3.1).

Submarine permafrost is known off arctic coasts (Figure 3·5; Mackay 1972*a*, 19–21), and it probably also occurs in the Antarctic. However, its distribution is poorly known.

The difference in Table 3.1 in the figure for the Northern Hemisphere ($22{\cdot}35 \times 10^6$ km²) and the total for Alaska, Canada, Greenland, and the USSR ($19{\cdot}04 \times 10^6$ km²) is due to the different methods of estimating the figures. If cementation is accepted as a criterion of permafrost, and

Table 3.1. Distribution of permafrost (*after Stearns, 1966,* 9–10)

	Continuous		*Discontinuous*		*Total*	
	km² × 10⁶	mi² × 10⁶	km² × 10⁶	mi² × 10⁶	km² × 10⁶	mi² × 10⁶
Northern Hemisphere	7·64	2·95	14·71	5·68	22·35	8·63
Antarctica	13·21	5·10	—	—	13·21	5·10
Mountains	—	—	2·59	1·00	2·59	1·00
Totals	20·85	8·05	17·30	6·68	38·15	14·73

	Land surface (per cent)	*Permafrost* km² × 10⁶	*Permafrost* mi² × 10⁶
Alaska	80	1·30	0·5
Canada	40–50	3·89–4·92	1·5–1·9
Greenland	99	1·68	0·65
USSR	50	11·14	4·3
		19·04 (max)	7·35 (max)

glaciers are excluded, a major reduction in the estimate for Antarctica and Greenland will apply if it turns out that the ice sheets here are extensively underlain by unfrozen material.[1] Excluding glaciers, Shumskii, Krenke, and Zotikov (1964), cited by Grave (1968*a*, 48; 1968*b*, 2), estimated that 14·1 per cent of the land surface is underlain by permafrost. R. J. E. Brown (1970, 7; cf. 27–

[1] Zotikov (1963) and Grave (1968*a*, 51–2; 1968*b*, 6) argued that the central part of the Antarctic is free of frozen ground, and recent drilling confirms the presence of water at the base of the Ice Sheet near Byrd Station (Gow, Ueda, and Garfield, 1968). However, the estimated PT melting point of $-1{\cdot}6°$ at the drilling site would still be indicative of permafrost according to Muller's definition. Assuming that the Antarctic Ice Sheet is in a steady state, Budd, Jenssen, and Radok (1970, 301–5; 1971, 117–26) calculated, contrary to Grave and Zotikov, that the central area would be frozen and only relatively small marginal areas would be thawed beneath the ice; their assumption is subject to question but Budd, Jenssen, and Radok concluded that the central area would be characterized by permafrost even if ice gain exceeded ice loss by a factor of two. According to similar calculations, most of the Greenland Ice Sheet has a basal temperature below $-5°$ (personal communication, W. Budd, 1971).

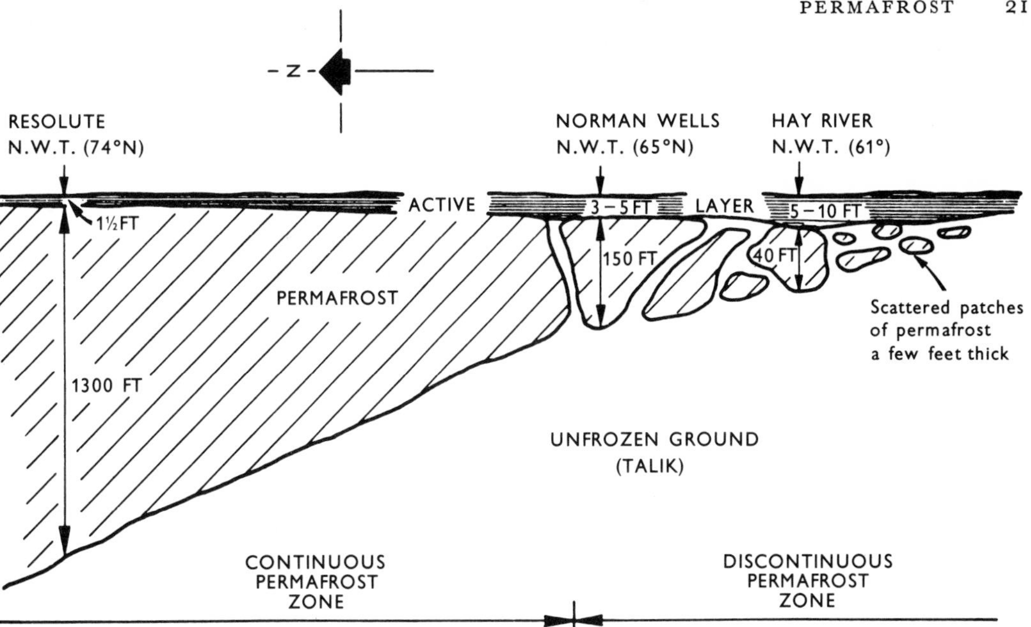

3.2 North-south vertical profile of permafrost in Canada showing decreasing thickness southward and relation to continuous and discontinuous permafrost zones (*R. J. E. Brown, 1970, 8, Figure 4*)

8) and Péwé (1971; personal communication, 1972) estimated 20 per cent (excluding glaciers), and Péwé in agreement with Stearns (Table 3.1) indicated that permafrost characterizes 80 per cent of Alaska and 50 per cent of Canada. He cited Jerry Brown for the estimate that the coastal plain of Arctic Alaska contained 907 km^3 (350 mi^3) of ground ice. According to Grave's report, the world's permafrost comprises 0·83 per cent of the total freshwater ice. Most of the remainder is in the Antarctic Ice Sheet (>90 per cent), the Greenland Ice Sheet, and other glaciers. Together this ice constitutes some 75 per cent of the freshwater resources of the earth. The amount represented by permafrost is very small but its significance far transcends its quantity.

Comparison with climatic maps shows that the southern limit of continuous permafrost is not necessarily parallel to air isotherms. It lies north of the −6° to −8° mean annual isotherm in Alaska (Péwé, 1966*a*, 78; 1966*b*, 68). In Canada it is common at *c.* −6·7° (20°F) (R. J. E. Brown, 1960, 171–2, Figures 6–7; 1969, 18; 19, Figure 2; 48), although in places discontinuous permafrost is still present at −9·4° (15°F) (R. J. E. Brown, 1969, 52). According to Crawford and Johnston (1971, 237), the discontinuous and continuous zones in Canada merge where the mean annual temperature is about −8·3°. Probably about −7° approximates the southern

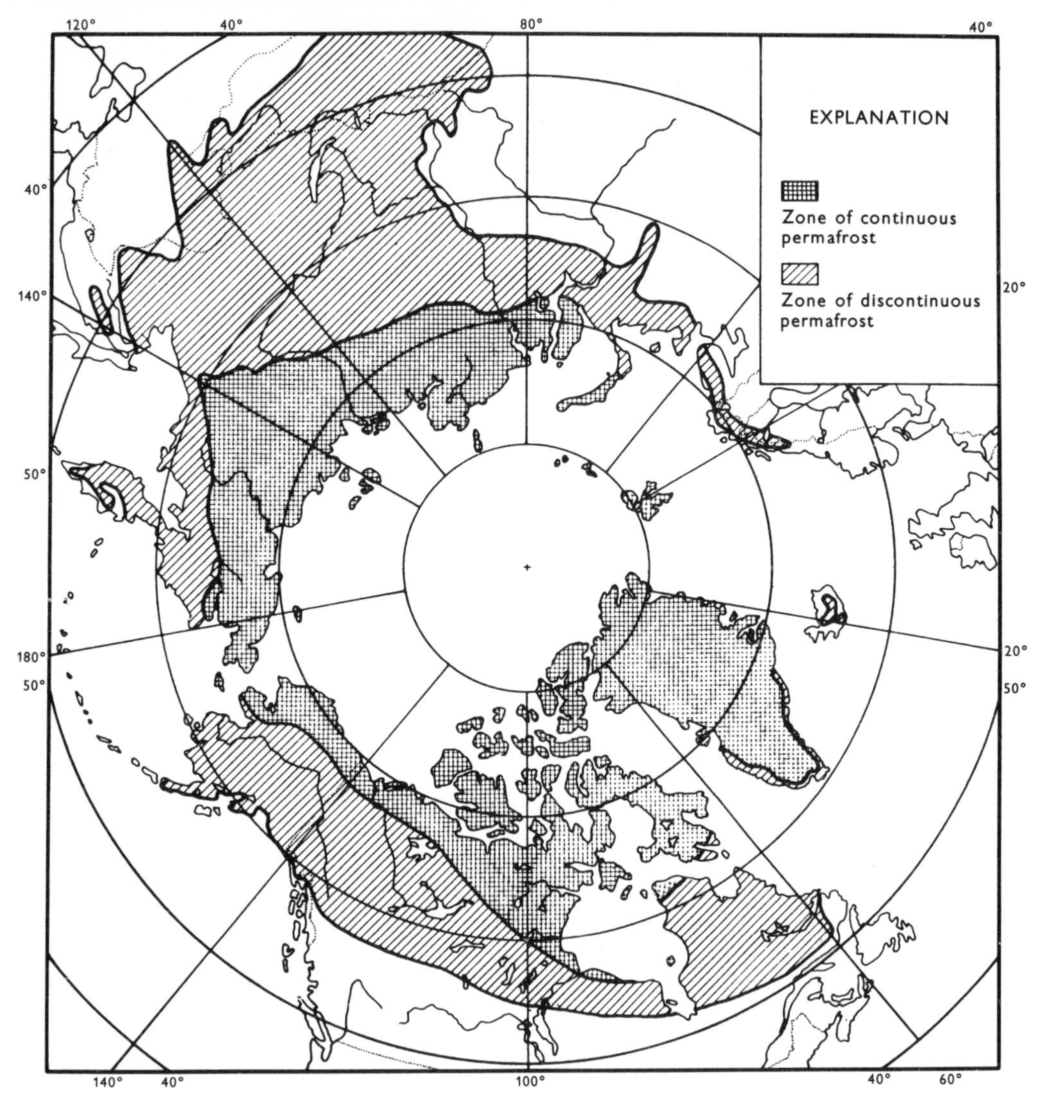

3.3 Permafrost in the Northern Hemisphere
(*Péwé, 1969, 5, Figure 1*)

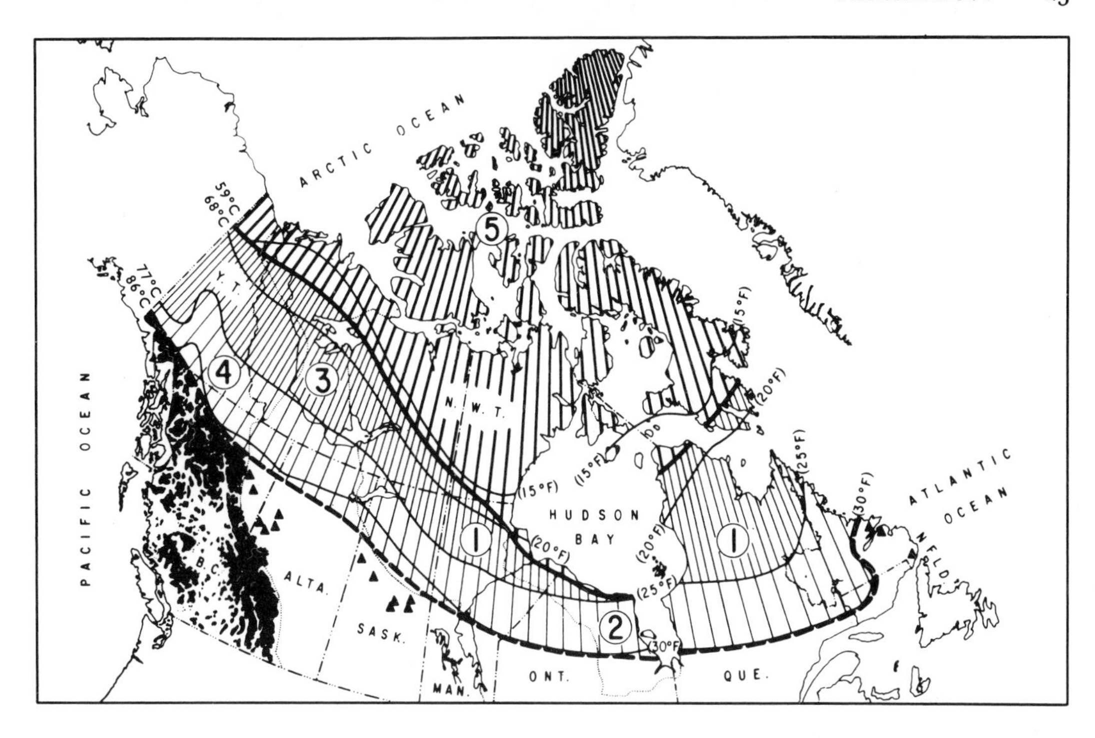

3.4 Permafrost in Canada (*R. J. E. Brown, 1970, Figure 3, opposite 8*)

boundary of continuous permafrost in the USSR (Baranov, 1959, Figure 24 opposite 201; 1964, 83, Figure 24; Akademiya Nauk SSSR, 1964, 234), but permafrost is absent in the Turukhansk region of USSR at −7° (Shvetsov, 1959, 79; 1964, 5–6; who cited Yachevskii (1889)[1] or −7·4° (Schostakowitsch, 1927, 396). R. J. E. Brown (1967*b*) has pointed out in an excellent review that there are significant differences as well as many similarities in environmental factors and the distribution of permafrost in North America and the USSR.

The distribution of permafrost can also be evaluated with respect to the temperature at the depth of zero annual amplitude – i.e. the depth to which seasonal changes of temperature extend

[1] Yachevskii (1889, 346) gave −8° (*c.*) for Turukhansk, which lies in an area of insular permafrost (Akademiya Nauk SSSR, 1964, Plate 234). Kendrew (1941, 212), referring to lack of permafrost at Turukhansk, gave a temperature of 17°F (i.e. also about −8°C).

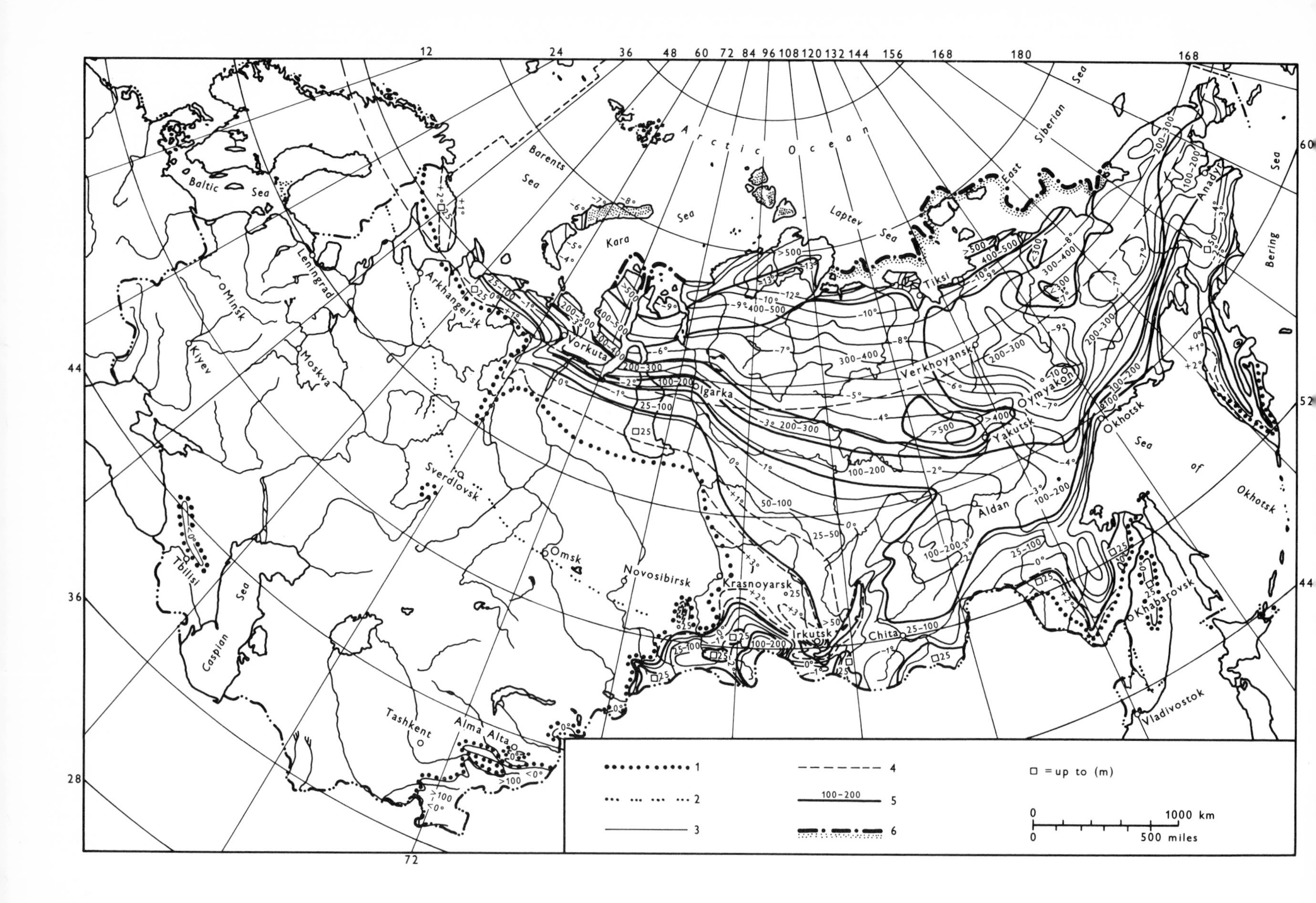
Arctic Ocean
Barents Sea
Kara Sea
Laptev Sea
East Siberian Sea
Bering Sea
Sea of Okhotsk
Baltic Sea
Caspian Sea
Leningrad
Minsk
Kiyev
Moskva
Arkhangel'sk
Vorkuta
Igarka
Tiksi
Verkhoyansk
Oymyakon
Yakutsk
Okhotsk
Anadyr
Aldan
Sverdlovsk
Omsk
Novosibirsk
Krasnoyarsk
Irkutsk
Chita
Khabarovsk
Vladivostok
Tbilisi
Tashkent
Alma Alta
1
2
3
4
5
6
100–200
□ = up to (m)
0 1000 km
0 500 miles

1 Boundary of permafrost area 2 Boundary of zone of frequent pereltoks 3 Minimum ground temperature at level of zero annual amplitude (in mountainous regions, shown for valleys) 4 Soil isotherms at 1–2 m depth under natural conditions 5 Maximum thickness of permafrost (m) 6 Permafrost zone under Arctic Ocean
Note: Temperatures in °C

3.5 (*Opposite*) Permafrost in the Soviet Union (*Baranov, 1959, Figure 24, opposite 201; cf. Baranov, 1964, 831, Figure 24*)

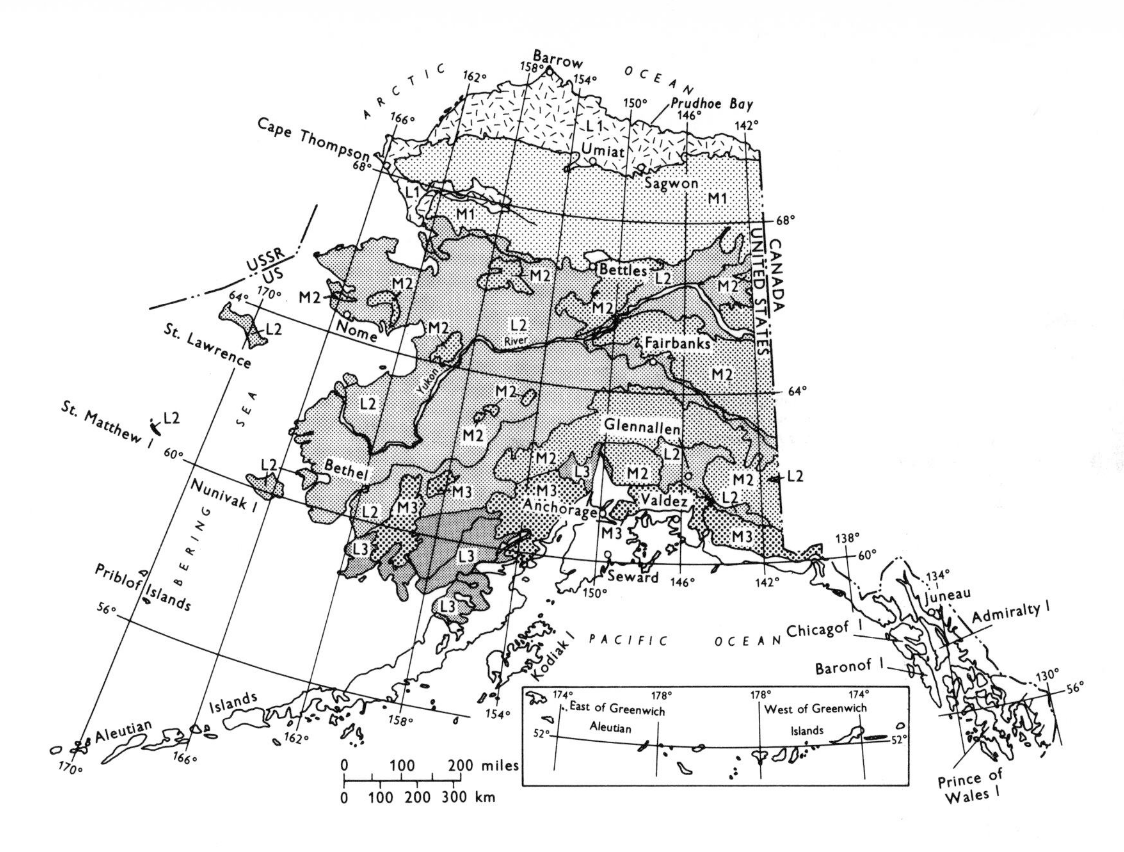

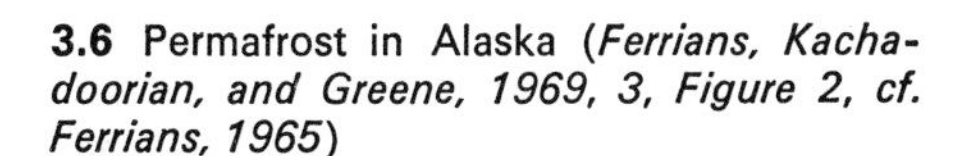

3.6 Permafrost in Alaska (*Ferrians, Kachadoorian, and Greene, 1969, 3, Figure 2, cf. Ferrians, 1965*)

AREAS WITHIN PERMAFROST REGION

Mountainous areas, generally underlain by bedrock at or near the surface

M1 Underlain by continuous permafrost

M2 Underlain by discontinuous permafrost

M3 Underlain by isolated masses of permafrost

Lowland areas, generally underlain by thick unconsolidated deposits

L1 Underlain by thick permafrost in areas of either fine-grained or coarse-grained deposits

L2 Underlain by moderately thick to thin permafrost in areas of fine-grained deposits, and by discontinuous or isolated masses of permafrost in areas of coarse-grained deposits

L3 Underlain by isolated masses of permafrost in areas of fine-grained deposits, and generally free of permafrost in areas of coarse-grained deposits

AREAS OUTSIDE PERMAFROST REGION

Generally free of permafrost, but a few small isolated masses of permafrost occur at high altitudes, and in lowland areas where ground insulation is high and ground insolation is low, especially near the border of the permafrost region

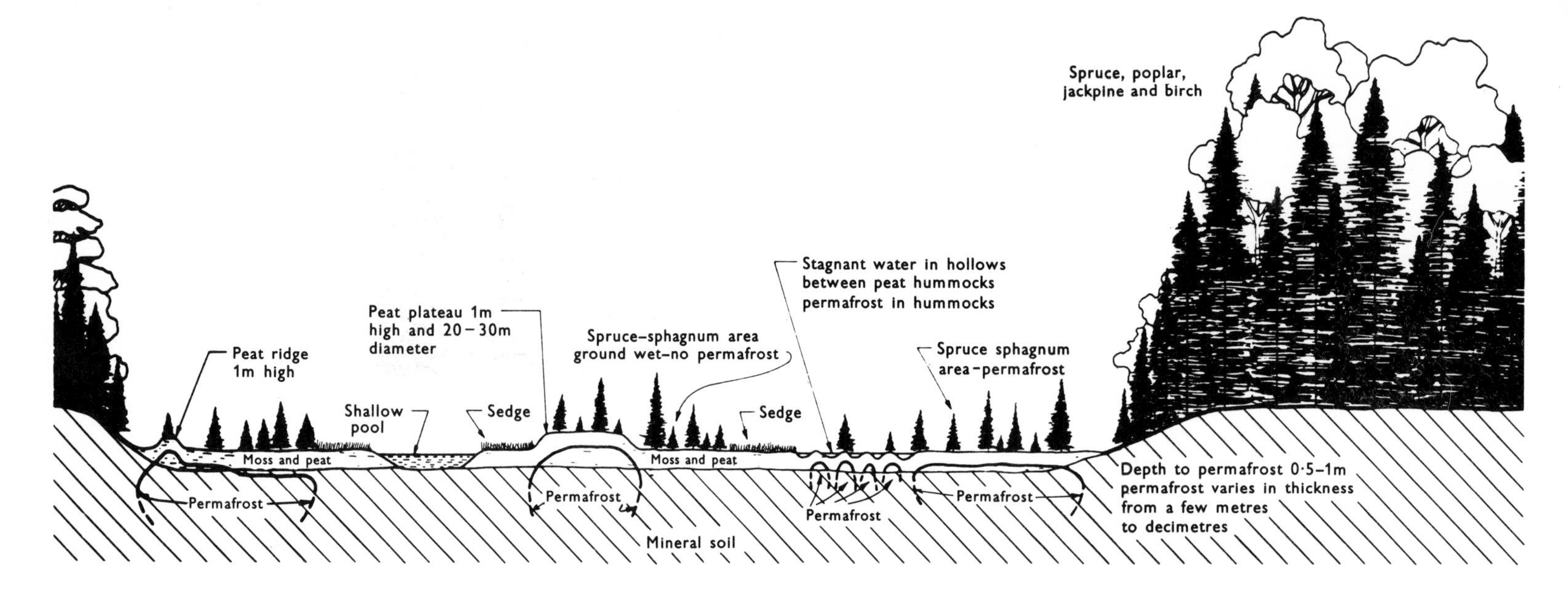

3.7 Profile through typical peatland in southern fringe of discontinuous permafrost zone in Canada, showing relation of permafrost to vegetation and topography (*R. J. E. Brown, 1968, 176, Figure 2*)

into permafrost. According to R. J. E. Brown (1967*a*; 1970, 8), the criterion of $-5°$ at this depth has been adopted in both North America and the USSR as an arbitrary boundary between continuous and discontinuous permafrost.

Permafrost is in a particularly delicate ecological balance in the Subarctic whose southern limit can also be considered as the southern limit of discontinuous permafrost (R. J. E. Brown, 1966*b*). Possibly much of the discontinuous zone is in disequilibrium with the present climate but detailed observations are needed. R. J. E. Brown (1967*a*; 1970, 21) reported that the southern limit of discontinuous permafrost in Canada coincides roughly with the $-1°$ isotherm for the mean annual air temperature. Discontinuous permafrost in Canada is in the belt of greatest peatland concentration, and the presence of permafrost here is controlled in part by the insulating qualities of the peat (Figure 3.7).

Permafrost is not confined to high latitudes but in highlands may extend far south of the limit as usually mapped. In the Canadian Cordillera where the southern limit is mapped at valley-bottom levels the lower altitudinal limit rises from about 1220 m (4000 ft) at lat. 54°30′N to an estimated 2134 m (7000 ft) at lat. 49°N (R. J. E. Brown, 1970, 12). In the United States permafrost occurs in the summit area of Mount Washington in New Hampshire (alt. 1917 m) (Goldthwait, 1969; John Howe, 1971; Thompson, 1962, 215) where the mean annual temperature is $-2{\cdot}8°$ (27°F), and it characterizes some areas above treeline in the Rocky Mountains (J. D. Ives and Fahey, 1971). Given high enough altitudes it can occur at even tropical latitudes. For instance, permafrost (possibly of Pleistocene rather than Holocene age) has been reported at a depth of about 0·5 m in the summit area of Mauna Kea, Hawaii (alt. 4206 m), where the mean annual temperature is about 3° (Woodcock, Furumoto, and Woollard, 1970; A. H. Woodcock, personal communication, 1970).

4 Depth and thickness

To date the maximum reported thickness of permafrost is 1400–50 m in the vicinity of the upper Markha River in Siberia (Grave, 1968*a*; 1968*b*, 3, diagram opposite 4). Bakker (*in* Büdel, 1960, 43) reported *c.* 1000 m in Siberia, and *c.* 600 m (2000 ft) occurs at Nordvik, Siberia (Stearns, 1966, 21, Table 1). In North America the maximum reported thickness is on the order of 610 m (2000 ft) at Prudhoe Bay, Alaska (Robert Stoveley, *in* Lachenbruch, 1970*b*, J2–J4). The maximum known in Canada is in the Arctic at Winter Harbour, Melville Island, where it is about 557 m according to observations of A. Jessop, although climatic considerations indicate the thickness may be almost twice as much in the interior of Baffin and Ellesmere Islands (R. J. E. Brown, 1972, 116–17). Table 3.2 shows approximate thicknesses as compiled by Stearns (1966, 21, Table 1). Regional variation in thickness in the USSR is shown in Figure 3.5.

5 Structure

The basic types of frozen ground in the CRREL classification (R. F. Scott, 1969, 17; Stearns, 1966, 73) are

Segregated ice not visible by naked eye: (1) poorly bonded; (2) well bonded.

Segregated ice visible by naked eye but ice 2·5 cm (1 in) or less in thickness: (1) individual ice crystals or inclusions; (2) ice coatings on particles; (3) random or irregularly oriented ice formations; (4) stratified or distinctly oriented ice formations.

Segregated ice greater than 2·5 cm (1 in) in thickness: (1) ice with soil inclusions; (2) ice without soil inclusions.

The thickness and orientation of ice segregations are especially critical aspects of the structure. The amount of ice tends to control the behaviour of permafrost upon thawing, and the amount and orientation of the segregations may indicate the origin of the ice.

Some characteristic structures of permafrost as determined by certain forms of segregated ice,

Table 3.2. Approximate depth to bottom[1] of permafrost at selected places in the Northern Hemisphere (*after Stearns, 1966, 21, Table 1*)

		M[2]	*Ft*
Alaska:	Cape Thompson	305–365	1000–1200
	Barrow	205–405	670–1330
	Cape Simpson	250–320	820–1050
	Umiat	320	1055 (deepest well)
	Kotzebue	75	238
	Fairbanks	30–120	100–400
	Nome	35	120
	Bethel	13–185	42–603
	Northway	65	207
Canada:	Aishihik	15–30	50–100
	Churchill	0–45[3]	0–140[3]
	Hay River	2	5
	Norman Wells	45–60	140–200
	Resolute	305–455	1000–1500
	Schefferville	15–60	50–200
	Uranium City	0–9	0–30
	Yellowknife	12–35	40–115
	Yukon and Klondike	60–120	200–400
Greenland:	Thule	520[4]	1700[4]
Spitsbergen:		240–305	790–1000
USSR:	Amderma	275	195 (measured)
	Kozhevnikov Bay	395	1300
	Nordvik	610	2000
	Taimyr	305–610	1000–2000
	Ust'-Port	455	1500
	Vorkuta	130	430
	Yakutsk	195–250	650–820

[1] Based on 0°
[2] Converted from feet, commonly to nearest 5 m
[3] Varies with distance from Hudson Bay and Churchill River
[4] Estimated from temperature gradient and from temperature at 306 m (1005 ft) depth

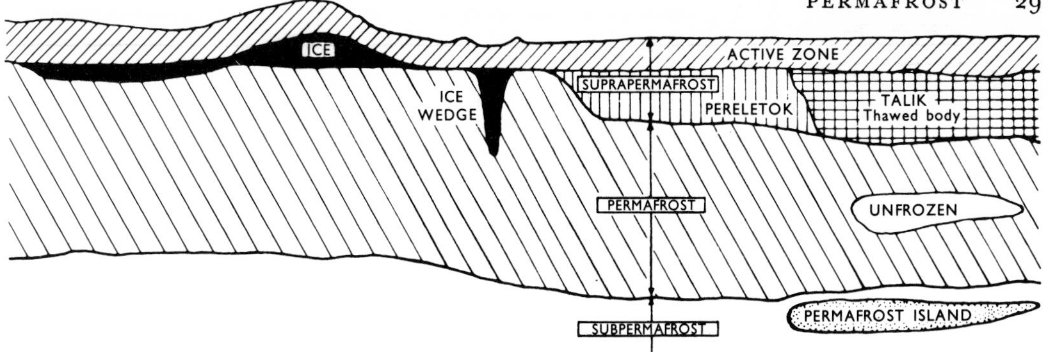

In the discontinuous zone, the permafrost breaks (both horizontally and vertically) into islands and separate masses. The unfrozen islands become more continuous

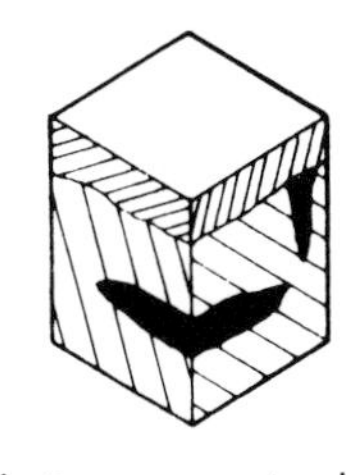

Active zone extends to continuous permafrost

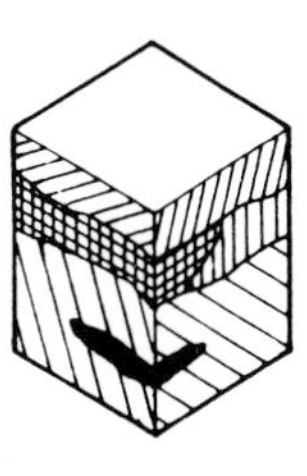

Transition zone with talik

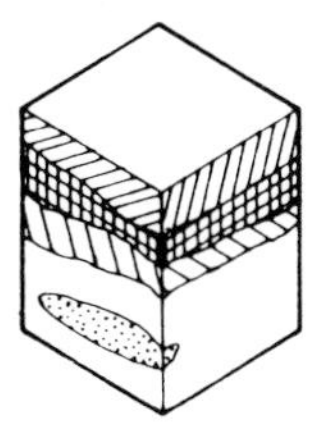

Discontinuous zone Talik, layers, island

3.8 Some structural features of permafrost (*Stearns, 1966, 17, Figure* 7)

including ice lenses and ice wedges described below, are shown in Figure 3.8, which also shows certain features described later, including thaw areas (taliks), temporarily frozen areas simulating permafrost (pereltoks), the upper limit of permafrost (permafrost table), and the overlying zone of seasonal freezing and thawing (active layer).

6 Forms of ice

a General Ground ice may take many forms. Some of the more common are Aufeis, glacier ice, pingo ice, ice lenses, ice veins, and ice wedges. A detailed classification, based on the origin of the water and on the transfer process involved, is shown in Figure 3.9.

b Aufeis Aufeis (sometimes called icings and naleds) is a sheet of freshwater ice (Figure 3.10)

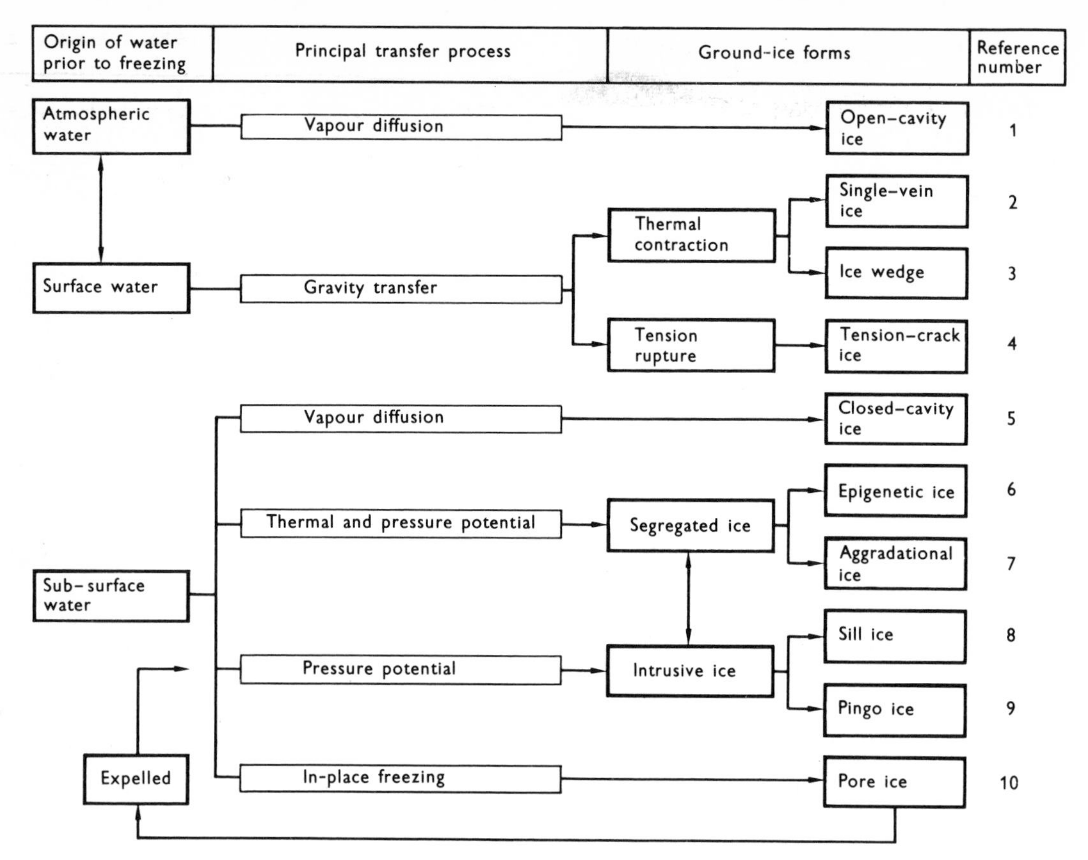

3.9 Classification of ground-ice forms (*Mackay, 1972a, 4, Figure 2. Courtesy, Assoc. American Geographers*)

usually formed by (1) a stream overflowing its banks during freeze-up, often under hydrostatic pressure due to continued flow of water beneath ice in shallow streams as freezing approaches the stream bed; or (2) ground water coming to the surface in natural seepages or springs, or in artificial cuts that intersect the groundwater table (Eager and Pryor, 1945). Development of Aufeis is favoured by underlying permafrost, and Aufeis can be preserved as permafrost by becoming buried under debris that insulates it from thawing. Where freezing is intense, Aufeis formed by seepages is most common at the toe of south-facing slopes (Kachurin, 1959, 396; 1964, 27).

According to Ferrians, Kachadoorian, and Greene (1969, 34–6), Aufeis accumulations are

3.10 (*Opposite*) Aufeis, Anaktuvuk Pass area, Alaska. Note smooth stone pavement between rougher stream bed and eroded Aufeis. Photo by S. C. Porter

seldom more than several hectares (a 'few acres') in area but are reported to have areas as large as 2185 hectares (5400 acres) and to be as much as 9·1 m thick. Mayo (1970, 310) reported flood-plain Aufeis in Arctic Alaska up to 4 m thick, 5 km wide, and as much as 50 km long. Shumskii (1964*b*, 194) cited accumulations up to 27 km long, 10 m thick, and containing up to 5×10^8 m³ of ice. Aufeis tends to be layered parallel to the surface of deposition (Shumskii, 1964*b*, 192–5), which provides a useful criterion of origin. Also, Aufeis (overflow ice) is reported to have a higher average extinction coefficient (0·43 cm^{-1}) than normal (clear) river ice (0·092 cm^{-1}) (Wendler, 1970).

c Glacier ice Like Aufeis, glacier ice may become buried and preserved as permafrost. The structure of glacier ice is rarely horizontal over large distances; rather it tends to be folded and faulted.

d Pingo ice Pingo ice, like buried Aufeis and glacier ice, is commonly a massive body of freshwater ice. It is usually formed by (1) progressive, all-sided freezing of a water body or water-rich sediments (closed system – Mackenzie type) or (2) injection of groundwater under artesian pressure into permafrost in the manner of a laccolith (open system – East Greenland type). In both cases, ice forms the core of a mound and tends to assume a planoconvex shape. Pingos are described in the chapter on Frost action.

The structure of pingo ice is not well known. Drilling of a pingo in the Mackenzie Delta area of northern Canada revealed ice with no apparent structure; the ice, some 9 m (30 ft) thick, was milky from small air bubbles (Pihlainen, Brown, and Legget, 1956, 1122). An exposure of pingo ice in the same area showed many crystals with diameters of 2·5–5·0 cm (1–2 in), the maximum crystal size noted being 20 cm (8 in); the optic axes had a preferred orientation towards the centre of the pingo's ice core (Mackay and Stager, 1966, 367).

e Ice lenses An ice lens is a dominantly horizontal layer of ice. It may be less than a millimetre to tens of metres thick and a few millimetres to hundreds of metres in extent (Figures 3.11, 3.12). The term is purely descriptive but many ice lenses form *in situ* by segregation of ice during the freezing process. The structure of an ice lens of segregation origin is dominantly horizontal, with any bubbles tending to be elongated and aligned normal to the horizontal layering. Soil fragments broken by freezing show a vertical separation (Mackay, 1971*b*, 411).

f Ice veins An ice vein, as the term is used by the present author, is a tabular, wedgelike, or irregular body of ice, hair thin to 5 cm (2 in) thick, whose longest dimensions lie in a dominantly vertical plane and may be several metres or more long. The term is purely descriptive. This usage follows that of T. L. Péwé (personal communication, 1970) in stressing the thinness of veins

3.11 Small ice lenses, Mamontova Gora, left bank Aldan River, 310 km above junction with Lena River, Yakutia, USSR. The lenses are in clayey silt and are mostly less than 1 cm thick

as opposed to ice wedges, discussed below. However, these terms tend to be used interchangeably in translations from the European literature.

g Ice wedges An ice wedge is 'a narrow crack or fissure of the ground filled with ice which may extend below the permafrost table' (Muller, 1947, 218). As defined, the term is purely descriptive, and this usage is followed here, but to many workers the term has come to imply an origin by frost cracking, discussed later. Ice wedges start as ice veins.

Ice wedges tend to be V-shaped (Figure 3.13) and to have a characteristic, dominantly vertical structure imparted by dirt and air bubbles oriented parallel to the wedge edges (Figure 3.14) as a result of the frost-cracking process. The air in the ice consists mainly of inert and biochemical gases almost devoid of oxygen because of oxidation of plant remains. The ice crystals tend to be columnar with C axis vertical in the upper part of a wedge and horizontal at depth, but a cataclastic texture may develop as an ice wedge grows (Shumskii, 1964*b*, 196–205). The shape and structure provide unambiguous criteria for well-developed ice wedges, and the ice texture and the composition of the included gases may assist in dubious cases.

7 Thermal regime of permafrost

a General The thermal regime of permafrost (Figure 3.15) is dependent on the quantity of heat affecting the permafrost and the overlying layer that freezes and thaws annually – the active layer, described later. The quantity of heat can be expressed as

$$Q_h = A k_h i_g$$

where Q_h = the quantity of heat flowing through an area at right angles to direction of flow per unit of time (cal/sec), A = area (cm^2), k_h = thermal conductivity (cal/cm sec °C), and i_g = geothermal gradient (°C/cm). The ultimate sources of heat are the earth's interior and the sun, and the geothermal gradient expresses their combined effect. These ultimate heat sources are quite stable but the distribution of heat is affected by so many variables that thermal conditions vary widely.

It is convenient to discuss first the relatively stable part of the permafrost, affected by mean annual air temperatures and long-term temperature trends, before discussing the less stable part above the zone of zero annual amplitude, where seasonal temperature changes dominate the thermal regime. In many respects it is the regime of this upper part that is the most critical for frost-action processes.

b Geothermal heat flow The geothermal heat flow inhibits freezing at the base of stable permafrost, and thaws the base of degrading permafrost. To illustrate (Terzaghi, 1952, 30–1): The rate of geothermal heat flow (q_i) varies somewhat from place to place but approximates 40 cal/cm^2 yr.

3.12 (*Opposite*) Large ice lens, right bank Lena River, 90 km north of Yakutsk. Lenses here are in bedded sand and up to 3 m thick

3.13 (*Far opposite*) Ice wedge near Brakes Bottom, Seward Peninsula, Alaska

3.14 (*Near opposite*) Ice wedge at Shamanskiy Bereg, left bank Lena River, 120 km north of Yakutsk, Yakutia, USSR. Scale given by tape numbered in decimetres

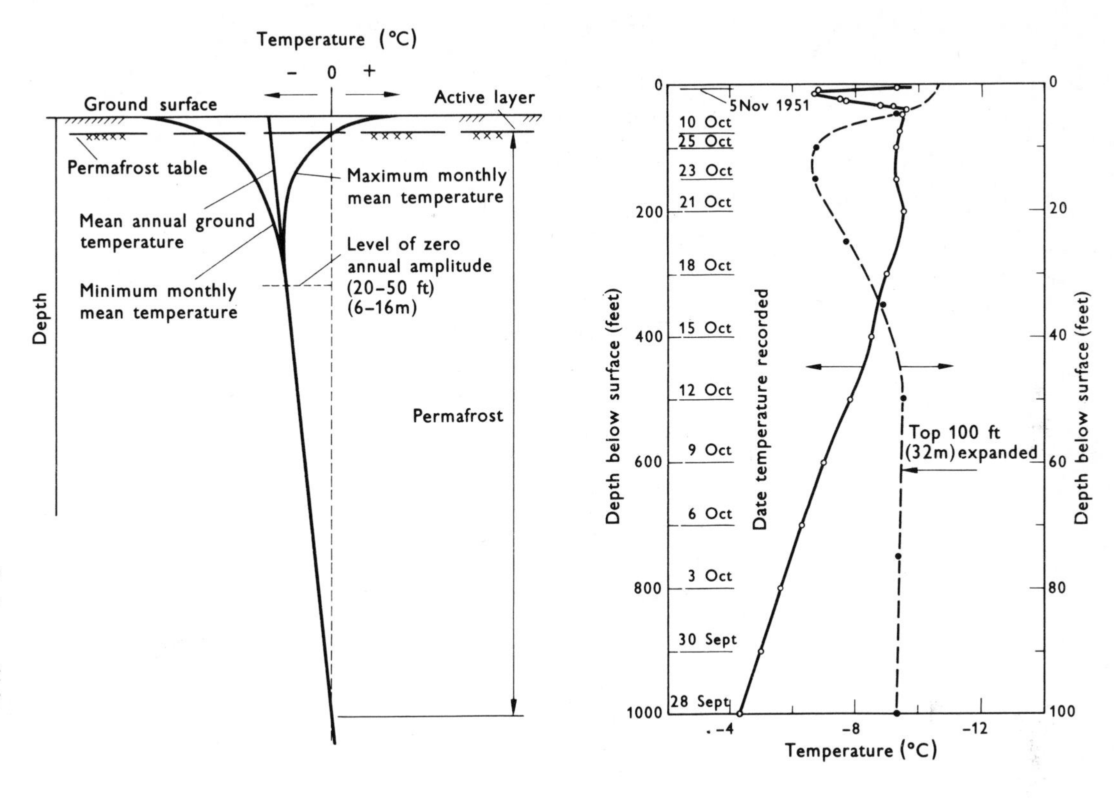

3.15 (*Near right*) Typical temperature regime in permafrost (*R. J. E. Brown, 1970, 11, Figure 6*)

3.16 (*Far right*) Ground temperatures at Thule, Greenland, September–November 1951 (*Stearns, 1966*, 34, *Figure 25*)

Considering the heat of fusion of ice (80 cal/gm or *c.* 70 cal/cm³), and assuming an ice content of 30 per cent, the heat (q_f) required to melt frozen ground at 0° would be $q_f = 0{\cdot}3 \times 70$ cal/cm³ = 21 cal/cm³. The rate of basal thawing (q_i/q_f) would be

$$q_i/q_f = \frac{40 \text{ cal}}{\text{cm}^2 \text{ yr}} \bigg/ \frac{21 \text{ cal}}{\text{cm}^3} = 2 \text{ cm/yr}$$

On this basis

$$t = \frac{H \text{ (cm)}}{2 \text{ (cm/yr)}}$$

where t = time required for thawing from below, and H = thickness of permafrost. If H = 200 m, t = 10,000 years. According to comparative calculations (Terzaghi, 1952, 31), a sudden rise in mean annual air temperature of 2° from −1° to 1° would result in less than 100 years in thawing 15 m of permafrost from the surface down while thawing from the base up would amount to only about 2 m.

c Geothermal gradient Where the geothermal heat flow is constant, the geothermal gradient is inversely proportional to conductivity. Thus if the mean surface temperature at two places is the same, permafrost extends deeper where conductivity is higher. The effect is startlingly illustrated by Robert Stoveley's observation that permafrost at Prudhoe Bay, Alaska, is 50 per cent thicker than at Barrow and 100 per cent thicker than at Cape Simpson, despite similarity of mean surface temperatures; the increasing thicknesses was ascribed to increasingly higher proportions of silicious sediments and consequently higher conductivity as between Cape Simpson, Barrow, and Prudhoe Bay (Lachenbruch, 1970*b*, J1).

Geothermal gradients in permafrost range from about 1°/22 m (1°F/40 ft) to 1°/60 m (1°F/110 ft) (Stearns, 1966, 33, Table v). The gradient at Thule, Greenland, which is about 1°/54 m (1°F/100 ft) is illustrated in Figure 3.16.

The term steep gradient commonly refers to a high °C/m ratio but gradients are also reported in m/°C, and depending on the axes chosen in plotting temperature, a high gradient may have a gentle slope in a graph.

Given sufficient time, the thickness of permafrost is related to temperature and geothermal gradient by the formula (Terzaghi, 1952, 27)

$$H_p = \frac{T}{-i_g}$$

where H_p = thickness (m), T = mean annual temperature (°C), and i_g = geothermal gradient (°C/m).

The geothermal gradient can provide proof of climatic change and indicate its amount rather precisely. For instance, Lachenbruch and Marshall (1969) described temperature profiles from three boreholes along the arctic coast of Alaska that show similar pronounced curvatures (Figure 3.17). The curvature is due to a climatic warming following establishment of thermal equilibrium represented by the linear gradient and its extrapolation to the surface. The extrapolated parts of the Barrow and Cape Simpson curves indicate a former mean annual temperature of −12·1°. The change to the present mean annual temperature (*c.* −9°) is about +3°. Because repeated observations at the Barrow borehole provided information on the rate of temperature change as a function of depth, Lachenbruch and Marshall were able to calculate that the mean annual *surface* temperature at Barrow must have risen about 4° since the middle of the nineteenth

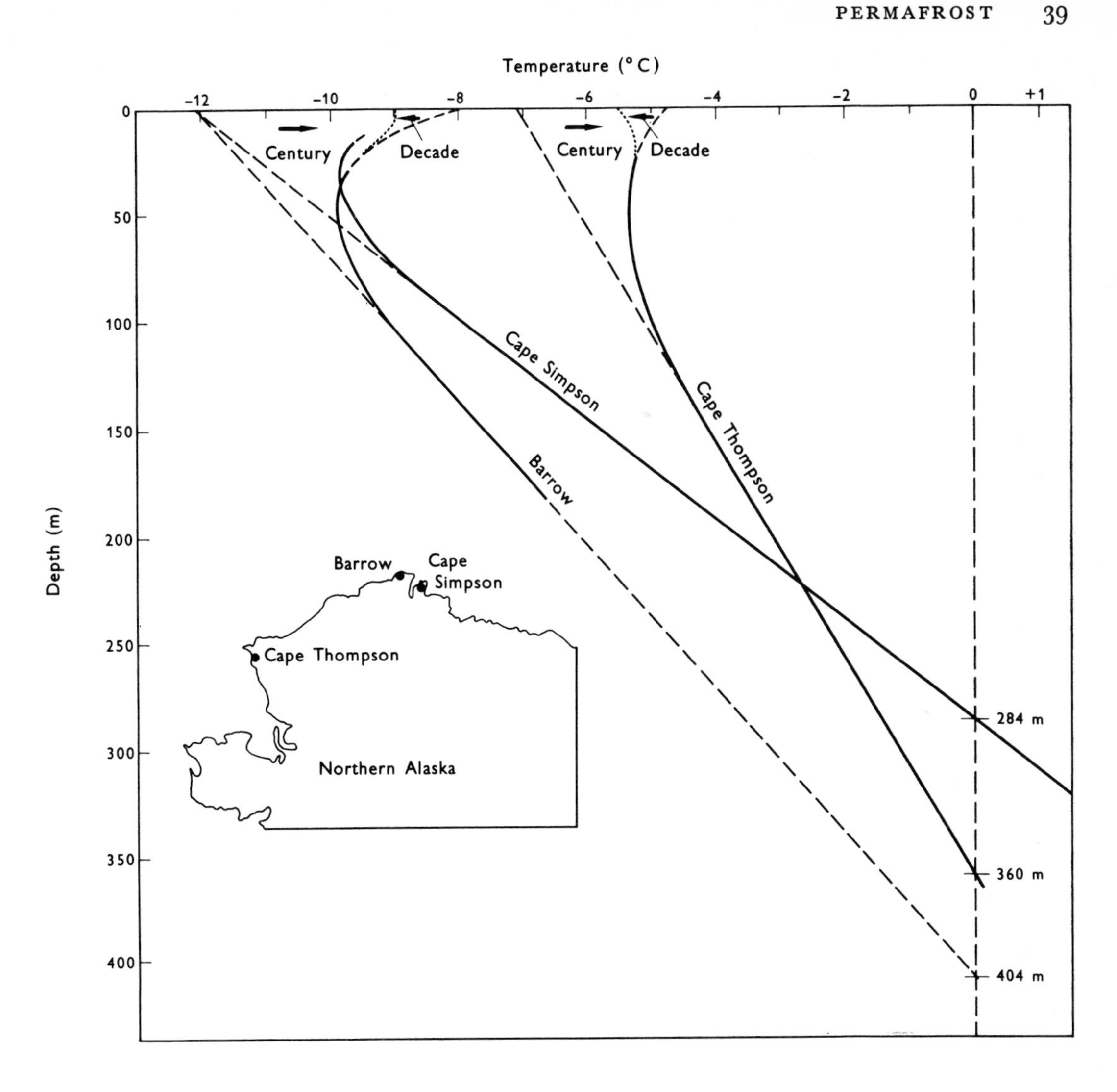

3.17 Geothermal gradients in three boreholes in Arctic Alaska. Extrapolations are shown as broken lines (*Lachenbruch and Marshall, 1969, 302, Figure 2*)

century. They concluded that the one degree difference from the 3° change with respect to the present surface temperature indicated a cooling within the last decade or so. The geothermal gradient at Cape Thompson revealed a similar history of climatic change.

d Depth of zero annual amplitude The depth of zero annual amplitude – the depth to which seasonal changes of temperature extend into permafrost – increases in direct proportion according to the equation (Terzaghi, 1952, 22)

$$z_1 = \sqrt{12\, a t_1}$$

where z_1 = depth, a = diffusivity, and t_1 = time. Diffusivity (cf. Terzaghi, 1952, 11) is defined as

$$a = \frac{k_h}{C_h \times W_t}$$

where a = diffusivity (cm²/sec), k_h = thermal conductivity (cal/cm sec °C), C_h = heat capacity (cal/gm°C), and W_t = unit weight (gm/cm³).

Temperatures at the depth of zero annual amplitude can be calculated according to the equation (cf. Grave, 1967)

$$t_{TA} = t_b + \frac{A_{cm}}{2}\left(1 - \frac{1}{f}\right)$$

where t_{TA} = temperature (°C) at depth of zero annual amplitude ('bottom of thermoactive layer'), t_b = mean annual air temperature, A_{cm} = annual amplitude of mean monthly air temperatures, and

$$f = e^{+z}\sqrt{\frac{\pi}{kT}}$$

where z = thickness of snow cover, k = thermal conductivity of snow, and T = period of oscillations equal to one year.

The temperature at the depth of zero annual amplitude is reported to be generally about 3° (6°F) warmer than the mean annual air temperature (R. J. E. Brown, 1967*a*; 1970, 20). However, there is considerable regional variation (1·5°–9·3°) in figures from various parts of the world as compiled by Grave (1967, 1341, Table 1).

The convergence of minimum and maximum temperatures to the level of zero annual amplitude is illustrated in Figure 3.18.

Above the depth of zero annual amplitude, the permafrost is subjected to seasonal fluctuations of temperature. However, a mean surface temperature can be deduced by upward extrapolation of the geothermal gradient from the level of zero annual amplitude, provided the measured gradient has achieved equilibrium and there are no recent climatic changes. As previously noted, it is this upper part of the permafrost and the overlying active layer that are especially critical for frost-action processes and effects.

8 Aggradation and degradation of permafrost

a General There are numerous ways in which permafrost can build up or aggrade, and thaw or degrade, all of which are controlled by the thermal regime. Changes that can lead to aggradation or degradation are climatic, geomorphic, and vegetational, and the ways in which they can affect permafrost are shown in Figures 3.19–3.20, following Mackay (1971*a*).

b Syngenetic and epigenetic permafrost Aggradational permafrost may be syngenetic or epigenetic depending on whether concurrent sedimentation is present or absent (Baranov and Kudryavtsev, 1966, 100–101). Thus, except for the ice, the rocks and soil of epigenetic permafrost are older

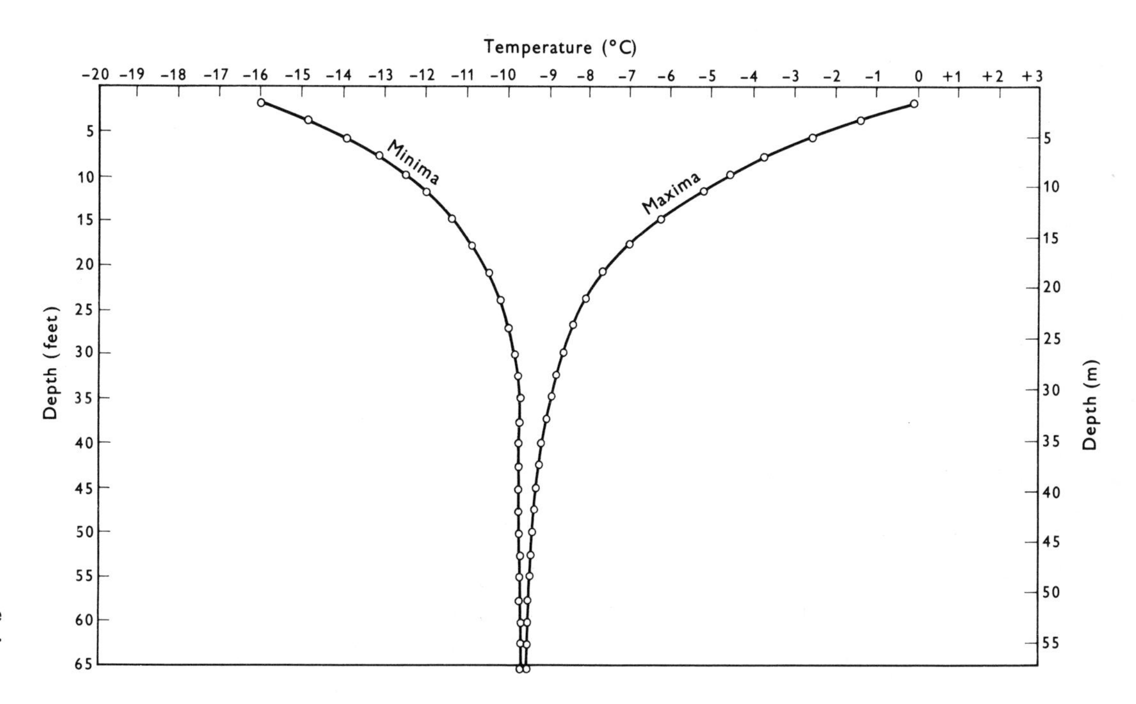

3.18 Amplitude of seasonal temperature changes at Barrow, Alaska, 1949–50 (*MacCarthy, 1952, 591, Figure 3*)

than the freezing, whereas those of syngenetic permafrost must be about the same age as the freezing. Epigenetic permafrost is characterized by a concentration of well-developed segregated ice in the upper horizons because of the water that is drawn upward from considerable depth during freezing, as described under Freezing process in the chapter on Frost action. In syngenetic permafrost the concurrent sedimentation and freezing favour accretion of thin layers without the possibility of significant moisture migration from depth.

c Thermokarst Thermokarst comprises 'karst-like topographic features produced by the melting of ground-ice and the subsequent settling or caving of the ground' (Muller, 1947, 223). As such, thermokarst records degradation of permafrost. The disturbance of the thermal regime may be local such as disruption of insulating vegetation, or more general such as climatic change. There are many cases of well-developed thermokarst features, due to local influences only, in the continuous permafrost zone. Thermokarst resulting from climatic change is more likely to be associated with the discontinuous zone where the thermal balance of permafrost is more delicate; however, by the same token, local disturbances in this zone are also more likely to cause thawing of permafrost. As discussed later the origin of thermokarst must be interpreted with care.

3.19 (*Above opposite*) Permafrost aggradation (*Mackay, 1971a, 29, Figure 1*)

3.20 (*Below opposite*) Permafrost degradation (*Mackay, 1971a, 30, Figure 2*)

d Taliks A talik is 'a layer of unfrozen ground between the seasonal frozen ground (active layer) and the permafrost. Also applies to an unfrozen layer within the permafrost as well as to the unfrozen ground beneath the permafrost' (Muller, 1947, 223).[1]

Soviet investigators recognize three kinds of talik but contrary to Muller they do not commonly designate the ground below the base of permafrost as talik. The three kinds are open (skvoznoy), closed (zamknuty), and interpermafrost (mezhmerzlotny) talik (Figure 3.21). Taliks within permafrost, even in the continuous zone, are a common feature of the thermal regime and may or may not indicate degradation of permafrost. Some are due to local heat sources, including ground-water circulation, and may be associated with quasi-equilibrium conditions. Other taliks result from a marked change of thermal regime and thus indicate permafrost degradation, in which case thermokarst is likely to be present.

9 Thermal regime of active layer

a General The active layer is the 'layer of ground above the permafrost which thaws in the summer and freezes again in the winter' (Muller, 1947, 213). Association with permafrost is inherent in this definition but some investigators (cf. Péwé, Church, and Andresen, 1969, 7) also apply the term to the layer of ground that freezes and thaws in a non-permafrost environment

[1] This definition is subject to the criticism cited on p. 44 (footnote).

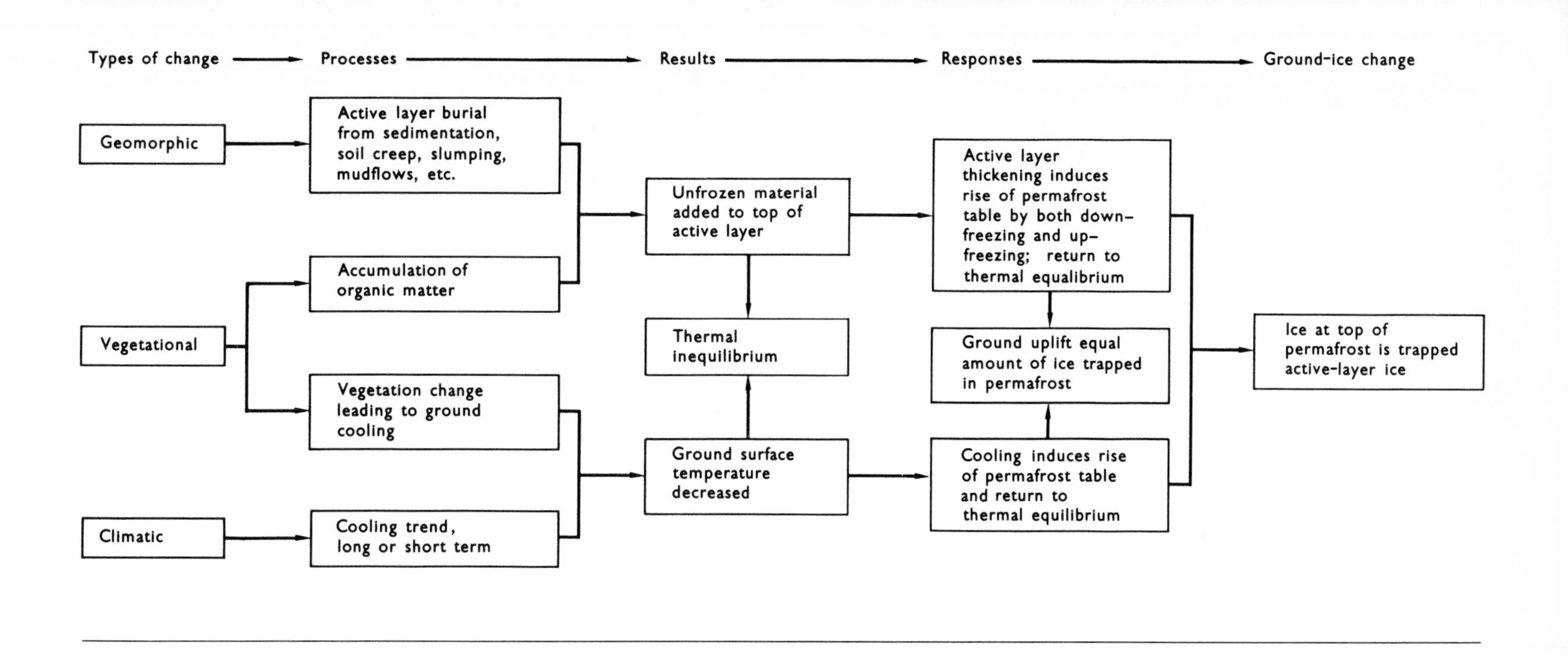
Types of change
Processes
Results
Responses
Ground-ice change
Geomorphic
Vegetational
Climatic
Active layer burial from sedimentation, soil creep, slumping, mudflows, etc.
Accumulation of organic matter
Vegetation change leading to ground cooling
Cooling trend, long or short term
Unfrozen material added to top of active layer
Thermal inequilibrium
Ground surface temperature decreased
Active layer thickening induces rise of permafrost table by both down-freezing and up-freezing; return to thermal equalibrium
Ground uplift equal amount of ice trapped in permafrost
Cooling induces rise of permafrost table and return to thermal equilibrium
Ice at top of permafrost is trapped active-layer ice

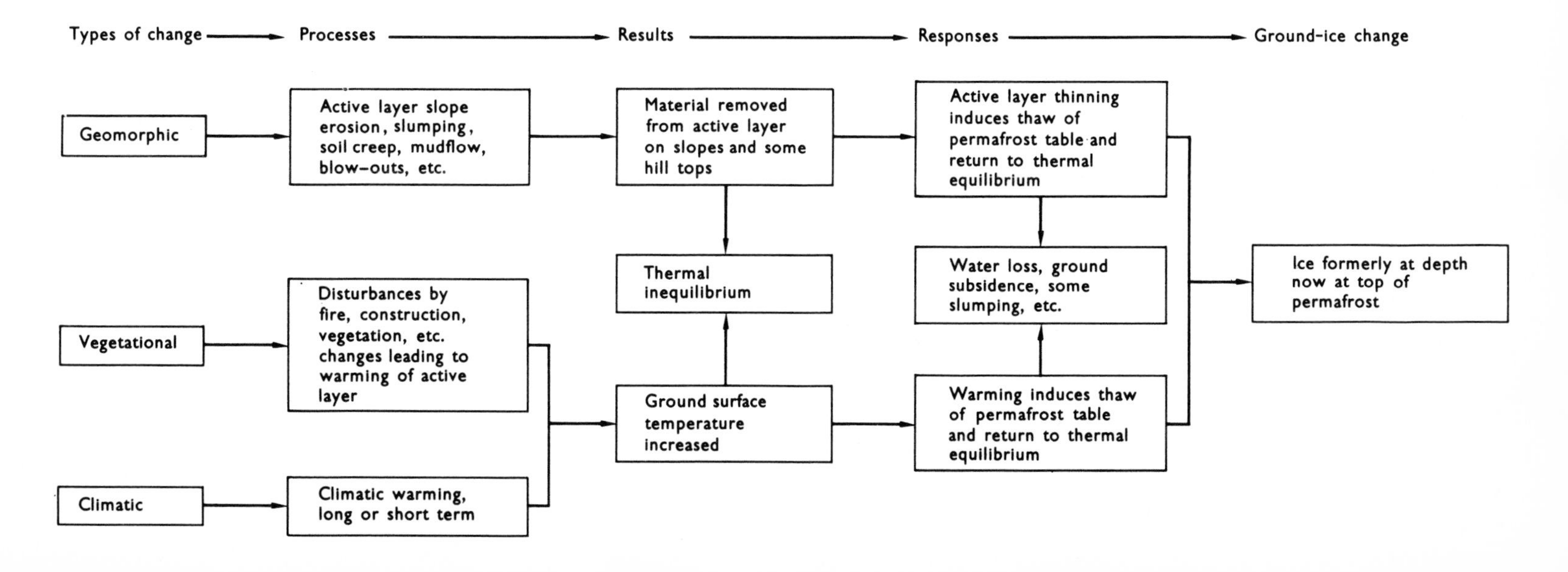
Types of change
Processes
Results
Responses
Ground-ice change
Geomorphic
Vegetational
Climatic
Active layer slope erosion, slumping, soil creep, mudflow, blow-outs, etc.
Disturbances by fire, construction, vegetation, etc. changes leading to warming of active layer
Climatic warming, long or short term
Material removed from active layer on slopes and some hill tops
Thermal inequilibrium
Ground surface temperature increased
Active layer thinning induces thaw of permafrost table and return to thermal equilibrium
Water loss, ground subsidence, some slumping, etc.
Warming induces thaw of permafrost table and return to thermal equilibrium
Ice formerly at depth now at top of permafrost

and a few others have used it in a still different sense (cf. Corte, 1969*a*, 130). However, the term is widely used as originally defined, and the present writer follows this usage.

The thermal regime of the active layer is controlled by the same environmental factors that control the development of frozen ground as discussed earlier. They include the basic factors of climate, topography, material, and time, and the dependent factors of snow cover, moisture, and vegetation. These interact in a complex fashion and determine such regime aspects of the active layer as its thickness, depth to the underlying permafrost table, upward freezing from the permafrost table, the zero-curtain effect, and the structure of the active layer as discussed below.

Mathematical models of thermal regimes, based on quantitative parameters representative of the controlling factors, are becoming increasingly sophisticated. Simulations derived from a model, utilizing values for a test site at Barrow, Alaska, have shown a remarkable similarity with the observed regime and help to evaluate the relative importance of selected factors (Nakano and Brown, 1972).

b Permafrost table The permafrost table is the 'more or less irregular surface which represents the upper limit of permafrost' (Muller, 1947, 219). As such it is the dividing surface between the permafrost and the active layer.

Any frozen surface in the active layer as it thaws downward towards the permafrost table is called a frost table and must not be confused with permafrost or the permafrost table. The term frost table is also applicable to a frozen surface in seasonally frozen ground in a non-permafrost environment.[1]

[1] Frost table is synonymous with tjaele, a frequently used term in the European literature. However, as pointed out by Bryan (1951), tjaele has also been erroneously applied to permafrost.

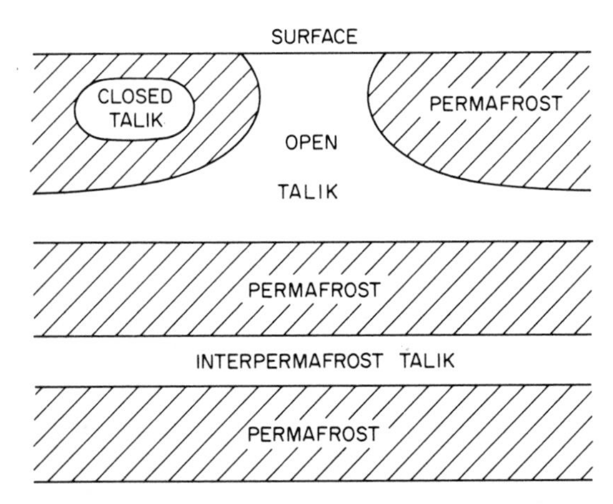

3.21 Diagram of open, closed, and interpermafrost talik (*courtesy, Jaromír Demek, 1969*)

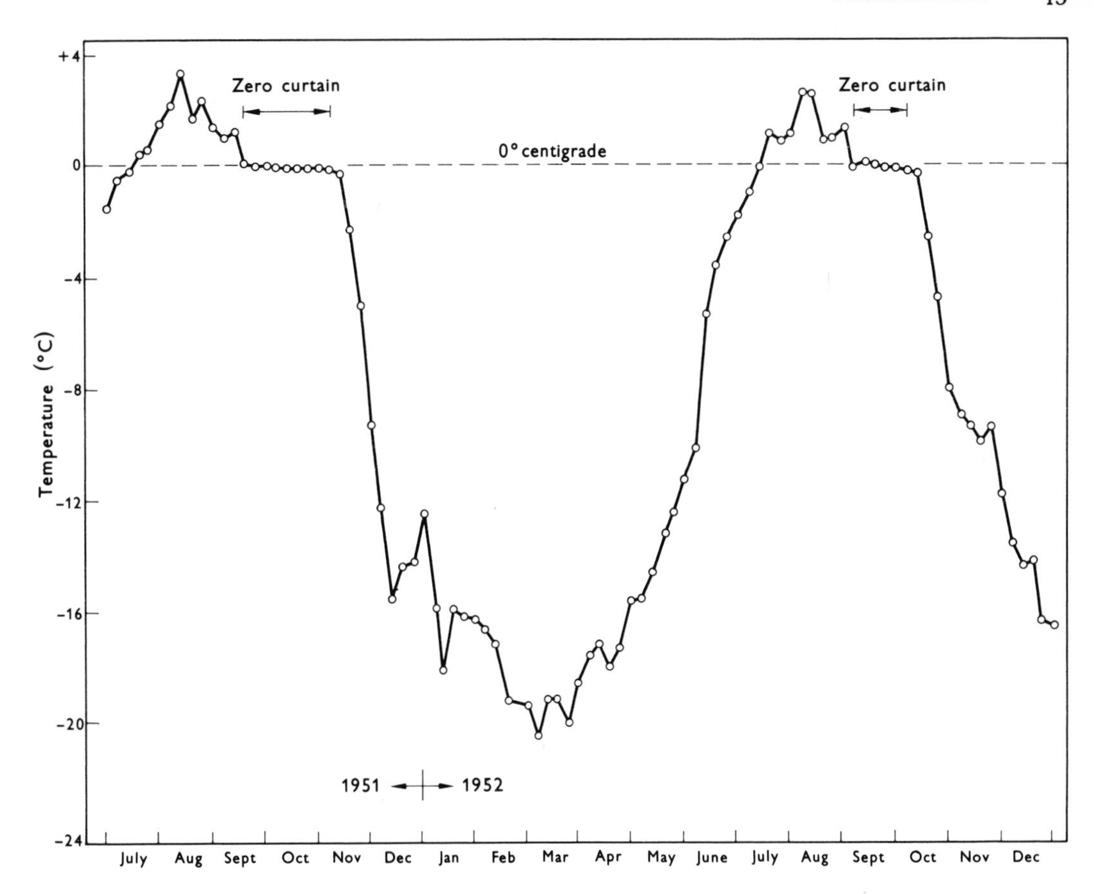

3.22 Time-temperature curve at a depth of 25 cm (10 in) near Barrow, Alaska (*Brewer, 1958, 22, Figure 4*)

Other conditions remaining constant, the thermal regime controls the depth to the permafrost table which in the zone of continuous permafrost usually coincides with the thickness of the active layer. However, the position of the permafrost table represents an average condition, and because of short-term fluctuations of regime, such as an unusually cold winter or warm summer, the permafrost table need not coincide with the maximum depth of thaw in a given year. Consequently there may be short-term anomalies that can cause a temporary talik (cf. p. 42) at the

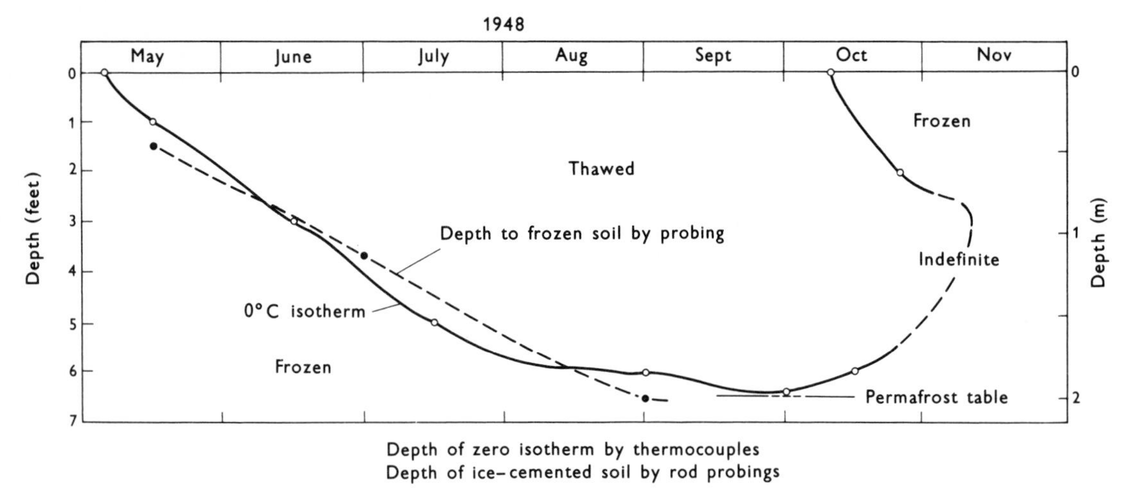

3.23 Ground temperatures at Fairbanks, Alaska. Note upward freezing from permafrost table in October (*Stearns, 1966, 24, Figure 13*)

average position, or a pereltok, defined as 'a frozen layer at the base of the active layer which remains unthawed for one or two summers (Russian term meaning "survives over the summer"). Pereltok may easily be mistaken for permafrost' (Muller, 1947, 219).

The permafrost table is a crucial boundary in several ways. It is commonly a rigid surface capable of bearing considerable loads without deforming. It also prevents moisture from seeping downward and it thereby favours a high moisture environment and the formation of ice lenses.

c Zero curtain The zero curtain can be defined as the zone immediately above the permafrost table 'where zero temperature (0°C) lasts a considerable period of time (as long as 115 days a year) during freezing and thawing of overlying ground' (Muller, 1947, 224).[1]

The zero curtain is caused by the latent heat of fusion of ice (80 cal/gm water), which delays freezing and thawing. The higher the moisture content at the permafrost table, the greater the delay. The zero curtain (Figure 3.22) plays a major role in some frost-action processes and is a notable aspect of the thermal regime of the active layer (cf. Nakano and Brown, 1972, 31–6).

[1] In this definition, the phrase 'zone immediately above the permafrost table' has been substituted for Muller's (1947, 224) phrase 'layer of ground between active layer and permafrost', because such a layer is inconsistent with Muller's definitions of permafrost table and active layer. The same caveat can be applied to his definition of talik (cf. p. 42).

d Upward freezing Because of the thermal regime of permafrost, freezing of the active layer may proceed upward from the permafrost table as well as downward from the ground surface (Drew *et al.*, 1958; cf. Meinardus, 1930, 40; Schmertmann and Taylor, 1965, 50; US Army Corps of Engineers (cf. Washburn, 1956*b*, footnote 19, 842). Figure 3.23 illustrates upward freezing of the frost table at Fairbanks, Alaska.

Upward freezing of a frost table in a non-permafrost environment is unlikely. Details as to the prevalence and significance of the process in a permafrost environment are meagre. However, it is probably an important factor in some frost-action processes as discussed later.

e Effect of thermal regime on structure The thermal regime contributes to the structure of the active layer. For one thing, there are generally smaller ice masses in the active layer than in permafrost. Also the active layer tends to have a threefold structure consisting of (1) an upper zone with small ice inclusions; (2) a middle zone that is desiccated because of withdrawal of moisture towards freezing fronts, one moving down from the surface, the other up from the permafrost table (upward freezing); and (3) a lower zone with a mixture of thin and thicker ice lenses immediately above the permafrost table (Khesthova *et al.*, 1961, 45; 1969, 5–6).

10 Origin of permafrost

Most permafrost may have originated during the Pleistocene. The evidence for this is that: (1) Tissues of woolly mammoths (*Mamuthus primigenius*) and other Pleistocene animals have been preserved in permafrost, indicating presence of permafrost at time of death (Gerasimov and Markov, 1968, 11), and there is no reliable evidence that the woolly mammoth survived into the Holocene (Farrand, 1961), although it appears to have lived as late as the Alleröd interstadial in northern Siberia (Heintz and Garutt, 1965, 76–7). (2) The upper boundary of some permafrost is considerably deeper than the present depth of winter freezing (Gerasimov and Markov, 1968, 12). This evidence merely indicates that some permafrost is not of present-day origin. (3) In places the temperature of permafrost decreases with depth, indicating residual cold (Gerasimov and Markov, 1968, 12). Again, this evidence does not necessarily prove that the residual cold records Pleistocene permafrost. (4) the thickest permafrost is commonly in areas that remained unglaciated, and therefore not insulated by ice, during the Pleistocene (Gerasimov and Markov, 1968, 14). (5) In the Soviet Union in the area covered by the early Pleistocene Kara Sea transgression, there are two layers of permafrost, the lower one dating from before the transgression (Grave, 1968*a*, 52–3; 1968*b*, i–ii, 6–8). (6) Some permafrost in the Mackenzie Delta area of northern Canada is believed to be of early Wisconsin age or older, since it is glacially deformed but occurs where there is no evidence of glaciation during the last 40,000 years (Mackay, Rampton, and Fyles, 1972).

In formerly glaciated areas, much permafrost is post-glacial. The argument for this is that: (1) Permafrost would have thawed beneath the Pleistocene ice sheets (Büdel, 1959, 305). The argument is supported by the finding of water beneath the Antarctic Ice Sheet (Gow, Ueda, and Garfield, 1968) but it need not apply to all parts of an ice sheet (Grave, 1968*a*, 51–2; 1968*b*, 6). (2) In polar regions permafrost is forming today in many areas where retreat of glaciers or recent emergence has exposed unfrozen material.

The conclusion is that (1) the upper part of most continuous permafrost is in balance with the present climate but the base may be slowly thawing, stable, or aggrading, depending on the present climate and the geothermal heat flow, (2) most discontinuous permafrost, both at its surface and base, is either out of balance with the present climate or in such delicate equilibrium that the slightest climatic or surface change will have drastic disequilibrium effects.

4 Frost action

I INTRODUCTION

Frost action, in the *Glossary of geology and related sciences* (Howell, 1960), is defined as 'The weathering process caused by repeated cycles of freezing and thawing.' However, the term is also used in a much broader sense, for instance in citing frost action as one of the causes of creep and patterned ground, in which the weathering aspect is commonly secondary. In the following, the term will be used broadly to designate the action of frost during freezing, thawing, and subfreezing temperature fluctuations. The last qualification is important because of the significance of temperature-dependent volume changes in ice as illustrated, for instance, by frost cracking. Frost action is really a collective term describing a number of distinct processes engendered mainly by freezing and thawing.

Parts of this chapter are based on previous discussions by the present writer (Washburn, 1956*b*; 1967; 1969*a*; 1969*b*), and some statements, like most of the preceding paragraph, are verbatim or only slightly altered from one of these earlier discussions.

II FREEZING PROCESS

1 General

The freezing of soil is a complicated thermodynamic process (Tyutyunov, 1964). Much remains to be learned about the process, including the movement of moisture towards a freezing plane in fine-grained soils, the expulsion of soil particles from developing ice and the consequent *in situ* segregation of clear ice masses in frozen ground, and many other facets of soil freezing (cf. Miller, 1966). The nature of phase-boundary water in freezing soils is far from established (D. M. Anderson, 1968; 1970).

There is some agreement that an electric double layer adjacent to a mineral surface, and the dipole nature of water are basic elements in the movement of water to a freezing plane. However, as reviewed by R. F. Scott (1969, 1–10), there are two primary hypotheses regarding the growth of an ice lens. According to Jackson and Chalmers (1956), the ice molecules are at a lower energy state than the water of the electric double layer, which in turn is at a lower energy level than the free water. Systems tend towards a lower energy level, so the free water flows towards the developing ice lens. The energy released by the freezing of supercooled water in fine-grained soils supplies the energy for frost heaving. This theory was developed and an experimental verification was attempted by Chalmers and Jackson (1970). On the other hand, according to Cass and Miller (1959) osmotic pressure, resulting from the concentration of cations in the electric double layer, causes a flow of relatively pure water to the ice-water interface. Supercooling is not necessarily involved because the ions in the double layer cause a reduction in the equilibrium temperature of the water there. Thus some of the ice in frozen ground can thaw at temperatures below 0°.

Whatever its complete explanation, the tendency for water to flow to a freezing front is influenced by a number of factors. Their combined effect can be regarded as a suction potential, or water potential ψ, in a freezing system (D. M. Anderson, 1971, 2–4). Table 4.1 shows ψ in various units and their relation to capillarity and freezing-point depression.

2 Closed *v.* open systems

The classic work on soil freezing was carried out by Taber (1929, 1930*a*; 1930*b*) who experimented with systems that were either closed or open with respect to water. He demonstrated that

Table 4.1. Energy state of soil water as expressed by the water potential ψ (*Anderson, 1971, 3, Table II*)

Appearance of soil	Soil water types and so called soil water constants	Water potential, ψ (all values are negative)				Relative humidity at 25 °C %	Freezing point depression °C	Equivalent capillary diameter mm	Common methods of estimating ψ
		cm of water	Atmospheres (approx.)	ergs/g $\times 10^8$	Cal/g				
	Oven dry	10^7	10^4	98000	235·2				
Dry	**Hygroscopic water** (unavailable to terrestrial plants and most micro-organisms)	10^6	10^3	9800	23·5	50	90		
						75			
		10^5	10^2	980	2·35	93		Colloidal	A
	Hygroscopic coefficient	31623	30·6			98			
	Wilting point	14125	13·6			99	1·12	0·0002	
		10^4	10	98	0·24				
							0·4	Coarse clay	B
Moist	**Capillary water** (Available to terrestrial plants and micro-organisms)								C D F
							0·2		H
							0·1	0·002	E
		10^3	1	9·8	0·024				
	Field capacity	501	0·5				0·04	Silt	
							0·01	0·02	
		10^2	10^{-1}	0·98	0·002				
	Aeration porosity limit	50	0·05					Fine sand	G
Wet	**Gravitational water** (subject to drainage)								
	(available to higher organisms)	10	10^{-2}	0·098	0·0002			0·2	
								Coarse sand	
								2·0	
	Saturation	1	10^{-3}	0·0098	0·00002				

A Psychrometry and water vapour pressure
B Freezing point
C Pressure membrane
D Bouyoucos resistance blocks
E Thermal units
F Centrifuge
G Tension meters
H Neutron thermalization and gamma ray attenuation

water is drawn towards a freezing plane and that in a closed system the build up of ice lenses tends to desiccate the still unfrozen soil. He found that heaving pressures are in the direction of crystal growth, normal to the freezing plane, and that heaving in a closed system is limited to the volume change of water to ice but in an open system is dependent on the amount of water drawn into it from outside. As a result heaving can be far greater than in a closed system; in fact heaving can be caused by freezing of liquids that contract rather than expand with the phase change.

3 Effect of temperature

To the extent that an open system prevails, it can favour build up of ice lenses and much heaving, depending on the temperature. Very rapid freezing reduces migration of moisture so that pore water tends to freeze *in situ*. However, if penetration of freezing is in equilibrium with movement of moisture towards the plane so the plane is stabilized, thick ice lenses can form as a result of the continued contribution of moisture. Much depends on pressure as well as temperature (Low, Hoekstra, and Anderson, 1967) and grain size.

Temperature also controls the amount of unfrozen water in fine-grained soils (Figure 4.1; Tables 4.1, 4.2; cf. also Tsytovich *et al.*, 1959, 111–12, Tables 2–3, Figure 5; 1964, 7, 96–9, Tables II–III; 105, Figure 5). The unfrozen water, which surrounds mineral grains as a film 3 to 50 Å or more thick, is readily measured to −10° and retains high proton mobility to below −50° (D. M. Anderson and Tice, 1970, 1).

Table 4.2. Per cent of unfrozen water as related to temperature and grain size (*Tsytovich, 1958*)

Temperature	*Per cent of unfrozen water*		
	Pure sand	Sandy clay	Clay
−0·2° (31·5°F)	0·3	18·0	42·0
−10° (14°F)	0·15	9·0	>20·0

It should be noted that the temperature at which the ice in a soil thaws may differ from that at which it freezes (Figures 4.2–4.3).

4 Effect of pressure

Pressure lowers the freezing point of water by a small amount (0·0073° per kg/cm²). Except in the case of large pressures, the effect is probably negligible compared to other factors affecting freezing. However, as noted under Permafrost in the chapter on Frozen ground, the pressure

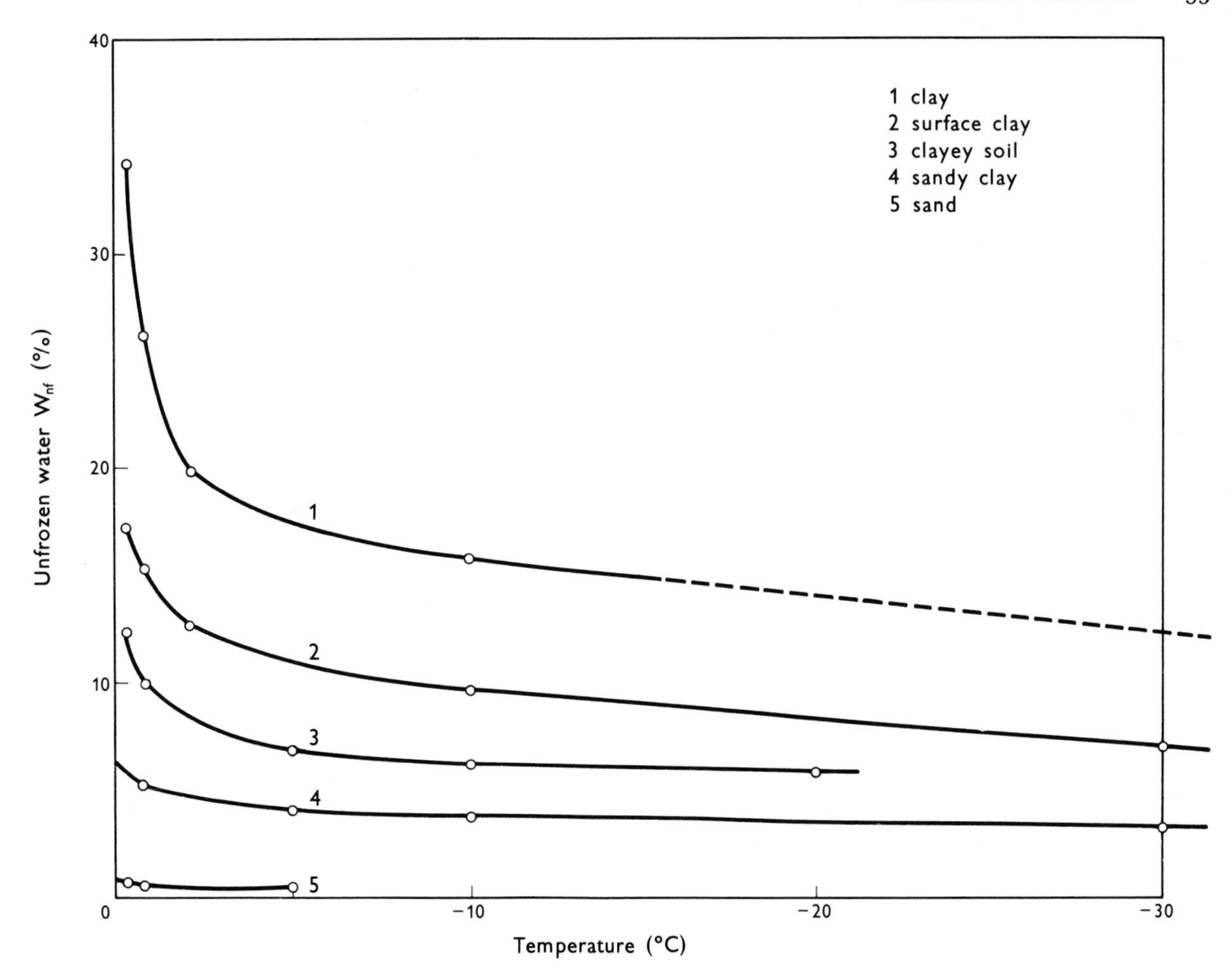

4.1 Unfrozen water content of frozen soils with changes of temperature (*after Tsytovich, 1957 116, Figure 1*)

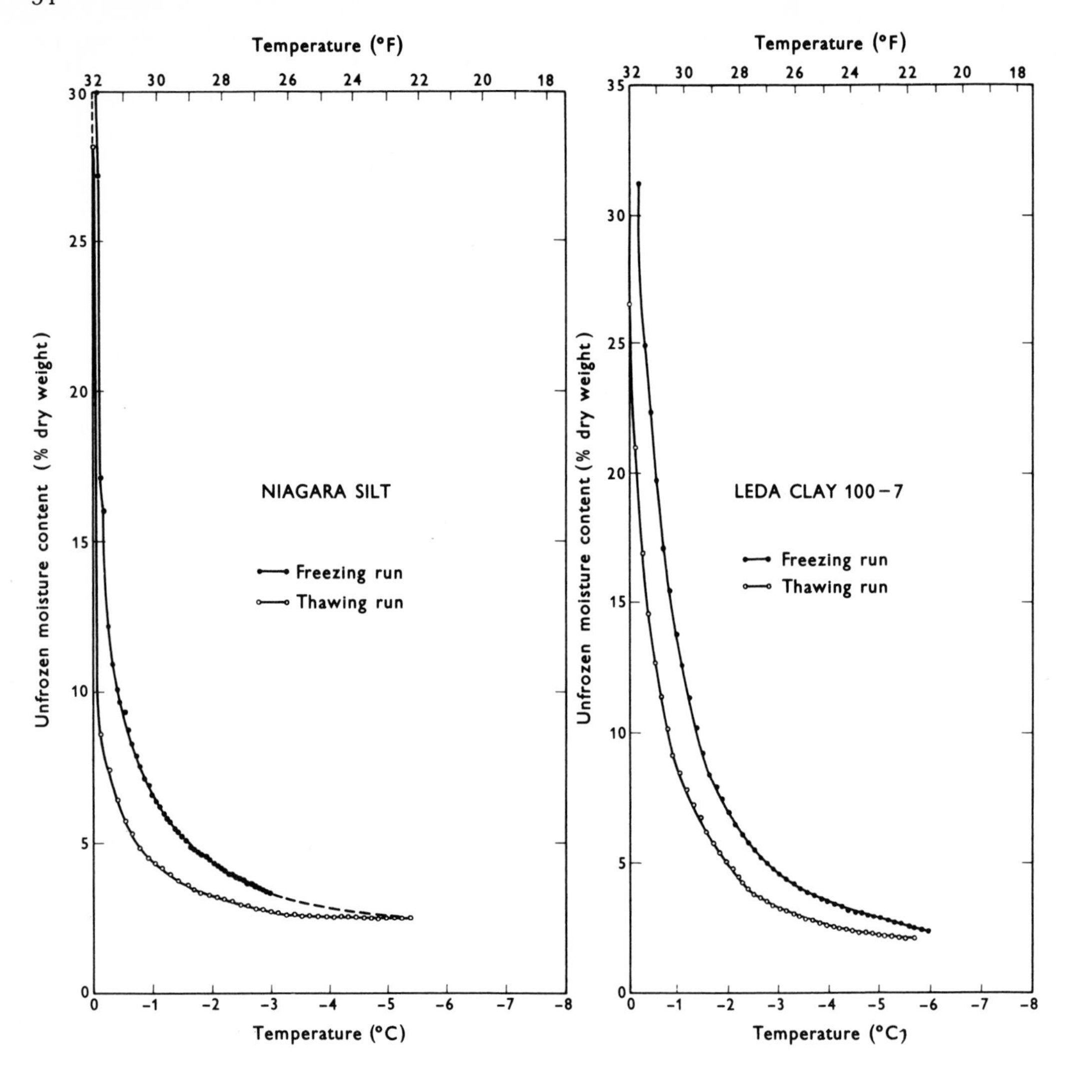

4.2 (*Left*) Unfrozen water content during freezing and thawing (*P. J. Williams, 1963, 121, Figure 5*)

4.3 (*Right*) Unfrozen water content during freezing and thawing (*P. J. Williams, 1963, 122, Figure 6*)

beneath the Antarctic Ice Sheet at the site of a drill hole reduced the freezing point to an estimated −1·6° and inhibited freezing.

Pressure also tends to counter the build up of ice lenses. As a result, other conditions being equal, ice lenses normally decrease in size with depth. Similarly, frost heaving can be artificially reduced by applying a large pressure to the top of freezing soil. The relationship between the freezing pressure or heaving pressure (Hoekstra, 1969), and the suction pressure or water potential ψ (D. M. Anderson, 1971, 2–4) remain to be adequately investigated.

5 Effect of grain size

Pore size and therefore grain size strongly affect the growth and form of ice in the soil by influencing (1) the freezing temperature of phase-boundary water and the water in fine capillaries, and (2) the movement of water to the freezing front.

It is well known that the amount of unfrozen water varies inversely with grain size, other conditions remaining equal (Figure 4.1). This accounts for the fact that silt and particularly clay freeze solid at a lower temperature than coarser material (Beskow, 1935, 13, 31–42; 1947, 5, 14–21). Some aspects of this are discussed below and others under Frost heaving and thrusting and under Mass displacement later in this chapter.

The basic factor accounting for the different freezing temperatures is the greatly increased contact area between solids and water as grain size is reduced, and the resulting increased tendency for surface effects to keep capillary and phase-boundary water from crystallizing. Thus in a closed system there would be less ice, the finer the grain size and the higher the freezing temperature. However, in an open system where water is continuously available for movement to the freezing plane, the water in material of fine grain size (i.e. with small pores) may remain mobile longer and hence build up larger ice lenses. Much depends on the rate of freezing as discussed previously.

Grain size influences movement of water to the freezing front because the potential for drawing water to the freezing front increases with decrease in grain size. This suction potential, or water potential ψ (Table 4.1), is absent in coarse sands and gravels lacking fines, so that the segregation of ice lenses as opposed to development of interstitial ice is inhibited. On the other hand, soils may be so fine grained as to become essentially impermeable, and consequently the movement of water to the freezing front to form ice lenses and cause heaving is at a maximum where the suction potential and permeability combine most effectively (Figure 4.4). As a result silt is particularly prone to heaving, since it permits relatively easy migration of moisture during freezing, and in an open system this factor favours a greater ice content in silt than in clay (other conditions remaining equal), as recognized long ago by Johansson (1914, 84, 93–4) and investigated in detail by Beskow (1935; 1947) and others. Linell and Kaplar (1959, 86, Figure 3; 88) found that silt and lean (slightly plastic) clays generally exhibit higher heave rates than fat (plastic) clays. ('Clays' in the engineering terminology they used includes all grain sizes <0·074 mm in diameter that produce plasticity – US Army Waterways Experiment Station, 1953, 3–4.) However, heaving pressures, which are also controlled by grain size, continue to increase beyond the point where heaving rates decline with decrease in grain size (Figure 4.4).

Even massive ice beds, 30 m or more thick and hundreds of metres in extent, can grow as segregated ice masses if porewater pressures are adequate to replenish ground water that is transformed to ice (Mackay, 1971*b*). As stressed by Mackay, such pressures would be generated by aggrading permafrost encroaching on permeable water-saturated sediments underlying fines, and he argued that such a system constituted a transition to the formation of injection ice as in pingos. The critical conditions for growth of beds of segregated ice as cited by Mackay (1971*b*, 411; following D. H. Everett, 1961; cf. P. J. Williams, 1967, 99) is expressed by

$$P_i - P_w = \frac{2\sigma}{r_{iw}} < \frac{2\sigma}{r}$$

where P_i = pressure of ice, P_w = pressure of water, σ = surface tension ice-water, r_{iw} = radius of ice interface, and r = radius of continuous pore openings.

As noted in the chapter on Environmental factors, the role of grain size has been stressed in the following 'rules of thumb'. According to these, a soil is frost susceptible (i.e. subject to build up of ice lenses and severe heaving) if it has (1) several per cent of particles <0·07 mm in diameter (passing through 200-mesh screen – *c.* upper limit of silt) (Terzaghi, 1952, 14), or (2) (the Casagrande criterion) contains more than 3 to 10 per cent particles <0·02 mm in diameter (*c.* middle of silt range), depending on the uniformity of the soil (Casagrande, 1932, 169; US Army Arctic Construction and Frost Effects Laboratory, 1958, 28–30). These 'rules of thumb' are subject to many factors (Dücker, 1956; 1958), and some soils heave that would be considered nonfrost susceptible by these standards (Corte, 1961, 10; cf. Corte, 1962*e*, 20). For instance, Kaplar (1971, 1) noted that sandy soils from Greenland showed substantial heaving when the content of particles finer than 0·02 mm was about 1 per cent, whereas the content in some other sandy materials could be up to 20 per cent before there was much heaving. More reliable criteria based on heaving rates are now being used by the US Army Corps of Engineers, and improved techniques are being investigated (Kaplar, 1971).

6 Effect of mineralogy

The mineralogy of the fines can influence the freezing process with respect to migration of water and heaving. Clays with expandable structure are able to hold more water but the water is

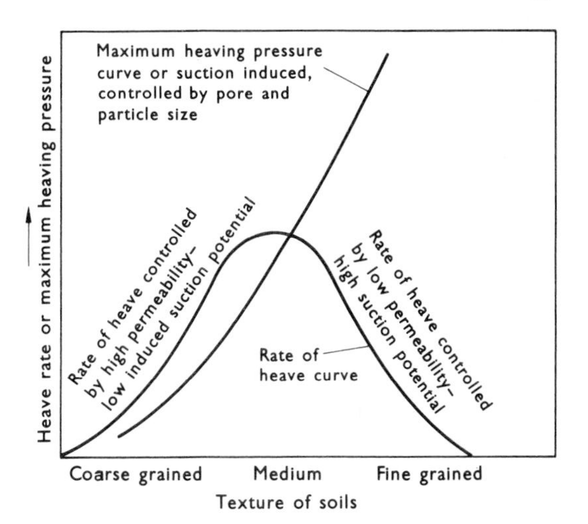

4.4 Diagram of relation between heaving rate, heaving pressure, and grain size (*Penner, 1968, 22, Figure 1*)

relatively immobile compared with non-expandable clays. Consequently, strong frost heaving is more likely to be associated with kaolinite than bentonite or montmorillonite (Dücker, 1940; Grim, 1952; Linell and Kaplar, 1959, 92–9). Salts such as NaCl and $CaCl_2$ influence freezing by depressing the freezing point significantly (Yoder, 1955) but other additives may be more effective in reducing the frost susceptibility of soils (Lambe and Kaplar, 1971).

III ESTIMATES OF FREEZING AND THAWING

1 Freezing and thawing indexes

Freezing and thawing indexes are measures of the heat balance at the ground surface (surface index) or at a height of 1·5 or 1·8 m (5 or 6 ft) above it (air index), usually given in degree days of freezing or thawing over an unbroken freezing or thawing period (Sanger, 1966, 253). For °C

$$I = \int_0^t T dt$$

where I = index, T = mean temperature for a day as represented by (maximum + minimum temperature)/2, and t = the period.

For °F, which is the most commonly used unit for freezing and thawing indexes

$$I = \int_0^t (T - T_0)\, dt$$

where $T_0 = 32°$ F.

Freezing and thawing indexes provide a measure of the severity of climate. They lend themselves well to mapping (Figure 4.5) and are useful in projecting depths of freezing and thawing (Sanger, 1966). On the other hand these indexes do not necessarily correlate well with mean annual temperature because of differences with respect to summer and winter temperatures in continental and maritime climates (P. J. Williams, 1961, 341, Table 1; 342, Figure 2), and for the same reason they give no information as to frequency of freezing and thawing.

2 Freeze-thaw cycles

The frequency of freeze-thaw cycles is an important control in the effectiveness of various kinds of frost action, including frost wedging. However, the purely climatic factor of the number of times the air temperature passes through the freezing point is not, in itself, an adequate measure of the effectiveness. The insulating effect of snow and vegetation, the nature of the rock material, and the rapid attenuation of temperature fluctuations with depth must all be taken into account in evaluating the frequency and effectiveness of freeze-thaw cycles in rock material.

Large discrepancies between air and ground-surface temperatures are possible as a result of insolation on dark surfaces. A difference of over 30° has been observed in Northeast Greenland, and similar occurrences have been noted elsewhere (cf. Washburn, 1969*a*, 45–6). In the Antarctic thawing of snow or ice adjacent to rock has been reported at air temperatures as low as −16·5° (Taylor, 1922, 47) and −20° (Souchez, 1967, 295), and freeze-thaw cycles in bedrock joints during negative air temperatures have also been observed there (B. G. Andersen, 1963, B142; Aughenbaugh, 1958, 168). On the other hand, because of outgoing radiation, it is also possible, for ground freezing to occur at positive air temperatures as reported from Iceland (Steche, 1933, 222, Figure 8).

Thus the frequency of freeze-thaw cycles at and beneath the ground surface can vary widely from the frequency in the air above. A climatic regime having many air cycles of small temperature range favours extreme examples; Meinardus (1923, 433; 1930, 50–1) reported that at Kerguelen Island in the Subantarctic no frost was observed at a depth of 5 cm in 1902–3, despite 203 air freeze-thaw cycles and 238 days with freeze-thaw cycles at the ground surface. At Cornwallis Island in the Canadian Arctic where the air cycles have a larger temperature range, Cook

4.5 (*Opposite*) Map of mean freezing indexes (°F) for Northern Hemisphere. Units explained in text (*after Corte, 1969a, Plate 6 opposite 161*)

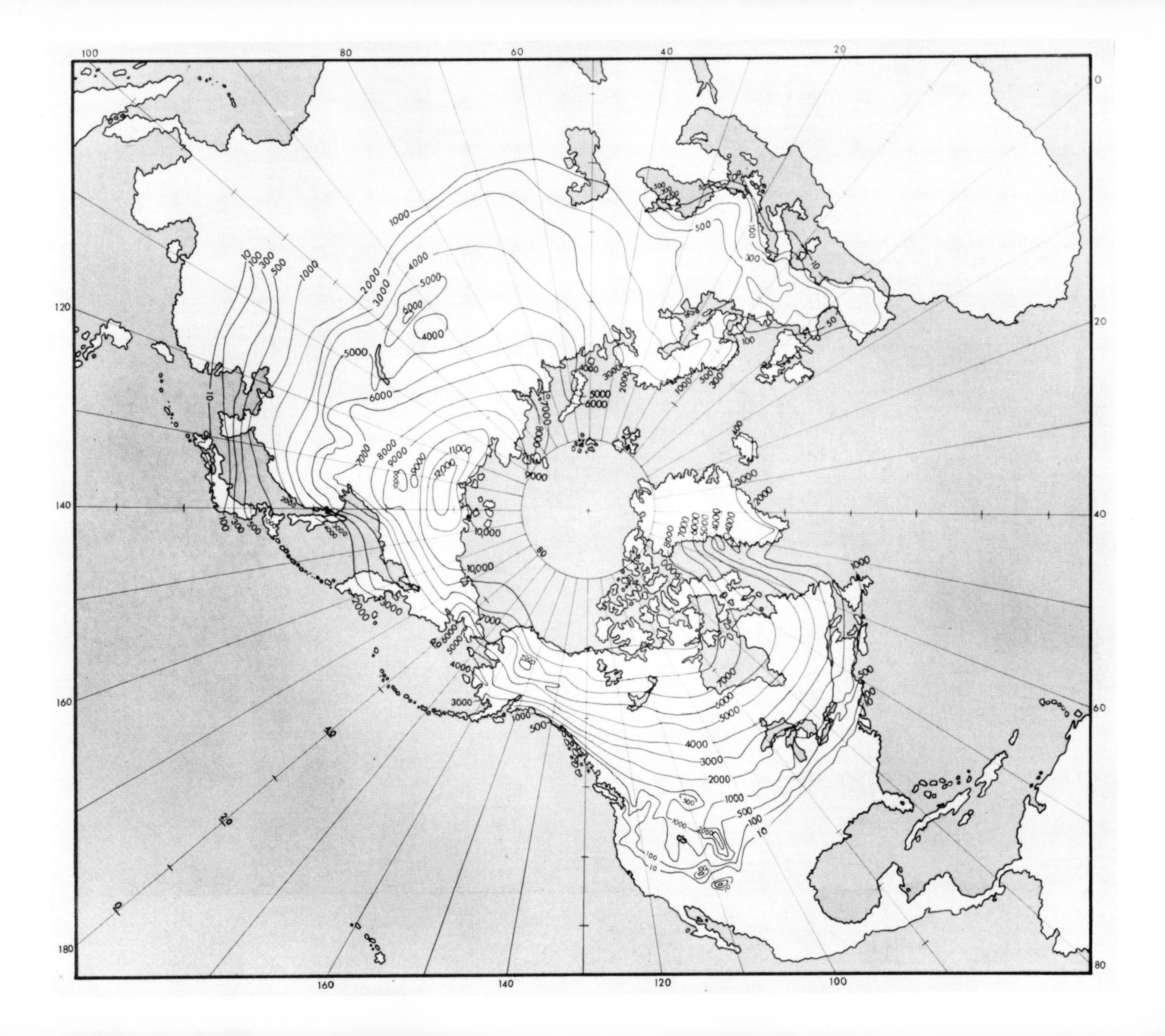

and Raiche (1962, 67, Table 1) found, from 1 May to 30 September 1960, 23 ground cycles in the range −2·2° to 0° (28° to 32°F) at the surface but only one cycle at a depth of 2·5 cm and none at depths of 10 cm and 20 cm. Observations in Arctic Canada convinced Cook (1966, 129) that 'there are no cycles, apart from the annual cycle, at depths below a few centimetres, and it follows that assumptions still widely held among some geographers and geologists (that frost cycles are a vigorous process producing frost splitting at depth in arctic countries today) are not valid'. Air freeze-thaw cycles become more common farther south in Canada (Fraser, 1959, 44–5) and the trend continues into the United States (Llewelyn Williams, 1964), and frost wedging is probably more frequent under such conditions. Whether it is also more effective with respect to size and quantity of resulting debris is less certain.

There are several methods for estimating the frequency of ground freeze-thaw cycles on the basis of air cycles. One method is to select a temperature range for an air cycle that supposedly also represents a ground cycle. There are many inherent uncertainties (cf. Matthews, 1962), and several workers employing the method have done so with reservations (Cook and Raiche, 1962; Fraser, 1959). One safeguard is to directly compare ground cycles with air cycles during a control period, and to work out a correlation from these data that permits estimates of ground cycles over a much longer interval (cf. Washburn, 1967, 133–8).

IV FROST WEDGING

1 General

Frost wedging is the prying apart of materials, commonly rock, by expansion of water upon freezing. It is synonymous with frost splitting and congelifraction (Bryan, 1946, 627, 640). The prying force is not necessarily confined to the 9 per cent volume expansion accompanying the freezing of water but in porous material may be due, in even greater measure, to the directional growth of ice crystals (Taber, 1929; 1930*a*; 1930*b*).

Some porous rocks may behave like certain soils in absorbing moisture, and frost wedging, although facilitated by foci of weakness, does not necessarily require pre-existing fractures (Taber, 1950; Willman, 1944). Based on his experiments at the US Army Cold Regions Research and Engineering Laboratory, Mellor (1970, 42–3) reported

> that the observed freezing strains come close to the strains for complete tensile failure of the rocks, while the potential freezing strains for completely saturated rock frozen rapidly far exceed the tensile failure strain for sandstone and limestone. It should also be noted that internal cracking of these rocks almost certainly occurs at tensile strains well below the ultimate tensile fracture strain.

The maximum pressure generated by the phase change of water to ice is 2115 kg/cm^2 under ideal laboratory conditions but is probably not attained in nature; for one thing the maximum tensile strength for rock is much less, being in the order of 250 kg/cm^2 (Grawe, 1936, 176–82).

2 Factors

Not only is the presence of moisture mandatory but the amount available can be critical, since the bulk freezing strain of a rock increases with increasing water content (Mellor, 1970, 50). Laboratory experiments show that limestone cubes originally 10 cm on a side break down much more readily in water 4 cm deep than 1 cm deep (Guillien and Lautridou, 1970), and that rocks half immersed generally disintegrate more rapidly than those that were saturated but surrounded by only a thin film of water (Potts, 1970, 112–13). Frost action has been reported to be particularly intense on shores where rocks are frequently wetted (Mackay, 1963*a*, 57; Taber, 1950).

Given adequate moisture, the nature of a rock is probably the most important factor determining its susceptibility to frost wedging. Sedimentary rocks such as siltstone and shale containing mica or illitic clays display, by virtue of the horizontal orientation of the micaceous minerals, planes of fissility through which water migrates preferentially. Thus shales break down more readily than igneous rocks (Potts, 1970, 114–22), although crystalline rocks rich in biotite or other mica may be also vulnerable to frost wedging. The process may become more critical if both biotite and muscovite are altered and the interlayer potassium has been replaced by hydrated calcium, magnesium, or sodium. Different behaviour may be anticipated in different minerals as the result of variations in hydration and expansion.

Temperature and time are other factors. Depending on the freezing rate, two different rocks can reverse their susceptibility to breakdown (Thomas, 1938, 63–8, 94). Mellor (1970, 31–43, 50–1) found that the bulk freezing strain of a rock increases not only with increasing water content but also, for any given water content, with increasing freezing rate, and he suggested that at water contents >50 per cent saturation, the freezing strain may be enough to cause internal cracking. Rapid freezing of saturated solid rocks or of rocks with cracks should favour frost wedging by sealing near-surface voids containing unfrozen water and thereby creating a closed system that would promote pressure effects. Thus frost wedging in cracks would be furthered by water freezing from the surface down and creating a solid plug of ice (Battle, 1960, 93–4; Pissart, 1970*a*, 44–5). On the other hand slow freezing may promote frost wedging of fine-grained uncracked rocks on wet soil by permitting flow of water to the freezing front to build up disruptive ice crystals, whereas rapid freezing would inhibit the flow of water (Taber, 1950).

Freeze-thaw cycles are a primary factor in frost wedging but the number of shifts of air temperature through the freezing point, which is sometimes used as an environmental measure of frost action, can be very misleading. Such shifts are by no means synonymous with freeze-thaw cycles in rock as discussed above under Estimates of freezing and thawing, since large discrepancies can occur between air temperature and ground temperature in polar climates. Furthermore, according to some experiments the intensity as well as the number and length of freeze-thaw cycles in rocks is important (Tricart, 1956*a*, 295–7), although other experiments are contradictory in part and indicate the primary importance of freeze-thaw cycles for a given rock type (Coutard *et al.*, 1970, 37; Lautridou, 1971, 69, 79; Potts, 1970, 113–14; Wiman, 1963, 116). Other data include the observation that 400 cycles resulted in a 10 to 20 per cent increase in shale and sandstone fragments having a size range of 0·1 to 10 mm (Legget, Brown, and Johnston, 1966, 25–6).

The multiplicity of factors affecting frost wedging – moisture, temperature, time, rock type – complicate the interpretation of the resulting products as stressed by Guillien and Lautridou (1970, 40–5).

3 Products

Frost wedging characteristically produces angular fragments that can be of widely varying size, ranging from huge, house-size blocks to fine particles.

'Frost weathering' has been cited as the mechanism by which unconsolidated sediments and shales disintegrate to granule-size particles by growth of segregated ice and subsequent thawing (Harrison, 1970), but the predominant size to which rocks can be ultimately reduced by frost action is much smaller and usually held to be silt (cf. Hopkins and Sigafoos, 1951, 59; Sørensen, 1935, 24–5; Taber, 1953, 330), although as stressed by Hopkins and Sigafoos the parent rock can exert a critical influence on the size of its products during disintegration. Laboratory experiments by Guillien and Lautridou (1970) showed that frost wedging of certain kinds of limestone can produce particles as fine as clay. Experiments by McDowall (1960) suggest that even clays may be further comminuted by frost wedging. On the other hand very little material finer than 0·06 mm was produced in experiments carried out by Potts, who concluded that 'processes other than frost shattering produce silt and clay particles which are found in solifluction deposits' (Potts, 1970, 120). Similarly, the significance of frost wedging in producing cold-climate loess remains to be fully established; probably there is considerable variability in view of other silt-producing processes such as glacial erosion.

Frost wedging resulting from repeated temperature cycles at negative temperatures appears possible, since water in minute rock openings should freeze and thaw at such temperatures like some of the water in fine-grained soils (Figures 4.1–4.3, Tables 4.1–4.2), as previously discussed in this chapter under Freezing process.

More experimental evidence is needed on frost wedging. In this connection it should be noted that when dealing with micro effects it is difficult to eliminate hydration as a factor in freeze-thaw experiments; furthermore, according to Dunn and Hudec (1965, 115–38; 1966) ordering and disordering of water molecules may simulate 'frost deterioration'.

4 Environmental aspects

In periglacial environments, thawing of snow or ice adjacent to dark rocks warmed by insolation is common at subfreezing air temperatures and must be a potent factor in frost wedging when meltwater seeps into joints and refreezes. Since subfreezing temperatures occur below the thawed layer throughout the year in a permafrost environment, refreezing of meltwater in lower-lying jointed bedrock or in cracks in unconsolidated material is not confined to the spring or autumn and to freeze-thaw cycles induced by changes of surface temperature. Nevertheless, the maximum effect of frost wedging as gauged by release of rock fragments probably occurs in the spring as shown by the frequency of rockfalls then. Products of rockfall accumulate on the snow during the spring thaw in many places and the process itself has been observed while the sun was thawing a cliff face, so it seems certain that the rockfall in such places is triggered primarily by frost wedging (cf. Rapp, 1960*a*, 104–9; 1960*b*, 17–23; Washburn, 1969*a*, 35). In many places this may be as much in response to an annual freeze-thaw cycle as to short-term cycles.

Coarse and angular, fresh rock debris mass-wasting from bedrock attests to the importance of frost wedging in polar and alpine regions. Some slopes without, or with only low, cliffs at their heads are formed mainly of material much like talus rubble. Where such coarse, angular debris is derived from the underlying rock without benefit of rockfall, frost wedging must be the predominant process responsible for detaching the fragments and forming block fields (on nearly horizontal surfaces) and block slopes (Figure 4.6), about which much remains to be learned as discussed later in the chapter on Mass-wasting.

Büdel (1969) suggested that frost wedging of bedrock in stream courses that dry up in the autumn accounts for especially rapid valley deepening in a permafrost environment. According to this view, thawing of the 'ice rind' of the top part of the permafrost during stream erosion releases frost-wedged products to fluvial transport, and by continually exposing more bedrock to such effects accelerates stream erosion where there is permafrost.

Salt wedging (Evans, 1970; Mortensen, 1933; Wellman and Wilson, 1965) may also break up rock material and produce silt-size particles. In fact ice and salts can cause many convergent phenomena as stressed by Tricart (1970) among others, and where both are present as in some arid cold climates it may be difficult to separate their effects.

V FROST HEAVING AND FROST THRUSTING

1 General

'The pressures generated by freezing water are exerted in all directions, but they are expressed in soil movements only upward and horizontally. The vertical expression has been termed heave; the horizontal, thrust' (Eakin, 1916, 76). Although Eakin's statement is open to the objections indicated below, the distinction between heave and thrust is important (cf. Hopkins and Sigafoos, 1954). Frost heaving is the predominantly upward, frost thrusting the predominantly lateral movement of mineral soil during freezing.

Taber (1929, 447–50; 1930*a*; 1930*b*, 116–18) demonstrated that the pressure generated by the growth of ice crystals is at right angles to the freezing isotherm and not necessarily in all directions; and because freezing extends downward from the ground surface he stressed the role of frost heaving as opposed to thrusting (cf. Taber, 1943, 1458–9; 1952). However, the complexities involved in the freezing of heterogeneous material can result in pressures in various directions as pointed out by Hamberg (1915, 600–3). Various investigators, including Schmid (1955, 92–5, 122, 130) and Corte (1962*c*, 14–17; 1962*d*, 58) have argued that varying conductivities in heterogeneous material influence the orientation of cooling surfaces and introduce lateral movement in places. Beskow (1935, 59; 1947, 30–1) emphasized resistance to expansion as well as orientation of cooling surfaces as an important factor controlling direction of expansion during freezing;[1] although the effect would normally be upward, this is not necessarily the case and frost thrusting cannot be neglected in considering frost action. Where thrusting as well as heaving may be involved, frost expansion is a useful term covering either direction or both.

Thermal expansion of ice upon temperature increase towards melting also causes lateral stresses referred to as thrusting (Laba, 1970). However, this process does not conform to Eakin's definition of *frost* thrusting, which specifies the pressure of freezing water and is a separate process.

The general process of frost heaving is determined by the thermodynamics of the freezing process and the growth of ice lenses, and is another way of looking at this complex and the controlling factors discussed in this chapter under Freezing process. Because of its practical significance in foundations for buildings, roads, and airfields in cold environments, frost heaving has been studied particularly by civil engineers, who have made many valuable contributions. A recent summary from the engineering viewpoint has been presented by Kaplar (1970). The practical implications and problems, especially in permafrost regions, are covered in numerous publications cited in the chapter on Frozen ground. The emphasis in the following is on certain specific frost-heaving processes that help to characterize the periglacial environment.

[1] Freezing may also cause contraction by desiccation as discussed later in this chapter under Patterned ground, and by withdrawing water from small pores to freeze in air-filled larger pores (Hamilton, 1966).

4.6 (*Opposite*) Block slope, Hesteskoen, Mesters Vig, Northeast Greenland (*cf. Washburn, 1969a, 34, Figure 18*)

2 Heaving of joint blocks

One of the striking results of frost action is heaving of joint blocks. Blocks, frost wedged from bedrock along joints, are raised well above the general surface in places, although the blocks are still tightly held by the surrounding bedrock (Figure 4.7).

Yardley (1951) has described such occurrences as 'frost thrust blocks', but this term is inappropriate because frost thrusting applies to lateral rather than vertical movements. Frost-heaved blocks are common in permafrost regions. For instance, they have been observed on Spitsbergen (Bertil Högbom, 1910, 41–2; 1914, 274–7), in the Canadian Arctic (Yardley, 1951), Northeast Greenland (Washburn, 1969*a*, 51–2) and elsewhere.

3 Upfreezing of objects

a Stones Ejection of stones from fines by upfreezing is commonly accepted because of prominent edgewise projecting stones in periglacial environments (Figure 4.8), the often-reported but rarely documented appearance of stones on previously cleared farmers' fields, the heaving of posts and other artificial structures, and other similar field occurrences. However, detailed observations and measurements relating to stones are conspicuously few. Morawetz (1932, 39) observed upward displacements of 4–7 mm for pea- to nut-size stones as a result of three to four freeze-thaw cycles in an alpine environment. Field experiments in Germany by Schmid (1955, 88–9, 130) were negative with respect to upheaving of stones not on edge; edgewise stones apparently heaved like posts but details were not reported. Laboratory experiments carried out and summarized by Corte (1966*b*) prove a vertical sorting involving an upward displacement of coarse soil particles relative to fine as the result of some types of frost action.

b Targets Czeppe (1959, 1960), working in Spitsbergen, found that wood pegs inserted to a depth of 15 cm were heaved out within a year, whereas the annual heave of targets originally at a depth of 35 cm was about 5 cm and of those at a depth of some 60 (or 65) cm was about 10 cm. Czeppe (1966, 70–1) also reported that targets inserted to depths of 20–35 cm heaved 5 cm in a year whereas those reaching depths of 40–50 cm heaved 10–11 cm (results from another site were inconsistent). Czeppe (1959, 1960) concluded that only the shallowest targets had been appreciably heaved by autumn freeze-thaw cycles and that the heave of the longer targets by the annual cycle was proportional to their depth of insertion.[1] However, the effective change of level between

[1] Although Jahn (1961, 9) reported that these and related experiments showed that depth of insertion had no essential effect on amount of target heaving, the apparent discrepancy is explained by the fact that Jahn was comparing the shallowest targets, which were affected by the spring heaving, with the deeper targets subject to annual heaving only.

4.7 (*Opposite*) Frost-heaved bedrock, Mesters Vig, Northeast Greenland (*cf. Washburn, 1969a, 51, Figure 25*)

some targets and the ground may have been consequent on spring thawing and collapse of mineral soil around the target pegs. Somewhat similar experiments were carried out by Chambers (1967, 18–19).

Experiments in Northeast Greenland (Raup, 1969, 21–6, 83–93, 122–3, 143–6; Washburn, 1969*a*, 58–81) utilized targets of two types – cone targets and dowels. The cone targets were mounted on wood pegs 1·5 cm in diameter. They were designed for theodolite observations of mass-wasting and were generally aligned across a slope at intervals of 2 m, and inserted in the ground to alternate depths of 10 and 20 cm. The dowels were short sticks 0·3 cm in diameter. They were commonly spaced at intervals of 10 or 20 cm and inserted to depths of 5 or 10 cm in lines or grid patterns to record frost-action and mass-wasting effects as measured with reference to strings or wires.

Several of the experimental sites and the heave data from them are illustrated in Figures 4.9–4.11 and Table 4.3.

The data clearly show that moisture content of the mineral soil, the vegetation, and depth of target insertion were critical variables controlling target heaving. Temperature conditions, although obviously important, were sufficiently similar at any one experimental site, and between most sites, to be considered constant. Grain size at most sites was also sufficiently similar to be considered constant. Exceptions were several sites that were relatively poor in fines and also characteristically dry so that the relative importance of grain-size and moisture conditions at these sites could not be well determined.

The critical importance of moisture was strikingly apparent in that the greatest target heaves were confined to characteristically wet places. However, vegetation was a controlling factor where moisture conditions were similar, since target heaves were consistently greater in thin tundra than in more richly vegetated areas.

Depth of target insertion influenced heaving in that, almost without exception, 10-cm targets heaved more than adjacent 5-cm targets, and 20-cm targets heaved more than adjacent 10-cm targets. Not all target heaves were progressive from year to year; in fact some targets tended to drop back slightly in some years, probably due to the weight of the target and/or that of overlying snow while the ground was thawed.

Most of the heaving probably occurred in the autumn as indicated by the following facts: (1) The greatest heaving was commonly in places that tended to remain wet throughout the summer because of lingering snowdrifts and were therefore particularly favoured with moisture during the autumn freeze-up; (2) most of these places tended to be protected from spring freeze-thaw cycles by lingering snow; (3) even places that were commonly snow free and wet in the early spring showed relatively little heaving; (4) low temperatures following thawing were more characteristic of autumn than spring.

4.8 (*Opposite*) Frost-heaved block in diamicton, Mesters Vig, Northeast Greenland. Scale given by trench shovel (*cf. Washburn, 1969a, 53, Figure 28*)

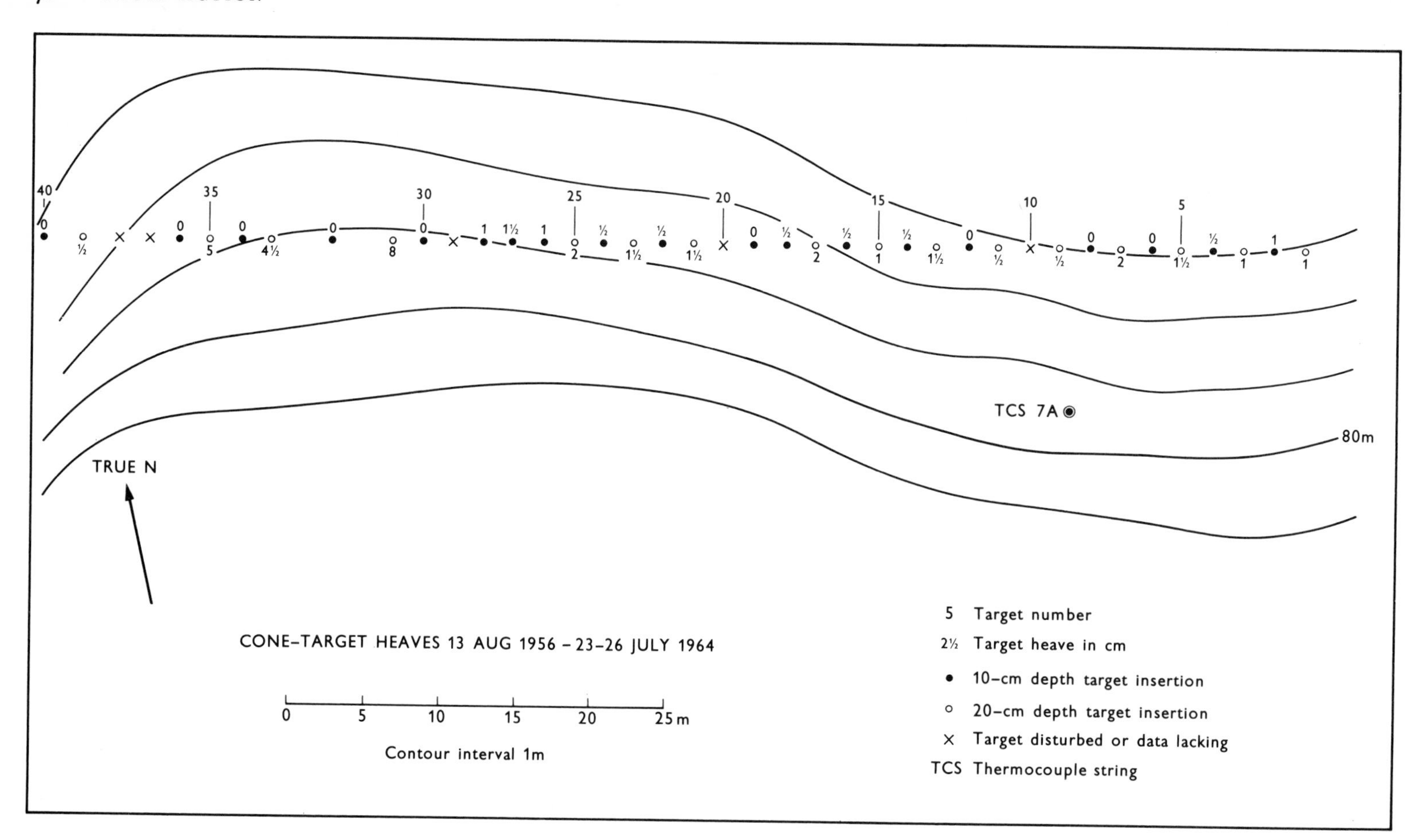

CONE–TARGET HEAVES 13 AUG 1956 – 23–26 JULY 1964
TRUE N
TCS 7A
80m
0 5 10 15 20 25 m
Contour interval 1m
5 Target number
2½ Target heave in cm
10–cm depth target insertion
20–cm depth target insertion
Target disturbed or data lacking
TCS Thermocouple string

The experiments of Chambers and Czeppe, and those from Northeast Greenland, show that progressively greater depths of target insertion correlate with progressively greater heaving, other factors remaining equal. These observations are consistent with Hamberg's (1915, 609) conclusion that the maximum upfreezing an object undergoes during a single freeze-thaw cycle is proportional to the object's 'effective height', which is the vertical dimension of the buried portion frozen to, and therefore heaved with, the adjacent material.

c Mechanisms The mechanics of upfreezing is poorly known, as was recognized by Schmid (1955, 23, 86–91). Bertil Högbom (1910, 53–4) suggested that when ground expanded during freezing it carried stones with it but, in contracting during thawing, fines adhered to each other and left the stones behind. Repetition of the process would thus lead to ejection of stones from fines, although Bertil Högbom (1910, 49) apparently did not specifically apply the hypothesis to upfreezing of stones as opposed to lateral sorting in patterned ground. Hamberg (1915, 603–10) argued that this explanation for stones not returning to their original position was inadequate. He elaborated the hypothesis that stones were pulled vertically with expansion of fines during freezing and did not return all the way during thawing because thawed material collapsed around them while their bases remained frozen. Beskow (1930, 626–7), adopting Hamberg's explanation, stressed the importance of growth of ice lenses in the process; he added the suggestion that stones, in addition to being hindered from dropping back by slumping in of material during thawing, would also be hindered because cavities left by stones as they were heaved would tend to be narrowed by frost thrusting during freezing (Figure 4.12). Vilborg (1955) stressed slumping in of sand and gravel during the heaving. Hamberg failed to discuss the formation and filling of such cavities, although his hypothesis logically entails these consequences; in any event the addenda cited strengthen this aspect of what can be called the frost-pull hypothesis. Time-lapse photography has demonstrated several stages of the frost-pull mechanism in the laboratory (Figure 4.13). Based on this mechanism the maximum distance a stone would be heaved in one cycle of freezing would be approximately (Kaplar, 1970, 36)

$$D = \frac{H_R L}{R_f}$$

4.9 (*Opposite*) Cone-target heaves, Experimental Site 7, Mesters Vig, Northeast Greenland (*cf. Washburn, 1969a, 65, Figure 38*)

where D = vertical distance (mm), H_R = heave rate (mm/day), L = vertical height of stone below its greatest horizontal diameter (mm), R_f = rate of frost penetration (mm/day).

Bertil Högbom (1914, 305), Nansen (1922, 117–18), Grawe (1936, 177), and Cailleux and Taylor (1954, 32) argued that upfreezing is explained by the greater heat conductivity of stones than fines, whereby ice would form around stones (Högbom) or at their base (Nansen, Grawe,

A
B

4.10 (*Opposite*) Experimental Site 11, Mesters Vig, Northeast Greenland. View west, 27 August 1956 when dowels installed (*cf. Washburn, 1969a, 69, Figure 43*)

4.11 Experimental Site 11, Mesters Vig, Northeast Greenland. View west, 20 July 1960 (*cf. Washburn, 1969a, 69, Figure 44*)

Table 4.3. Target heaves, Experimental Site 7, Mesters Vig, Northeast Greenland (*cf. Washburn, 1969, table E IV*)

Target no. and dry or wet	13 Aug 1956 *Depth cm*	1957 *Heave (noted)*	8 Aug 1958 *Heave cm (estimated)*	21 Aug 1959 *Heave cm*	25 Aug 1960 *Heave cm*	21 Aug 1961 *Heave cm*	23–26 July 1964 *Heave cm*	Remarks
1 D	20		3·0	1·0	1·5	2·0	1·0	
2 D	10		2·5	1·5	1·0	1·0	1·0	0·5-cm void at base
3 D	20		0·5	1·0?	2·0?	2·0?	0·5	
4 D	10	x	1·0	0·5	0·5	1·0	1·5	
5 D	20	x	1·0	0·5	1·0	1·5	3·5	
6 D	10		0·0	1·0?	1·5	0·5?	0·0	
7 D	20		1·0	0·5	1·0	1·0	2·0	
8 D	10		0·5	0·0	0·0	0·5	0·0	
9 D	20		0·5	0·5?	0·5?	1·0?	0·5	
10 D	10		1·5	0·5	1·0	1·0		Disturbance noted July '64
11 D	20		0·0	0·5?	0·5?	0·5?	0·5	
12 D	10		1·5	1·0?	2·0?	1·5?	0·0	
13 D	20	x	3·0	1·5	1·0	1·5	1·5	
14 D	10		0·5	0·5	1·0	1·0	0·5	0·5-cm void at base
15 D	20		0·5	0·5	0·5	0·5	1·0	
16 D	10		0·0	0·5?	0·0	0·0	0·5	
17 D	20	x	1·5	0·5	1·5	1·5	2·0	Small pebble at base
18 D	10		1·0	0·5	0·5	0·5	0·5	
19 D	10		0·0	?	0·0	0·0	0·0	1·9 cm void at base
20 D	10		0·0	0·0	0·0	0·0		Disturbance noted July '64
21 D	20		0·5	0·5	0·5	1·0	1·5	
22 D	10		1·0	0·5	0·5	0·5	0·5	
23 D	20	x	2·0	2·0	2·0	1·5	1·5	
24 D	10		1·5	0·5	0·5	0·5	0·5	
25 D	20		0·5	0·5	0·5	1·0	2·0	
26 D	10		1·0	0·5	0·5	1·0	1·0	
27 D	10		0·5	0·5	0·5	0·5	1·5	
28 D	10		0·5	0·0	0·0	0·5	1·0	
30 W	10		0·0	?	?	?	0·0	Large pebble at base
31 W	20	x	3·0	1·5	0·0	2·5	8·0	6-cm void at base
32 W	10		1·5	1·0	1·5?	1·0	0·0	
33 W	20		0·0	1·0?	?	1·0?	4·5	
34 W	10		2·0	0·5?	0·0	0·0	0·0	Well vegetated
35 W	20	x	3·0	3·0	3·0	4·0	5·0	
36 W	10		0·0	0·5?	0·0	0·5	0·0	Well vegetated
38 D	10		0·0	?	0·0?	?		Disturbance noted July '64
39 D	20		0·0	?	?	?	0·5	Well vegetated
40 D	10		0·0	?	?	0·0?	0·0	
Mean (with standard error)								
Questioned occurrences excl.								
10-cm targets								
Dry			0·7 (0·2)	0·5 (0·1)	0·5 (0·1)	0·6 (0·1)	0·6 (0·1)	
Wet			0·9 (0·5)	1·0 (0·0)	0·0 (0·0)	0·5 (0·3)	0·0 (0·0)	
20-cm targets								
Dry			1·1 (0·3)	0·8 (0·2)	1·1 (0·2)	1·3 (0·1)	1·4 (0·2)	
Wet			2·0 (1·0)	2·3 (0·8)	1·5 (1·5)	3·3 (0·8)	5·8 (1·1)	

Cailleux and Taylor) and force them up, and they thought that seeping in of fines during thawing would prevent stones returning to their original position. The above concept can be called the frost-push hypothesis; it has been recently proposed again, apparently independently, by Bowley and Burghardt (1971). Their laboratory experiments showed a correlation between rate of upfreezing and number of freeze-thaw cycles, but they did not discuss the possibility that the frost-pull mechanism caused the heaving, as it demonstrably did in Kaplar's (1965; 1970) experiments.

Streiff-Becker (1946, 154–5) argued that stones are kept from sinking back by preservation of ice at their base as the result of their poor conductivity, but Schmid, like Bertil Högbom and Nansen, pointed out that stones are better conductors than finer material; similarly, it has been

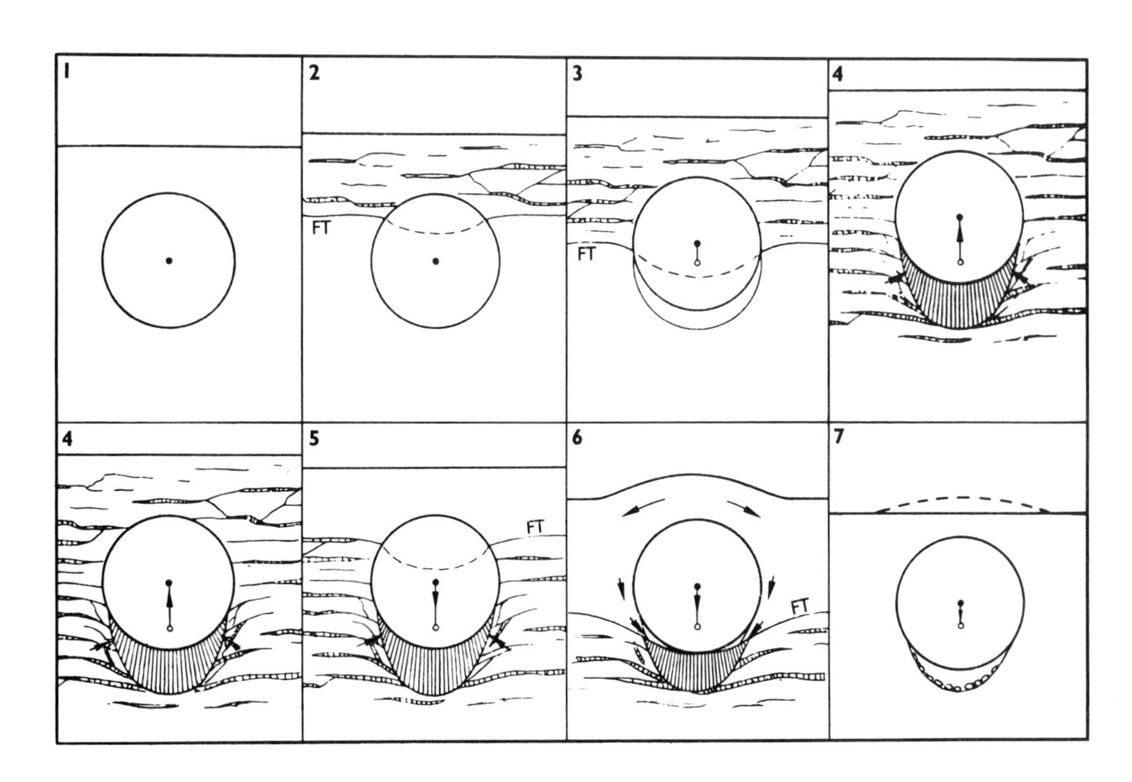

Top row = During freezing
Bottom row = During thawing
FT = Frost table
Hachures = Ice lenses

4.12 Upfreezing of stones according to frost-pull hypothesis (*Beskow, 1930, 627, Figure 4*)

stressed that diffusivity, which takes account of heat capacity as well as conductivity, also leads to stones being better conductors (cf. Bowley and Burghardt, 1971; Washburn, 1956*a*, 808; 1956*b*, 855–6). Only to the extent that a sizeable accumulation of ice would delay thawing because of its latent heat of fusion would there be a tendency for such accumulations to persist. The result pictured by Streiff-Becker is therefore misleading. Ice and frozen ground may support a stone at its base while thawing proceeds near its top, but this would be despite, not because of, differences in conductivity between stones and finer material.

Following lifting of a stone by the frost-pull mechanism, melting of ice above the bases of stones could lead to collapse of the ground around their upper portions and thus contribute to their upfreezing, and if very near the surface would leave the upper portions of stones projecting. However, where edgewise stones project in an area where the tops of many non-edgewise stones are at the surface without projecting markedly, such an explanation is dubious. Furthermore, vegetation- and silt-capped stones (Washburn, 1969*a*, 55, Figures 30–31) indicate a comparatively rapid forcing up of individual stones. Measurements show that the average upward movement in places can be as little as 0·6 cm (¼ in) due to autumn and winter freezing but that the total annual amount may be up to 5·1 cm (2 in) for individual stones as the result of a later movement of uncertain nature (L. W. Price, 1970, 107–8). In any event, a movement that breaks the vegetation cover in the way described above suggests appreciable accretions of ice at the base of the stone and supports the frost-push concept. That ice does develop preferentially adjacent to stones in material containing fines, and especially beneath them as argued by Nansen (1922, 117), has been frequently observed (Müller, 1954, 35, Figure 76; 128, Figure 52; Washburn, 1950, 41). In line with the frost-push hypothesis, this should occur because the better diffusivity of stones than fines would lead to earlier freezing of fines in contact with stones than of fines at the same level in the ground farther away, and consequently moisture should be preferentially drawn to stones during freezing, particularly to their base from below where the supply of moisture would tend to be uninterrupted. Also stones provide discontinuities that of themselves would favour localization of ice in heterogeneous material (Beskow, 1935, 41–2; 1947, 21). Thawing as well as freezing would first occur adjacent to stones and result in a tendency for them to drop back, but material thawed around their upper portions while their bases were still frozen would tend to fill in around the stones and prevent return to their original position.

The shape of stones can strongly influence their behaviour (Figure 4.14). For instance, frost-heaved, wedge-shaped stones with their narrow end down would not readily sink back upon thawing of the ground. Shape also influences the upfreezing of tabular stones in that their long axes tend to become oriented normal to the freezing surface. Suggested mechanisms call upon application of Stokes Law (modified), lateral compression during thawing (Cailleux and Taylor, 1954, 33, 54), or force couples set up during freezing (Figure 4.15) (Pissart, 1969; Schmid, 1955, 98).

4.13 (*Opposite*) Time-lapse photography of upfreezing stone. Scale in inches (*Kaplar, 1965, 1520, Figures 1–4*)

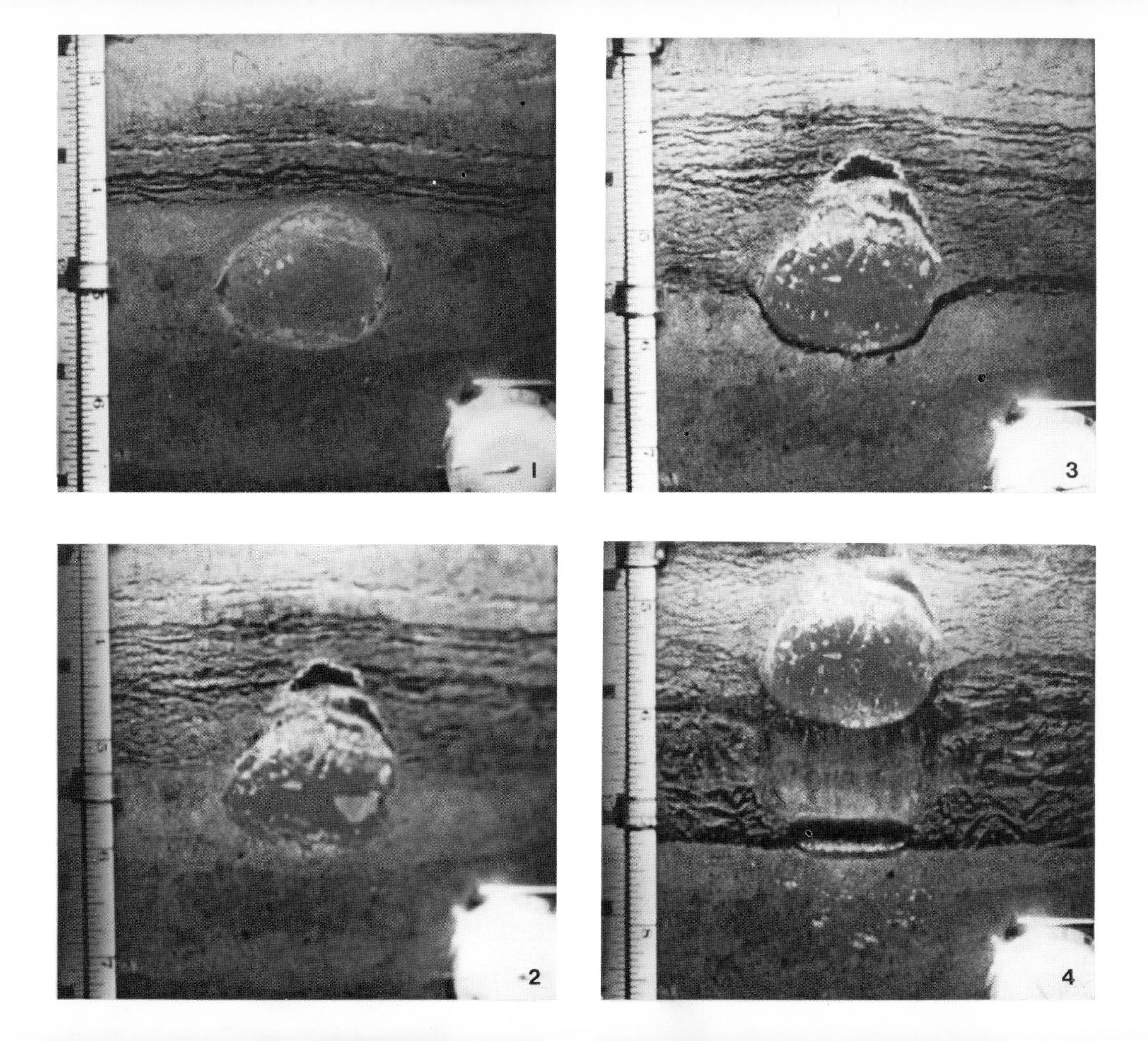
1
3
2
4

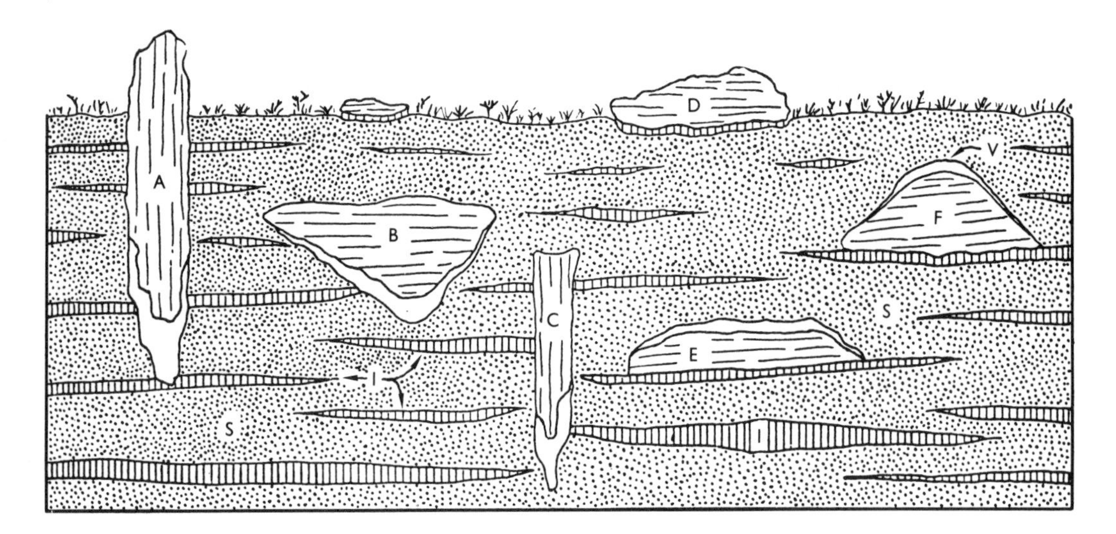

4.14 Diagram of upfreezing stones (*after Taber, 1943, 1453, Figure 3*)

Schmid (1955, 90–1) referred to an upward pressing of stones into still unfrozen ground by formation of ice at their base but omitted details as to how the freezing isotherm could reach the base of a stone without having already passed the level of the top and frozen the adjacent material. Because Taber (1943, 1455) believed that stones cannot be forced through overlying frozen soil, he rejected the view that growth of ice beneath stones could explain their upfreezing. However, the possibility of a buried stone being forced into still unfrozen fines is favoured by (1) the fact that freezing is partly a function of grain size so that silt and clay freeze solid at a lower temperature than coarser material (cf. Table 4.2 and discussion under Freezing process in this chapter), and (2) upward freezing from the permafrost table.

Ice lenses in soft clay have been observed in laboratory experiments (Taber, 1929, 432; 1930*a*, 311), and similar occurrences in the field have been reported in material with temperatures of −4° to −5° (Holmquist, 1898, 418) and as low as −15° (Terzaghi, 1952, 14–15). In frozen clay there can be 15 to 20 per cent unfrozen water at −10° (Tsytovich *et al.*, 1959, 111–12, Tables 2–3, Figure 5; 1964, 7; 98–9, Tables II–III; 105, Figure 5; cf. Tsytovich, 1958). In general, unfrozen water in fine-grained mineral soil at temperature ranges to −25° can cause appreciable variations of strength with temperature (Lovell, 1957), although the relationship is complex (Yong, 1966). Partial thawing as well as freezing is frequently cited in the Russian

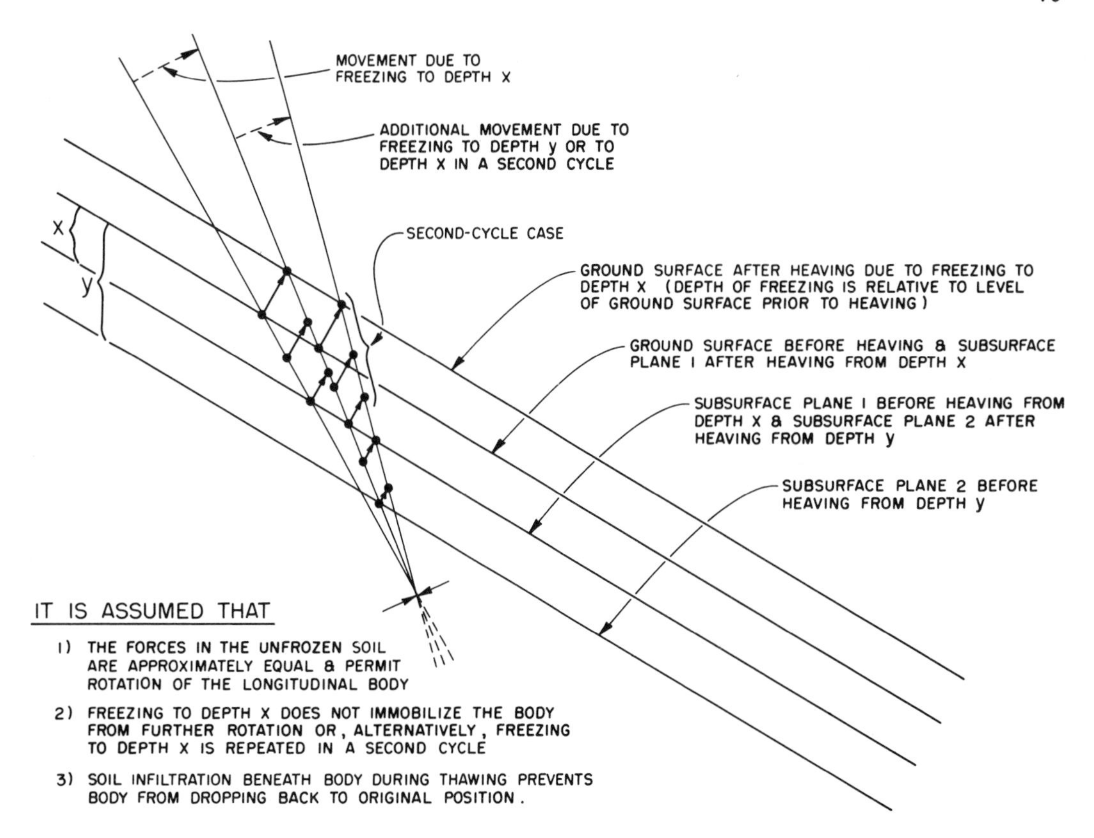

4.15 Frost heaving tending to rotate a longitudinal body towards a vertical plane

literature (cf. Savel'ev, 1960, 163–6). It may be significant that frost-heaved stones are commonly in materials having an appreciable content of fines. To the extent that freezing takes place from a permafrost table upward, as well as from the surface downward, it would also promote frost heaving of stones into unfrozen material. Upward freezing of the frost table occurs in some permafrost areas as reported long ago by Domrachev (1913, cited in Shvetsov, 1959, 81–2; 1964, 8–9), and it is supported by recent observations as reviewed elsewhere (Washburn, 1967, 138–40), but the magnitude of the phenomenon is not well known. Müller (1954, 130–1) reported freezing at depth to be more rapid than at the surface at Ella Ø, Northeast Greenland.

Even if material immediately above or adjacent to a stone is frozen, it would not necessarily prevent heaving of the stone, since the force of ice crystallization beneath the stone might lift overlying frozen material along with the stone. Also, if the stone were near the surface, localized shearing could lead to relative displacement of the stone with respect to the surrounding material. However, even without this shearing of frozen material the stone would work its way to the surface if the ice beneath it became replaced by thawed material. The effect then would be the same as the infilling of a void left by a stone that had been lifted by frost heaving of overlying material as depicted in the frost-pull hypothesis.

As reviewed by Corte (1966*b*), his detailed laboratory research demonstrates that repeated freezing from the surface downward can cause an upward displacement of coarse soil particles relative to fine. However, the exact mechanism is not clear and the experiments are hardly applicable to the forceful and rapid upfreezing of some stones and to the accompanying soil shearing (Washburn, 1969*a*, 55, Figures 30–1).

The above analysis is focused on the upfreezing of stones from depths below the influence of needle ice, which is essentially a surface or near-surface phenomenon as discussed under Work of needle ice in this chapter. It is well known that needle ice can lift sizeable stones, as illustrated by Troll (1944, 582, Figure 10; 1958, 29, Figure 10), and it may be a factor in the upfreezing of stones whose bases are below the surface but near it. Schmid (1955, 87) maintained that the only stones affected by needle ice would be those lying on the surface.

In the writer's opinion two mechanisms are primarily responsible for upfreezing of objects: the frost-pull and the frost-push mechanisms. The frost-pull mechanism seems particularly applicable to the upfreezing of targets. However, where there is evidence of rapid upfreezing as in the case of vegetation- and silt-capped stones, it would appear that the frost-push mechanism is more important.

The discussion has dealt primarily with frost heaving and the upfreezing of individual stones but frost thrusting could operate in a similar manner. This could happen wherever freezing isotherms are appropriately oriented in heterogeneous material or adjacent to contraction or dilation cracks.

It should also be noted that objects may move upward through a shallow layer as the result of cycles of wetting and drying (Cooke, 1970, 571–8; Jessup, 1960, 194; Springer, 1958, 64–5). How significant and widespread the process is does not appear to be known but in an arctic environment it is unlikely to approach the importance of upfreezing.

4 Local differential heaving

Local differential heaving is frost heaving that is significantly greater in one spot than in the area around it because of less insulation or other factors. Taber (1943, 1458–9) cited local differential heaving as an important cause of some forms of patterned ground. Under differential heaving, Taber (1943, 1452–6) included various aspects of frost heaving such as upfreezing of stones. The formation of dilation cracks by doming or irregular arching of the ground is another aspect (Benedict, 1970*a*). Thus local differential heaving does not specify the exact heaving process but emphasizes its local effect.

5 Work of needle ice

a General Needle ice (Taber, 1918, 262), also known as pipkrake, is an accumulation of slender, bristle-like ice crystals practically at, or immediately beneath, the surface of the ground. Elongation of the crystals is perpendicular to the cooling surface (i.e. parallel to the temperature gradient) commonly producing a nearly vertical structure on a horizontal surface except for long needles that tend to curve. Anomalously, the optic axis of crystals tends to be perpendicular rather than parallel to the temperature gradient (Corte, 1969*a*, 131; Steinemann, 1953, 502–5; 1955, 4–6).

Needles can range in length from a few millimetres to as much as 35–40 cm (Krumme, 1935, 39) depending on temperature, moisture, and soil conditions. Needles 0·5–3 cm long are quite common. There may be several tiers of needle ice separated by a thin layer of mineral soil, each tier sometimes representing a different freezing regime. In many cases tiers represent diurnal cycles but shorter or longer periods can be represented (Beskow, 1947, 7). Two tiers have been formed in the laboratory during a single freeze (Soons and Greenland, 1970, 585–7, 591–2). Needle ice can lift stones as large as cobbles, and thawing and collapse of needles can be a significant factor in sorting (Gradwell, 1957; Hay, 1936; Troll, 1944, 586–8, 669; 1958, 32–4, 92) and in frost creep.

It is noteworthy that needle ice often occurs as if raked. Troll (1944, 585–6; 1958, 30–2, 36–7) reported that the direction of 'raking' is parallel to the wind. On the other hand, J. R. Mackay and W. H. Mathews on the basis of experiments and repeated observations since 1958 in Canada (British Columbia) and New Zealand concluded that it is parallel to the late-morning sun and is a shadow effect developed during thawing rather than being a freezing effect as indicated by Troll (J. R. Mackay, personal communication, 1972; cf. Outcalt, 1970, 88). The fact that the phenomenon can occur on the equator (Mt. Kenya) where Troll observed a northwest-southeast orientation coinciding with nightly winds suggests that the wind hypothesis requires further research before being eliminated, even though the sun hypothesis can be demonstrated in places and may well be the general explanation.

Grain-size analyses of mineral soil in which needle ice has been seen appear to be rather few. Gradwell (1954, 245, Table 3; 249–52, 256–7) found that needle ice was best developed in loamy soil where fines exceeded 30 per cent, although other factors were also critical. Observations in Northeast Greenland (Washburn, 1969*a*, 82–5) support the views of Beskow (1947, 6) and several earlier investigators that a very high percentage of fines may inhibit needle ice under certain conditions. Depending on the rate of freezing, water movement through a 'tight' soil can be too slow for needle growth, in the same way that it can inhibit other forms of frost heaving (cf. Washburn, 1967, 102–3).

According to Outcalt (1971), three conditions must be fulfilled for needle ice to form: (1) a surface equilibrium temperature at least as low as $-2°$ for ice nucleation;[1] (2) a soil-water tension, $\gamma < \frac{2\sigma}{r}$, where σ = interfacial energy ice-water and r = effective radius of soil pores; and (3) the rate of heat flow from the freezing plane is balanced by the latent heat of fusion of the soil water supplying the growing ice crystal.

Needle ice may be important in the origin of some forms of patterned ground (cf. Hay, 1936; Troll, 1944, p. 586–8, 669; 1958, 32–4, 92) but its role here is not completely clear (Brockie, 1968). Occurrences of needle ice in Greenland (Boyé, 1950, 132; Washburn, 1969*a*, 82–5), in Arctic Alaska (K. R. Everett, 1963, 50–1, 135–7), and in the Canadian Arctic (Washburn, 1956*b*, 847, footnote 23) show that it is fairly widespread in arctic regions. It is probably much more widespread elsewhere (cf. Troll, 1944, 575–92; 1958, 24–37), and its disruptive work, including the effect on vegetation (Brink *et al.*, 1967; Schramm, 1958), seems to be far more common than suggested by the relatively few observations of needle ice itself, the difference presumably resulting from the fact that the effects long outlive the process. Nubbins, discussed below, are probably also due to needle ice.

b Nubbins A nubbin, as used here, is 'a small round-to-elongate earth lump, one to several centimeters in diameter' (Washburn, 1969*a*, 85–6). Nubbins include Feinerdeknospen (fine-earth buds) amid stony accumulations (Furrer, 1954, 233, Figures 15–16; 1955, 149; Höllerman, 1964, 101) but they are not restricted to such features.

Nubbins observed in Northeast Greenland (Washburn, 1969*a*, 85–8) tended to be elongate downslope, were commonly 1–2 cm wide and 2–6 cm long, and gave the surface a broken and very irregularly wrinkled aspect (Figure 4.16). Some nubbins were bare of vegetation, others were covered with black organic crust; in several places only the crust itself had the nubbin shape. The mineral soil of a representative nubbin was gravelly$_9$–clayey$_{18}$–silt$_{36}$–sand$_{37}$ in the top 5 cm and showed no significant difference at the 5–10 cm depth. There were a few stones at the surface

[1] The undercooling does not imply an air temperature $< -2°$ but can result from evaporation and thermal radiation; it is a momentary phenomenon, since once ice nucleation starts the temperature of the freezing water rises to 0° because of the latent heat of fusion. In nature, freezing usually begins close to 0° (cf. P. J. Williams, 1967, 93).

but the nubbins occurred whether or not stones lay in inter-nubbin depressions. Frozen nubbins were quite porous and contained small granular ice masses. Although needle ice was not seen in the nubbins, some nubbins with black organic crust were disrupted as if by needle ice. Moreover, needle ice was observed in the same central areas that had nubbins but not in areas where nubbins were largely lacking.

Nubbins or forms very similar to them were ascribed to the work of needle ice by Furrer (1954, 233–4; Figures 15–16; 1955, 149), Müller (1954, 135, 198), Mohaupt (1932, 32), and Troll (1944, 612–13; 1958, 52–3). Furrer and Mohaupt observed needle ice in the forms but Müller and Troll did not. Although it seems probable that most nubbins owe their origin to needle ice, Stingle (1971, 28–9) reported other forms.

6 Other surface effects

a Gaps around stones Narrow gaps encircling stones at the surface of the ground are another manifestation of frost heaving (Figure 4.17). They resemble features discussed by Bertil Högbom (1914, 302–3), Behr (1918, 102–10), and Schmid (1955, 89–90). Investigations in Northeast Greenland (Washburn, 1969*a*, 88–90) confirm Behr's and Schmid's suggestion that these gaps can be explained by the heaving and lifting of the immediate surface material away from the stones, since freshly frozen crusts were observed to have done just that. No ice was present in the resulting gaps, so that ice wedging between the stones and the adjacent mineral soil was not a factor in these instances. The top 3 cm of the soil had become so porous after freezing and heaving that irregular voids extended right through it. The ice responsible for the heaving was in two forms – small spots of needle ice with needles up to 0·5 cm long and, more prevalently, numerous tiny irregular lenses up to 0·1 cm thick. The frost-heave explanation for the gaps is probably generally valid with respect to stones that are large enough and of such shape that they would escape being pushed up by needle ice or pulled up by adjacent soil at the very start of freezing.

Many of the stones had shallow encircling gaps that were 0·3–1·0 cm wide and had a washed appearance; some of these gaps contained granules and small pebbles. Very probably these gaps originated as described and were then modified by meltwater run-off and rainwash.

b Surface veinlets Surface veinlets are 'sliver-like ice crystals that grow in a thin surface layer of the ground and intersect it vertically or at large angles so that their exposed sections appear needlelike' (Washburn, 1969*a*, 90). Individual veinlets generally taper towards both exposed ends but feathery forms occur, and branching at angles of 30° and 60° is described as common. Descriptions include those of T. M. Hughes (1884, 184), Marbut and Woodworth (1896, 992), and Behr (1918, 110–12; cf. 115–17). Hughes, and also Marbut and Woodworth, referred to

4.16 Nubbins, Mesters Vig, Northeast Greenland. Scale given by 17-cm rule (*cf. Washburn, 1969a, 87, Figure 62*)

4.17 (*Opposite*) Gaps around stones, Mesters Vig, Northeast Greenland. Scale given by 17-cm rule (*cf. Washburn, 1969a, 89, Figure 63*)

needles (among other designations) but the term surface veinlets avoids confusion with needle ice.

Surface veinlets observed in Northeast Greenland (Washburn, 1969*a*, 90) were in clayey silt and in silty sand. Upon thawing they left characteristic slits commonly a fraction of a millimetre up to 0·3 cm wide, 1–9 cm long, and 0·5–1·0 cm deep (Figure 4.18). Drying may contribute to their width with the result that the slits become desiccation cracks as described by Marbut and Woodworth. Although the needle-like aspect is commonly emphasized because of the surface appearance, the slits left by thawing are considerably deeper than wide. The growth of the veinlets must exert a lateral stress, however small, and therefore produce a kind of frost thrusting.

VI MASS DISPLACEMENT

1 General

The term mass displacement is defined 'as the *en-masse* local transfer of mobile mineral soil from one place to another within the soil as the result of frost action' (Washburn, 1969*a*, 90). It will be used mainly for upward movements of one kind of mineral soil into another but includes lateral movements. Among possible causes of mass displacement are cryostatic pressure, changes in density and intergranular pressure, and artesian pressure.

2 Cryostatic pressure

Cryostatic pressure was originally described as the hydrostatic pressure set up in pockets of unfrozen material trapped between the downward-freezing active layer and the permafrost table when the active layer becomes irregularly anchored to the permafrost table by freezing to it in some places sooner than others (cf. Washburn, 1956*b*, 842–5). An impermeable and resistant surface such as bedrock could have the same effect as the permafrost table. The variable freezing temperatures resulting from variations in grain size and moisture content are among the critical factors in irregular freezing and maintenance of plasticity. Cryostatic pressure can produce heaving (Muller, 1947, 21, 68) and probably thrusting. Direct field evidence is meagre but pressure effects have been measured in the laboratory (Pissart, 1970*a*, 35–6, 40). The process has been invoked by Philberth (1960; 1964, 142–7, 169–90) to explain the regularity of some forms of patterned ground, and by Corte (1967) to explain the formation, in the laboratory, of small soil mounds and intrusive phenomena. However, subsequent reports by Corte (1969*b*; Higashi and Corte, 1972), extending the same experiments, emphasized the role of frost heaving rather than cryostatic pressure.

Pissart (1970*a*, 40–4) demonstrated that pressures set up within a freezing mass, presumably as

4.18 (*Opposite*) Slits left by thawing of surface veinlets, Mesters Vig, Northeast Greenland. Scale given by 17-cm rule (*cf. Washburn, 1969a, 89, Figure 64*)

the result of some areas within the mass remaining unfrozen longer than others, are up to 4 kg/cm^2. These pressures are also basically cryostatic in being caused by confined freezing in a closed system but the system can be local as part of a larger open system. Thus, as stressed by Pissart, pressure effects are to be expected in seasonal freezing in a non-permafrost environment as well as in a permafrost environment.

3 Changes in density and intergranular pressure

In soil mechanics it is well known that

> The stresses that act within a saturated mass of soil or rock may be divided into two kinds; those that are transmitted directly from grain to grain of the solid constituents, and those that act within the fluid that fills the voids. The former are called *intergranular pressures* or *effective stresses*, and the latter *porewater pressures* or *neutral stresses* . . . only the intergranular pressures can induce changes in the volume of a soil mass. Likewise, only intergranular pressures can produce frictional resistance in soils or rocks. (Peck, Hanson, and Thornburn, 1953, 58.)

Where rock particles in frozen ground are separated by ice lenses and do not touch, density relationships are changed and melting of the ice can radically alter the pressure relationships. Whereas formerly only intergranular pressures existed, significant porewater pressures can be generated. If more meltwater is produced than can be accommodated in the available pore spaces, porewater pressures may keep some of the rock particles separated, reduce shear resistance caused by friction, and cause liquefaction of the material that was associated with the ice lenses (cf. R. F. Scott, 1969, 58). Under these circumstances the overlying material, if sufficiently heavy, will squeeze the excess water and probably some of the mobilized material associated with it towards areas of less porewater pressure. Fines and possibly even small stones could be displaced towards the surface, whereas large stones would be likely to sink. The process would be repetitive year after year as ice lenses redeveloped and thawed again, and the upward movement of material could be by small increments. In addition to melting of ice lenses, other less general changes such as slumping and drainage modifications could also lead to increased porewater pressures with similar effects. How delicate the equilibrium can be between stones and underlying fines is illustrated by stones sinking into mud 5 m from where a person was disturbing the ground by walking (Dybeck, 1957, 144). Disturbance of the ground by walking has also led to upwelling of mud that formed a nonsorted circle, a variety of patterned ground discussed later in this chapter.

> I heard a sound behind me. When I looked I observed mud spurting up from one spot along the path we had just taken. The mud continued to pour out for at least 15 minutes until it covered an area about 30 cm in diameter, forming a nonsorted circle essentially identical to the others nearby. (J. R. Reid, 1970*a*.)

On a slope such hydrostatic pressures would be aided by a hard desiccated surface and artesian effects, as suggested by Reid in reporting the above case.

Mortensen (1932, 421–2) was among the first to suggest the hypothesis that moisture-controlled density differences are important in the origin of patterned ground but he thought of the process as convection. Related significant contributions include those of Sørensen (1935, 32–53) and Jahn (1948*a*, 36–41, 89–100; 1948*b*, 52, 54, 57–8). The present concept (cf. Washburn, 1956*b*, 855) is consistent with (1) observations of plugs of finer material, characterized by vertically oriented elements, intruding coarser material (Bunting and Jackson, 1970, 200–8; Corte, 1962*b*, 12, 16, 25; Figures 100*a*, 100*b*; 1963*c*, 18, 20–1, 36; Figure 11; Mackay, 1953, 34–6; Washburn, 1956*b*, 844), (2) experiments by Dżułyński (1963; 1966, 16–21), Anketell, Cegła, and Dżułyński (1970), and Cegła and Dżułyński (1970) who outlined an essentially similar concept but with less emphasis on melting of ice, and (3) experiments by Jahn and Czerwiński (1965) who stressed equilibrium disturbances introduced by freezing and thawing. The hypothesis has the advantage over the cryostatic concept in that intrusion is into unfrozen material, and it accounts for displacement of fines into coarse material such as beach gravels that tend to remain free of ice and unconsolidated, however low the temperature.

The hypothesis is particularly applicable to some types of patterned ground (cf. Rohdenburg and Meyer, 1969) and to interpenetrating beds (involutions), discussed later in this chapter. Field testing of the hypothesis presents major difficulties, and the ideas may be more amenable to detailed laboratory investigation first.

Some investigators, for instance Kostyaev (1969), stressing the role of a high-moisture environment whether or not it is ice-induced, have argued that density-controlled mass displacements explain a variety of patterned ground forms and therefore that such forms are azonal and have no necessary periglacial significance. This argument in an extreme form can be countered by pointing out the lack of typical periglacial-type forms of patterned ground in other environments. On the other hand, pending further research, differences of opinion will continue as to the significance of density-controlled mass displacements as a periglacial process.

4 Artesian pressure

By definition, artesian pressure arises from differences in hydrostatic head. It differs from cryostatic pressure and changes in density and intergranular pressure in being slope induced, although it may aid these processes where they operate on slopes. Artesian pressure of water-saturated soil beneath a frozen surface is perhaps a more common mass-displacement process than is usually recognized.

Selzer (1959) observed the growth of isolated earth mounds up to 60 cm high, consisting of unfrozen stony mineral soil that was apparently injected upward from beneath a frozen surface layer some 50 cm thick. Eye-witness accounts of such soil movements are very rare, and detailed field and laboratory data are lacking. On the other hand, there is abundant evidence that artesian flow of water beneath frozen ground is a very common process. Its role in growth of pingos is discussed later.

VII FROST CRACKING

1 General

Frost cracking is fracturing by thermal contraction at subfreezing temperatures. This definition conforms to general usage in excluding dilation cracking by differential heaving, although other usages (cf. Benedict, 1970*a*) occur. Frost cracking was suggested as early as 1823 (Figurin, 1823, 275–6) in explanation of large polygonal cracks in Siberia. However, not all cracking is polygonal (Figure 4.19). Following other pioneer Russian investigations (cf. Shumskii, 1959, 276–9; 1964*a*, 5–9), the process was elaborated in considerable detail by Leffingwell (1915; 1919, 205–14) on the basis of work in Arctic Alaska. Although attacked by several later workers (Pissart, 1964; Schenk, 1955*a*, 64–8, 75–6; 1955*b*, 177–8; Taber, 1943, 1519–21), the process is accepted by most investigators including Pissart (1970*a*, 10–19). Taber's criticism was effectively answered by Black (1963; cf. Washburn, 1956*b*, 851) who measured seasonal changes in crack widths, and by the work of Berg and Black (1966) and Mackay (1972*c*) who demonstrated repeated cracking and the growth of ice wedges. A sound theoretical basis for the process was elaborated by Lachenbruch (1961; 1962; 1966) and supported by field observations of Kerfoot (1972). As discussed below, frost cracking is not restricted to a permafrost environment but it rarely forms well-defined and persistent features elsewhere. Frost cracking, and wedge structures described later in this chapter, have been comprehensively reviewed by Dylik (1966; cf. also Dylik and Maarleveld, 1967, for a review of the literature).

However, even though frost cracking is a valid process, other cracking processes such as desiccation and dilation leave features simulating fossil frost-crack phenomena (Dżułyński, 1965; Johnsson, 1959; Wright, 1961, 941–2).

2 Characteristics

In frozen ground, ice is the critical material. Pure ice has a coefficient of linear contraction of about $50 \times 10^{-6}/°C$, varying somewhat with temperature. If the ice has a salt content, the coefficient also varies with the salinity (D. L. Anderson, 1960, 310–15). Contraction calculations,

4.19 (*Opposite*) Linear frost crack, Mount Pelly, Victoria Island, Northwest Territories, Canada (*cf. Washburn 1947, Figure 2, Plate 31*)

however, are complicated by the fact that the ice content of frozen ground is highly variable and that ice does not deform purely elastically but follows a power law, which according to Glen (1958) has the form

$$e = \left(\frac{\sigma}{A}\right)^n$$

where e = strain rate (per cent/yr), σ = stress (bars. 1 bar = 10^6 dynes/cm^2 $\approx$ 14·5 pounds/in^2), A = a constant (bars·yr$^{-(n+1)}$) at a given temperature, and $n = 2 - 4$.

As indicated by Lachenbruch (1966, 65–6), frost cracking is more dependent on the rate of temperature drop than on the actual subfreezing temperature at time of cracking. The first time cracking occurs it starts at the surface and extends to a depth of 3 m (10 ft) or more (Lachenbruch, 1970*b*, J1) in a pattern determined by temperature conditions and the rheological behaviour of the frozen ground, and open cracks as deep as 4 m have been observed by Mackay (1972*c*). Most frost cracks are in unconsolidated material, and in such material in a permafrost environment the crack pattern is primarily imprinted in the permafrost, rather than in the active layer where thawing tends to destroy the pattern. Therefore subsequent cracking starts at the permafrost table and is propagated upward as well as downward. Where frost cracking occurs in bedrock, as reported for some nonsorted polygons (Arthur *in* Corte, 1969*a*, 140), or takes place in unconsolidated material where permafrost is lacking, the same analysis does not apply.

The rate of temperature drop needed for cracking varies with conditions. Black (1963, 264; 1969*b*, 228) cited a rapid temperature drop of 4° as being adequate to crack frozen ground having a thermal coefficient of contraction approaching that of ice; a change of about 2° would be adequate to propagate the crack downward. Where the coefficient is near that of rock (*c.* 1/5–1/6 that of ice), field data suggest that a temperature drop of about 10° is required to start the crack, and a change of some 4° is needed to propagate it downward.

Frost cracking in a permafrost environment is commonly accompanied by growth of ice wedges or sand wedges. Ice wedges, whose ice characteristics are described under Permafrost in the chapter on Frozen ground, grow as the result of surface water, ground water, or water vapour entering a fissure in permafrost and ice growing there. Ice wedges start as ice veins, which are the focus of subsequent cracking. The addition in any one year may be very small – 1–20 mm according to Shumskii (1964*b*, 198) – but repeated cracking at the same place over a period of years leads to well-developed ice wedges (Figures 4.20–4.22), some being massive in suitable material. Exposures parallel to the trend of an ice wedge can show continuous ice and give the impression of a continuous ice layer (Figure 4.23). Other things remaining equal, large ice wedges tend to develop in fines or in peaty soil rather than sandy material, the width of wedges in pure sand being limited to 0·1–0·5 m (Shumskii, 1964*b*, 198). Massive wedges can extend to depths of

30 m and more, and can grow so large laterally, squeezing and deforming the intervening soil into columns, that the volume of ice exceeds the volume of soil (Shumskii, 1964*b*, 199). Adjacent wedges may even form essentially continuous stratiform masses (Figure 4.24), which may be up to 80 m thick in the New Siberian Islands and along the Arctic Coast of Siberia (Gerasimov and Markov, 1968, 14; Grave, 1968*a*, 53; 1968*b*, 9). However, some stratiform ice masses described as of ice-wedge origin may be massive beds of segregated ice such as those described by Mackay (1971*b*; 1972*a* 15–19).

Ice wedges may be syngenetic or epigenetic in the same sense that permafrost may be one or the other – that is, depending on the presence or absence of continued sedimentation. Wedges growing in the absence of sedimentation are epigenetic. They increase in thickness but not in the vertical dimension (Figure 4.25A). Growth is limited by the compressive stresses generated by repeated cracking and crack filling. Wedges growing concurrently with sedimentation are syngenetic and increase in both thickness and vertical dimension (Figure 4.25B). Assuming a constant depth of cracking, the width of syngenetic wedges is limited by the expression (Dostovalov and Popov, 1966, 103)

$$m = \frac{\bar{h}\bar{c}}{d}$$

where m = thickness of ice wedge, $\bar{h}$ = mean depth of cracks, $\bar{c}$ = mean width of crack at surface, and $\bar{d}$ = mean thickness of sediment deposited during one cycle, or rate of sedimentation. Thus the width is inversely proportional to rate of sedimentation. Because of concurrent development of ice lenses, syngenetic ice wedges are more likely than most to intersect lenses in a honeycomb-like arrangement.

Sand wedges, formed by repeated frost cracking and infilling with dry sand or loam, occur in the arid parts of the Antarctic (Berg and Black, 1966, 70–3; Péwé, 1959). Attempts have been made to use the width of ice wedges or sand wedges as an age indicator of the surface in which they occur (Berg and Black, 1966), but the method must be used very cautiously (Calkin, 1971, 386).

Ice wedges require certain temperature conditions for growth and preservation. In Alaska, for instance, the southern limit of presently active wedges roughly coincides with a mean annual air temperature of $-6°$ to $-8°$ (Péwé, 1966*a*, 78; 1966*b*, 68). Upon thawing of permafrost, the sites of wedges are filled in by collapsed material that becomes a cast of the wedge. Such ice-wedge casts (Figures 4.26, 4.27) are among the few acceptable criteria for former permafrost.

The infilling of a thawed ice wedge may be sand and form a so-called sand wedge. However, as noted above, sand wedges can form instead of ice wedges under arid conditions such as those in the Antarctic dry valleys today. Thus as fossil features of former permafrost, sand wedges may record such an origin or be an ice-wedge cast associated with a moister climate. The difference in

4.20 Ice wedge, Ostrov Chagen, Lena River near Shamanskiy Bereg, 130 km north of Yakutsk, Yakutia, USSR. 15-cm rule as scale

4.21 (*Opposite*) Ice wedge at Kurankh, right bank Aldan River, 170 km above junction with Lena River, Yakutia, USSR

4.22 (*Above left*) Ice wedge at Mamontova Gora, left bank Aldan River, 310 km above junction with Lena River, Yakutia, USSR

4.23 (*Above right*) Continuous ice exposure parallel to an ice wedge, left bank Aldan River, 130 km above junction with Lena River, Yakutia, USSR

4.24 (*Left*) Stratiform ice mass interspersed with baydjarakhs, Novosibirskiye Ostrova (*Toll, 1895, Plate III*)

origin might be hard to establish. Péwé (1959) suggested that some of the loess or loam wedges of Europe may be similar to Antarctic sand wedges rather than being ice-wedge casts formed by the thawing of ice wedges and collapse of the adjacent or overlying material. Subsequent reports (Black, 1969*b*, 229) indicate that original soil wedges formed in dry polar climates differ from ice-wedge casts in being (1) thinner, (2) vertically foliated, and (3) limited to material small enough to enter narrow cracks up to 1–2 cm wide. New cracks tend to form between older

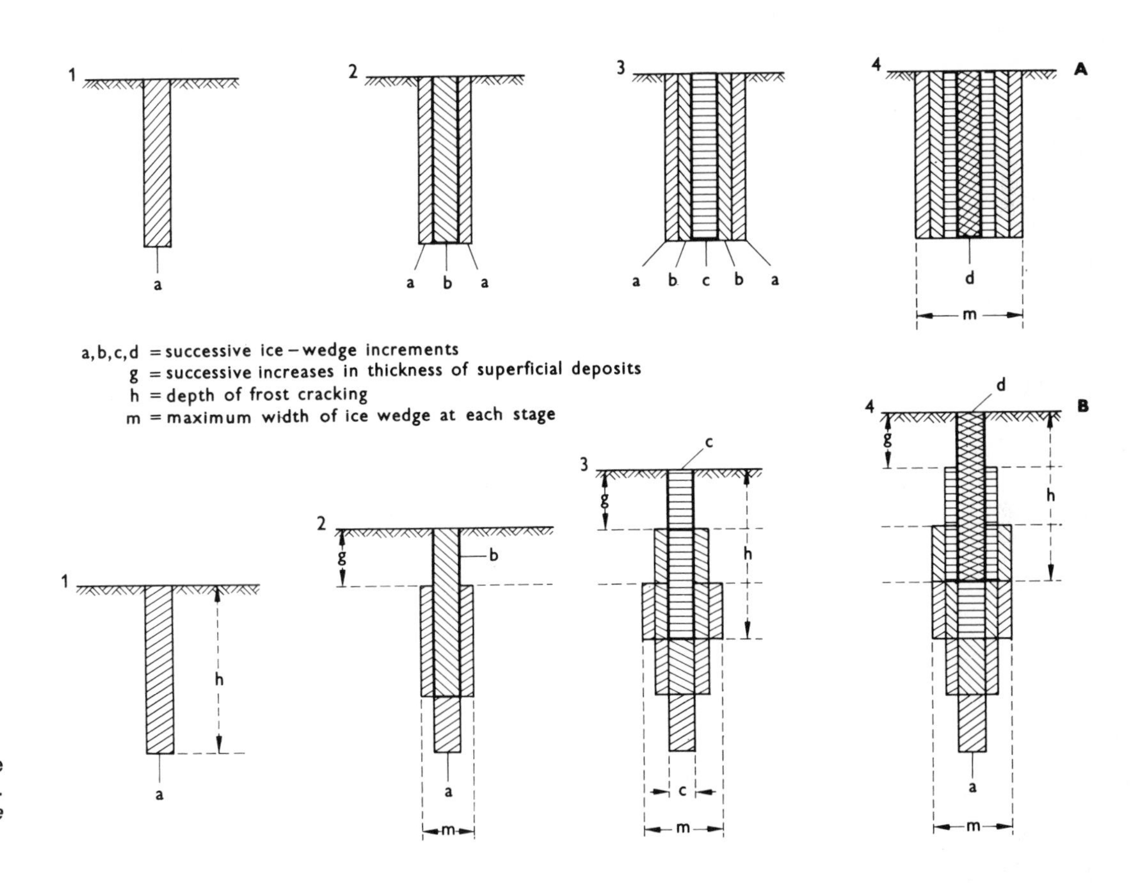

4.25 Diagram of growth of an epigenetic ice wedge (A) and syngenetic ice wedge (B). *(Dostovalov and Popov, 1966, 103, Figure 2)*

ones and with time result in the development of many vertical narrow wedges. Ice-wedge casts on the other hand, as the result of collapse of material into voids left by melting ice, tend to (1) be broader, (2) have foliation that is less distinctly vertical, and (3) contain larger stones if the overlying or adjacent material is stony.

Frost cracking confined to a seasonally frozen layer – seasonal frost cracking – has been reported from both a permafrost environment (cf. Pissart, 1968, 177–9; 1970*b*, 15–16) and from non-permafrost areas (cf. Washburn, Smith, and Goddard, 1963). However, frost cracks demonstrably formed in a non-permafrost environment rather than occurring in one formerly characterized by permafrost appear to be either temporary or features formed under conditions of such deep freezing as to approach permafrost conditions. Where seasonal frost cracking is accompanied by wedge growth, the wedges are of the sand or loam variety and not ice-wedge casts.

Pissart (1970*a*, 18–19) reviewed seasonal frost cracks as follows. (1) The cracks are developed in sands; (2) the wedge bedding is more or less vertical, and if a range of grain sizes is present there is sorting with the largest grain sizes at top; (3) the sediments enclosing the wedge tend to be downwarped along the contact, whereas in ice-wedge casts they are upwarped; (4) the structures are shallow, being confined to the seasonally frozen layer; (5) when fully developed the wedges may be very broad, approaching equilateral triangles. Pissart pointed out that the first three characteristics also apply to some wedges where permafrost occurs but that the great vertical extent of these cracks (>2m) can be used as evidence of permafrost.

While agreeing that caution needs to be used and that wedges (or cracks) with great vertical extent indicate permafrost, the present writer questions whether it has been demonstrated that well-developed sand- or soil-wedges form in a non-permafrost environment. The situation described by Danilova (1956; 1963) and Bobov (1960) and cited by Pissart (1970*a*, 17–18) as a demonstration is not convincing because, even if present cracking occurs, the wedges may not have formed under present conditions. In fact Danilova (1956, 117–18; 1963, 94) thought they had formed under former colder conditions and based the hypothesis that the cracks were confined to a seasonally frozen layer on an interpretation of wedge features. Moreover, even though development of some wedges is limited to the active layer above permafrost (Katasonov and Solov'ev, 1969, 12–18; Katasonov, 1972), soil wedges may still be indicative of a permafrost environment unless it is proved that comparable wedges can form elsewhere.

4.26 (*Far opposite*) Ice-wedge cast, Zakrecie, Poland

4.27 (*Near opposite*) Ice-wedge cast, Chodecz, Poland

VIII SORTING BY FROST ACTION

A number of frost-action processes contribute to size sorting of mineral soil. Upfreezing of stones, work of needle ice, mass displacement due to moisture-controlled changes in intergranular pressure, and probably cryostatic pressure, which have been discussed above, are among them.

These and other processes related to the origin of sorted forms of patterned ground have been reviewed elsewhere (Washburn, 1956*a*; 1956*b*, 836–60).

More recently the work of Corte (1961; 1962*a*; 1962*b*; 1962*c*; 1962*d*; 1962*e*; 1963*a*; 1963*b*; 1966*a*; 1966*b*) has resulted in new data on sorting accompanying the movement of a freezing front. These data show that under certain laboratory conditions there is a general tendency for finer material to migrate ahead of the freezing front whatever its direction of movement, while coarser material is more readily entrapped by freezing and left behind. The factors are complex and include the orientation of the freezing front, rate of freezing, moisture content of the mineral soil, and shape, size, and perhaps density of particles. All these enter into what Corte termed horizontal and vertical sorting. In addition he reported a mechanical sorting involving downward movement of fines in an unfrozen upper layer under the influence of a freezing plane that simulated upfreezing from a permafrost table (Corte, 1962*c*, 8–9; 1962*d*, 54, 65; 1966*b*, 193–7, 232). That sorting in nature occurs by migration of fines ahead of a freezing front has been challenged by Inglis (1965) but supported by Jackson and Uhlmann (1966) and Pissart (1966*a*). Kaplar (1969, 35–6) argued that contrary to Corte's conclusion, the process would not apply to a downward freezing front because of the difficulty of pushing fines into the underlying soil. The upfreezing of individual stones, described in this chapter under Frost heaving and frost thrusting, is probably related to some aspects of the sorting investigated by Corte but it is distinct in involving, in places, a dynamic heaving through fines. It has been shown that particles imbedded in ice can migrate under a temperature gradient (Hoekstra and Miller, 1965, 6–8; Römkens, 1969), but whether this process has significance for sorting in nature is not known. In fact all the sorting processes related to frost action are poorly understood and much remains to be learned from both field and laboratory work. The possibility that rise of capillary moisture, or movement of moisture to loci of ice development, may carry clay-size particles and contribute to sorting merits investigation (cf. Bakulin, 1958 – cited *in* Kachurin, 1959, 374; 1964, 10; Cook, 1956, 16–17; Johansson, 1914, 19–20, 22; Thorodssen, 1913, 253–4; 1914, 260; Washburn, 1956*b*, 841–2). Laboratory work of R. Brewer and Haldane (1957, 303, 308) supports the possibility.

Finally, as noted under Frost heaving and frost thrusting, and as is also true of some other periglacial phenomena, processes other than frost action can result in convergent sorting effects.

IX PATTERNED GROUND

1 General

'Patterned ground is a group term for the more or less symmetrical forms, such as circles, polygons, nets, steps, and stripes, that are characteristic of, but not necessarily confined to, mantle subject to intensive frost action' (Washburn, 1956*b*, 824). Much of the following discussion is based on the publication cited.

Although various active forms of patterned ground occur in a variety of environments as reviewed by Troll (1944), periglacial forms are the most widespread. Some cartographic syntheses relate patterned ground to environment (Figure 4.28), but there are so many forms of patterned ground to consider that much detailed work will be required for accurate syntheses. For instance, Hövermann (1960) argued that Troll's (1944, 661) contention that the lower limit of patterned ground rises towards the continental interior away from sources of moisture is not valid, but on the contrary has an opposite trend to treeline and snowline and is lower in Eurasia than in European mountains at comparable latitudes. Detailed regional overviews by Hollerman (1967) for the Pyrenees, the eastern Alps (cf. also Stingl, 1969; 1971), and the Apennines (cf. also Kelletat, 1969), by Jan Lundqvist (1962) for Sweden, and the comparison by Furrer (1965) of altitudinal ranges of various kinds of patterned ground in the Swiss Alps and in the Karakoram Range of northern India indicate the kind of studies needed. Also helpful are regional inventories such as the one for the mountains of Snowdonia, North Wales (Ball and Goodier, 1970), and detailed local studies such as those of Rudberg (1970; 1972).

Patterned-ground forms tend to have a characteristic sequence related to altitude. Excluding forms controlled by local factors adjacent to glacier fronts, the sequence from lower to higher altitude as reported by Furrer (1965, 72) for the Swiss Alps and the Karakoram Range comprises the following zones: (1) zone of ploughing blocks (Wanderblöcke) – i.e. stones sliding downslope with a ploughing action, leaving a linear depression upslope and forming a frontal ridge downslope. Most earth hummocks and turf hummocks (Erdbülten) lie in this zone; (2) zone of garlands (Girlanden) – i.e. terrace-like features; (3) zone of earthflows (Erdströme); (4) zone of miniature patterned-ground; (5) zone of sorted or stony lobate forms (Steinzungen); (6) zone of large sorted patterned ground. These zones overlap and the average altitudes of zones 5 and 6 are practically identical. Furrer (1965, 72) cited Jäckli (1957, 21) for the view that zone 6 tends to be characterized by permafrost.

Many patterned-ground forms are known and many different terms have been employed for them. A purely descriptive terminology in common use is based on geometric form and presence or absence of prominent sorting[1] (Washburn, 1950, 8–9). However, it should be recognized that some forms are gradational in both pattern and sorting as pointed out by Black (1952, 125) and others.

[1] The terms well sorted and poorly sorted are used in the geologic sense to indicate, respectively, well-developed and poorly developed uniformity in grain size. The engineering usage is the opposite with well sorted indicating a wide range of grain sizes.

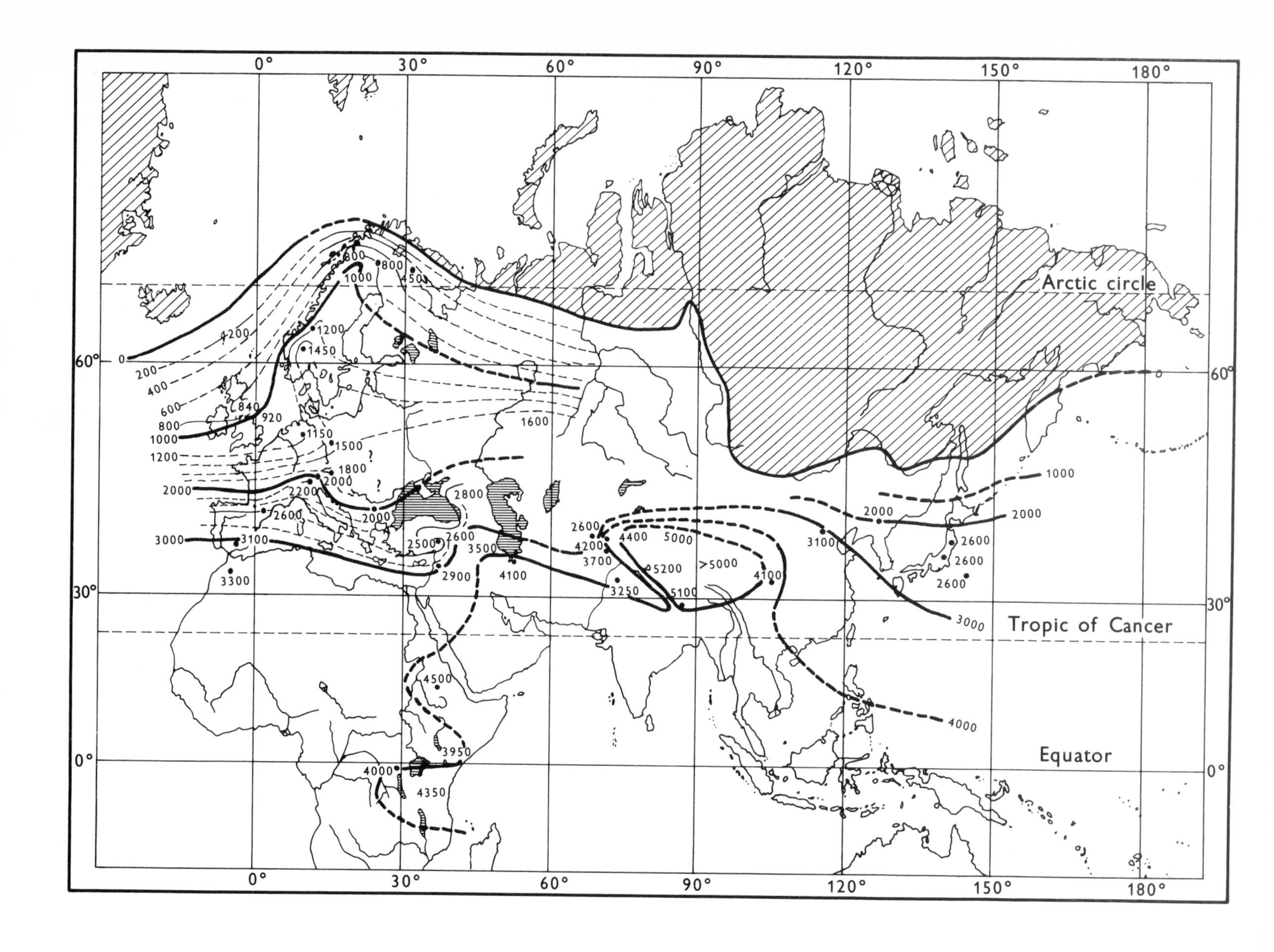

0°
30°
60°
90°
120°
150°
180°
60°
30°
0°
Arctic circle
Tropic of Cancer
Equator
0
200
400
600
800
1000
1200
2000
3000
1200
1450
840
920
1150
1500
1800
2000
2200
2600
3100
3300
800
1000
800
450
1600
2000
2800
2500
2600
2900
3500
4100
2600
4200
3700
3250
4400
5000
5200
>5000
5100
4100
3100
2000
1000
2000
2600
2600
2600
3000
4000
4500
3950
4000
4350

The following forms of patterned ground are recognized:

Circles	nonsorted sorted
Polygons	nonsorted sorted
Nets	nonsorted sorted
Steps	nonsorted sorted
Stripes	nonsorted sorted

Circles, polygons, and nets characteristically occur on nearly horizontal surfaces. Their unit component, termed a mesh or cell, tends to become elongated over a transition gradient of 2°–7°, depending on conditions, and in sorted forms to turn into stripes (cf. Hussey, 1962; Washburn, 1956*b*, 836–7). All stripes are confined to slopes but the sequence of slope relationships is less well established for nonsorted stripes than sorted, and both nonsorted and sorted stripes can occur without merging upslope or downslope into mesh forms.

Inactive and fossil patterned ground can be of great value in environmental reconstructions and paleoclimatic studies. Inactive (also called relict) forms have features similar to active forms (except for evidence of inactivity) and might become reactivated through some environmental change. Fossil forms are commonly less obvious, may be revealed in cross-section only, and are more clearly irreversible in activity. In periglacial literature, the term fossil does not imply former organic life.

4.28 (*Opposite*) Lower limit (m) of patterned ground in the mountains of Eurasia and Africa. The lower limit in eastern Siberia lies much farther north than shown (*Troll, 1969, 233*) (*Troll, 1947, 164, Figure 1; key translated*)

2 Nonsorted circles

Nonsorted circles are patterned ground whose mesh is dominantly circular and lacks a border of stones (Figures 4.29–4.30). They are characteristically margined by vegetation, and occur singly or in groups. Common diameters are 0·5–3 m. The central areas tend to be slightly dome shaped and cracked into small nonsorted polygons. Most nonsorted circles occur on nearly horizontal surfaces.

The mineral soil normally has a high content of fines and may or may not contain stones. Cross-sections indicate that the material of the central areas in some forms has risen from depth (Figures 4.31, 4.32).

4.29 Nonsorted circle (Experimental Site 3), Mesters Vig, Northeast Greenland (*cf. Washburn, 1969a, 108, Figure 69*)

4.30 (*Opposite*) Nonsorted circles joined by narrow neck, Mesters Vig, Northeast Greenland (*cf. Washburn, 1969a, 109, Figure 70*)

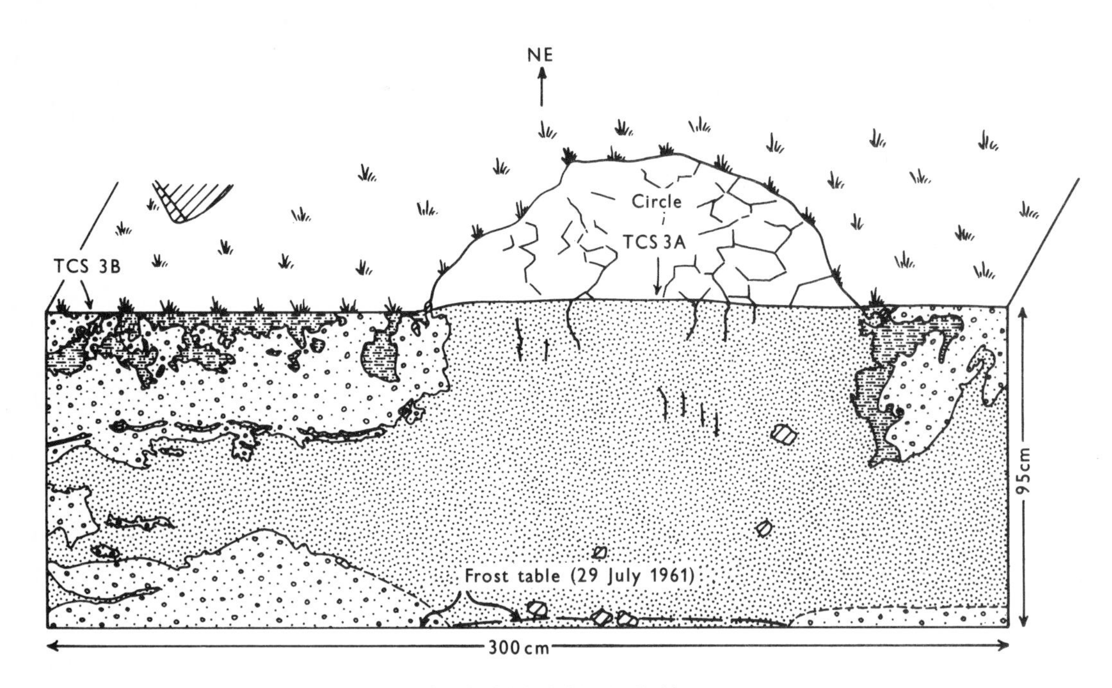

From field sketches made by J. Scully and F. Ugolini and from tracing of photograph

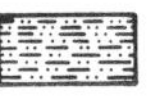
Silty–gravelly sand with humic material Very dark grayish brown (10YR 3/2) to moderate yellowish brown (10YR 5/4)

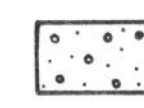
Gravelly sand with shells Brown (10YR 5/3) to light yellowish brown (10YR 6/4)

Sandy–silty clay with shells Dark gray (10YR 4/1)

Cracks

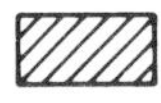
Stones

Vegetation

TCS Thermocouple string

4.31 Cross-section of nonsorted circle (Experimental Site 3), Mesters Vig, Northeast Greenland (*cf. Washburn, 1969a, 112, Figure 72*)

Nonsorted circles are prominent in polar, subpolar, and highland environments. Their distribution in northern Eurasia is shown in Figure 4.33. Similar forms known as 'normal gilgai' and characterized by a high clay content, occur in sub-humid and semi-arid areas in Australia (Costin, 1955; Hallsworth, Robertson, and Gibbons, 1955), and may be more widespread than is commonly recognized (cf. E. M. White and Agnew, 1968).

Recognition of cold-environment, inactive nonsorted circles that are out of balance with their present surroundings is dependent on proof of inactivity. Surface features such as lichen-covered stones and vegetation-covered central areas provide evidence. Fossil forms in stratigraphic sections would be hard to recognize but may be represented by some involutions (cf. Hopkins and Sigafoos, 1951, 98; Poser, 1947*a*, 12; 1947*b*, 233).

4.32 (*Opposite*) Parallel cross-sections of nonsorted circles joined by narrow neck, Mesters Vig, Northeast Greenland (*cf. Washburn, 1969a, 114–15, Figure 74*)

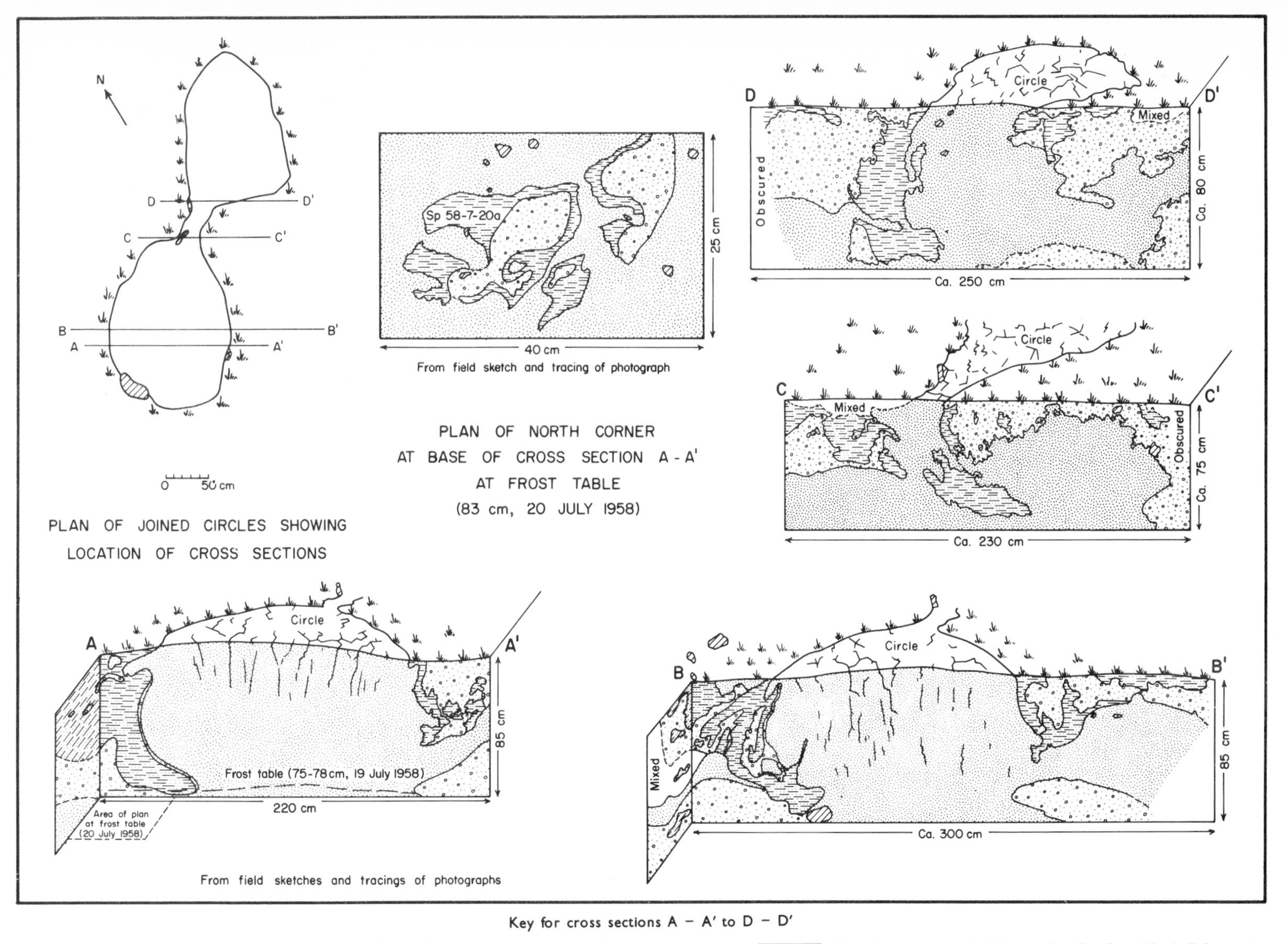

Key for cross sections A – A' to D – D'

Gravelly sand with humic material
Dusky brown

Gravelly sand with shell fragments
Moderate yellowish brown (10YR 5/4)

Gravelly-clayey sand-silt to sandy-silty clay with shell fragments
Light to dark gray with brownish to greenish tints in places

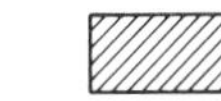
Cracks

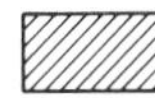
Stones

Vegetation

SKETCHES OF EXCAVATION 58-5-19 12 m EAST OF EXPERIMENTAL SITE 3

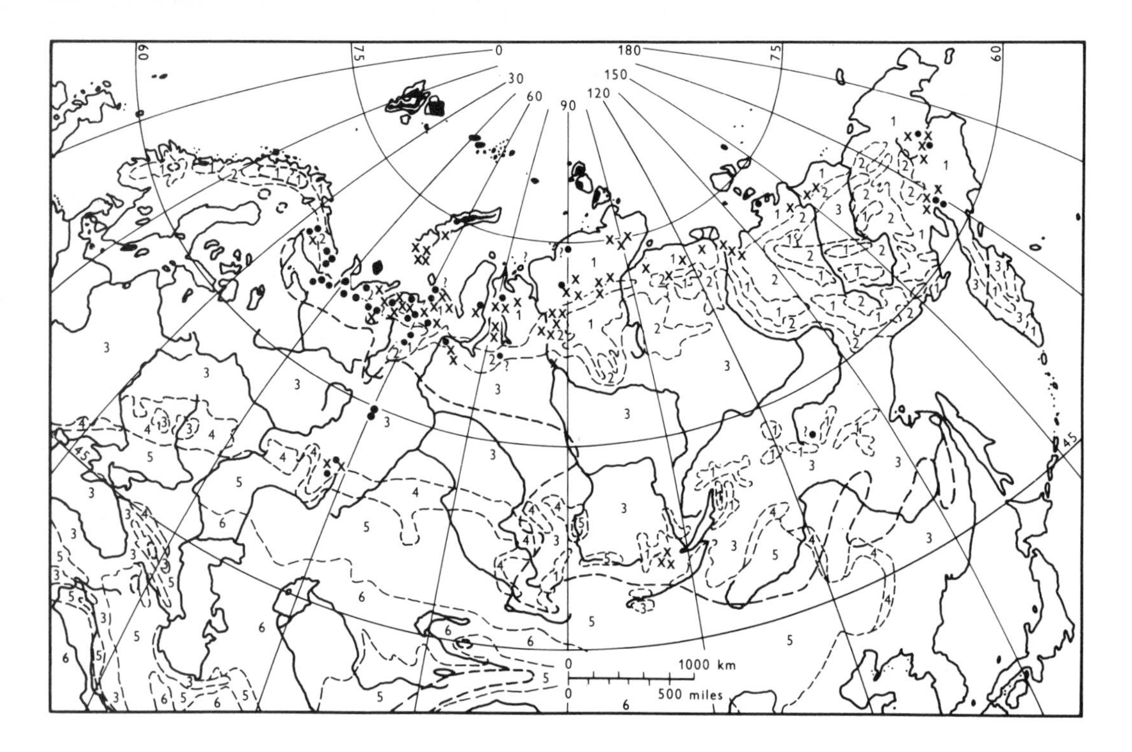

Glaciers
Limit of permafrost, including discontinuous zone

1 Tundra
2 Forest tundra
3 Forest
4 Forest steppe
5 Steppe
6 Desert

4.33 Distribution of active nonsorted circles and related forms with flat (•) and domed (×) surfaces in northern Eurasia (*Frenzel, 1960a, 1011, Figure 8*)

3 Sorted circles

'Sorted circles are patterned ground whose mesh is dominantly circular and has a sorted appearance commonly due to a border of stones surrounding finer material' (Washburn, 1956*b*, 827) (Figures 4.34–4.36).[1] Debris islands are sorted circles amid blocks or boulders. Like nonsorted circles, sorted ones occur singly or grouped, and the size range is also similar. Sorted circles are common on nearly horizontal surfaces but debris islands may occur on gradients as steep as 30°.

[1] The qualification 'commonly' was introduced because of certain stone-centred and other forms surrounded by edgewise slabby stones. Subsequent reports indicate that in places these forms developed by wave action rather than frost action (Dionne and Laverdière, 1967, Dionne, 1971b). Whether this is true of all stone-centred forms remains to be determined.

4.34 (*Opposite*) Sorted circles, Hornfund-fiord, Vest Spitsbergen. Photo by Alfred Jahn

4.35 Sorted circle (debris island), Mesters Vig, Northeast Greenland. Scale given by 16-cm rule (*cf. Washburn, 1969a, 155, Figure 96*)

4.36 (*Opposite*) Sorted circle (debris island) on 31° gradient, Hesteskoen, Mesters Vig, Northeast Greenland. Scale given by hammer at centre (*cf. Washburn, 1969a, 157, Figure 98*)

Such slope forms tend to have central areas that are considerably less steep than the general gradient but can be as high as 25°.

The central areas have a concentration of fines, either with or without stones. The stones of the borders surrounding the central areas tend to increase in size with the size of the circles. Tabular stones tend to be on edge with their long axis in the vertical plane parallel to the border, the next most common long-axis orientation being at right angles to it.[1] This fabric is reported to characterize the central areas as well as the borders, with the former having most of the stones dipping at angles >45° (Furrer, 1968; Furrer and Bachmann, 1968, 9–12).

Sorted circles are characteristic of polar environments. However, they also occur in subpolar and highland regions. Although best developed where there is permafrost, they are present in Iceland in places where it is lacking (Steche, 1933, 209).

Inactive sorted circles, identified as such by lichen-covered stones in the borders and central areas and by the vegetated nature of the central areas, are more easily recognized than inactive nonsorted circles. Fossil forms in stratigraphic section should also be easier to identify but well-documented reports are rare.

4 Nonsorted polygons

Nonsorted polygons are patterned ground whose mesh is dominantly polygonal and lacks a border of stones. It is convenient to classify nonsorted polygons into small forms (diameter < 1 m) and large forms (diameter > 1 m), since in many places they have distinct origins (Washburn, 1969*a*, 123–49). Meshes of small forms (Figures 4.37–4.38) measure as little as 5 cm across, those of large forms (Figures 4.39, 4.40) can exceed 100 m (Black, 1952, 130). Nonsorted polygons are group forms whose mesh (or borders between polygons) is commonly but not always delineated by a furrow and a crack. Where vegetation is sparse, the vegetation is generally concentrated along the furrow and emphasizes the pattern. Nonsorted polygons are most frequent on nearly horizontal surfaces but are by no means confined to them. Small forms have been noted on gradients as high as 27°, and large ones are also known to occur on steep slopes in polar regions.

The mineral soil of nonsorted polygons can be well-sorted fines, sand, or gravel, or it can be a diamicton. Ice-wedge polygons have an ice wedge coincident with the borders and their borders tend to be depressed or raised with respect to the central areas, reflecting, respectively, growth or decay of the ice wedges. During the thaw season, the low-centred polygons often contain ponds in

[1] As used here 'on edge' does not necessarily imply that the long axis is dominantly horizontal in the vertical plane. The prevailing orientation of *a* and *b* axes in the vertical plane in the bordering stones of sorted forms of patterned ground still remains to be established (Watson and Watson, 1971, 113–14).

4.37 (*Opposite*) Small nonsorted polygons, DeSalis Bay, Banks Island, Northwest Territories, Canada (*cf. Washburn, 1950, 26, Figure 1, Plate 6*)

4.38 Small nonsorted polygons on a 27° gradient, Mesters Vig, Northeast Greenland. Scale given by 35-mm camera case at centre (*cf. Washburn, 1969a, 133, Figure 84*)

4.39 (*Opposite*) Large nonsorted polygons, Arctic Coastal Plain, Alaska

4.40 (*Opposite*) Large nonsorted polygons, Mesters Vig, Northeast Greenland. Polygon marked by ice axe is 3 m in diameter (*cf. Washburn, 1969a, 148, Figure 92*)

the central areas, the high-centred ones water in the bordering depressions. (Hussey and Michelson, 1966, 165–70). Some ice-wedge polygons enclose small pingo-like, ice-cored mounds (French, 1971*a*). Similarly, sand-wedge polygons have coincident sand wedges instead of ice wedges. Some nonsorted polygons occur essentially in bedrock (Arthur *in* Corte, 1969*a*, 140; Berg and Black, 1966, 69–70; W. E. Davies, 1961*a*; Washburn, 1950, 47–9). Nonsorted polygons in cross-section are illustrated in Figures 4.41–4.42.

Small nonsorted polygons due to desiccation are ubiquitous. Large forms occur mainly in two contrasting arid environments – cold and warm. The large, cold-environment forms are characteristically polar, and ice-wedge polygons and probably most sand-wedge polygons are necessarily associated with permafrost. In the continuous permafrost zone of Alaska this implies (Péwé, 1966*a*; 1966*b*) a mean annual air temperature ranging from approximately $-6°$ to $-8°$ in the south to $-12°$ in the north, and a mean annual degree-days (°C) freezing range of 2800 to 5400 (Figures 4.43–4.45). Minimum temperatures at the top of the permafrost here range from some $-11°$ to $-30°$. Annual precipitation is about 20 cm as rain and less than 140 cm as snow. In the

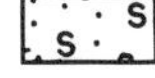

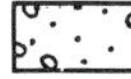

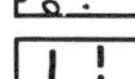

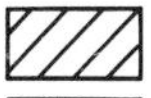

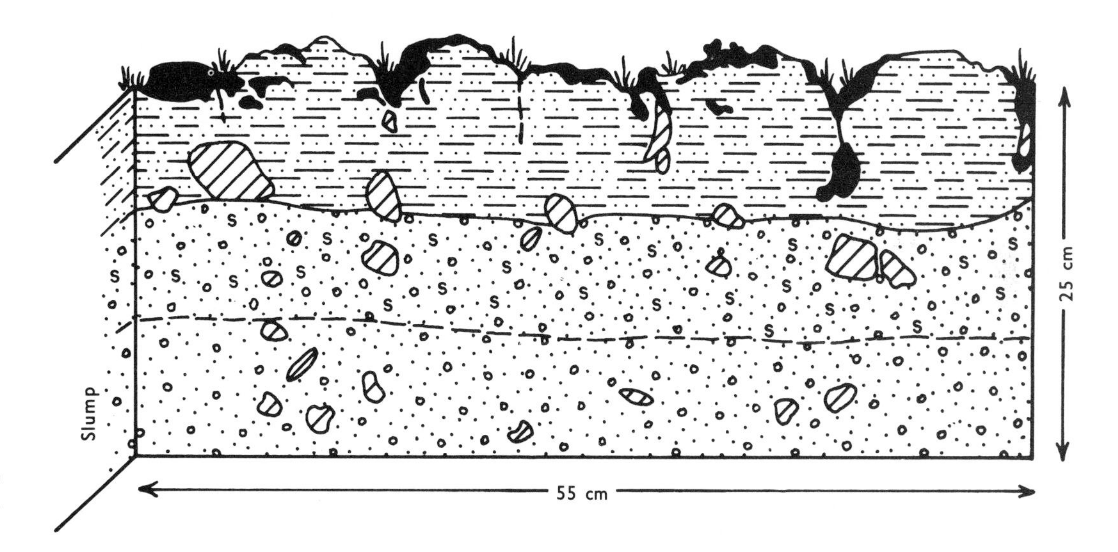

4.41 Cross-section of small nonsorted polygons, Mesters Vig, Northeast Greenland (*cf. Washburn, 1969a, 135, Figure 86*)

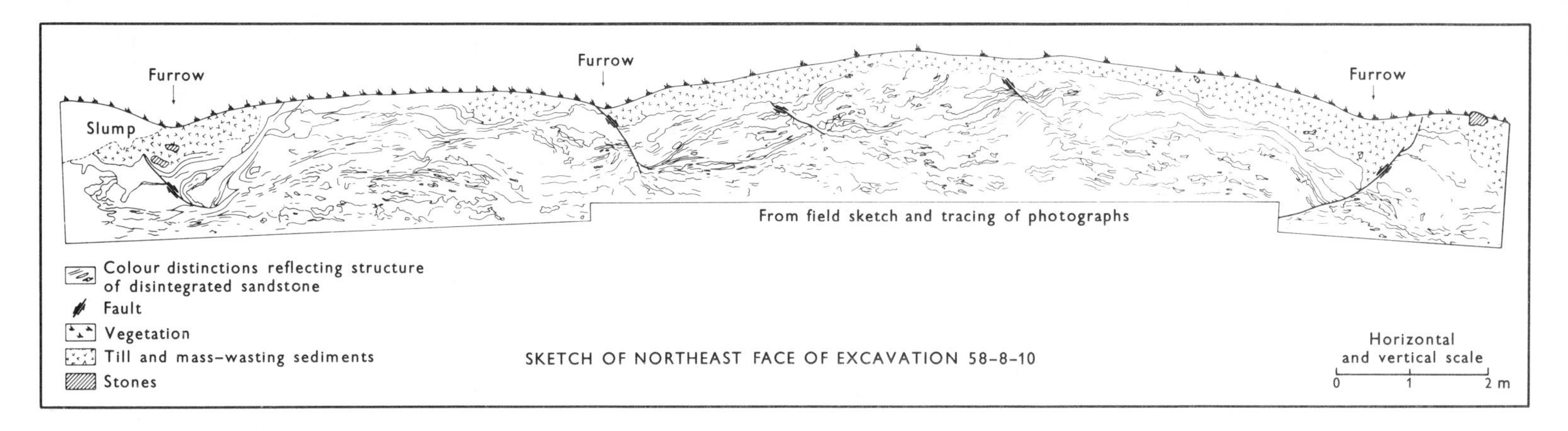

4.42 Cross-section of five large nonsorted polygons, Mesters Vig, Northeast Greenland (*Cf. Washburn, 1969a, 144, Figure 90*)

discontinuous permafrost zone of Alaska where there are inactive ice wedges, the mean annual air temperature ranges from about −2° in the south to −6° to −8° in the north, and the mean annual degree-days freezing range is 1700 to 4000. Precipitation is greater in the west than in the east, and the snowfall is 100 to 200 cm. The distribution of ice-wedge polygons and related forms in northern Eurasia is shown in Figure 4.46.

Small nonsorted polygons are likely to be passing features following an environmental change, and inactive forms and fossil forms are rarely reported. However, large inactive and fossil nonsorted polygons are among the most reliable evidences of environmental change, although inactive forms may be difficult to distinguish from weakly active forms that crack rarely. In the case of ice-wedge polygons, ice wedges can be present in both the inactive and the weakly active forms and both occur in the discontinuous permafrost zone of Alaska.

In places, fossil ice-wedge polygons can probably be recognized from their surface appearance as suggested by Henderson (1959*a*, 48–57), although the cases he cited are troublesome because of their large size and central depressions, some of the depressions being suggestive of collapsed pingos. Most fossil ice-wedge polygons commonly manifest themselves in section through ice-wedge casts, formed by the melting out of the former wedges and the slumping in of adjacent or overlying sediments. Some of the variations that may occur during infilling were discussed by Brüning (1964). If the infill is sand, fossil ice-wedge polygons might be difficult to distinguish from fossil sand-wedge polygons (Péwé, 1959). However, as previously noted, sand wedges are likely to be thinner, vertically foliated, and devoid of stony material (Black, 1969*b*, 229).

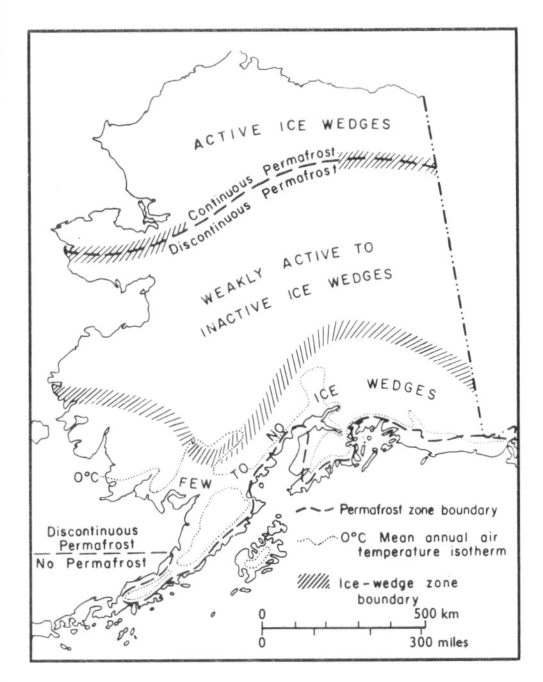

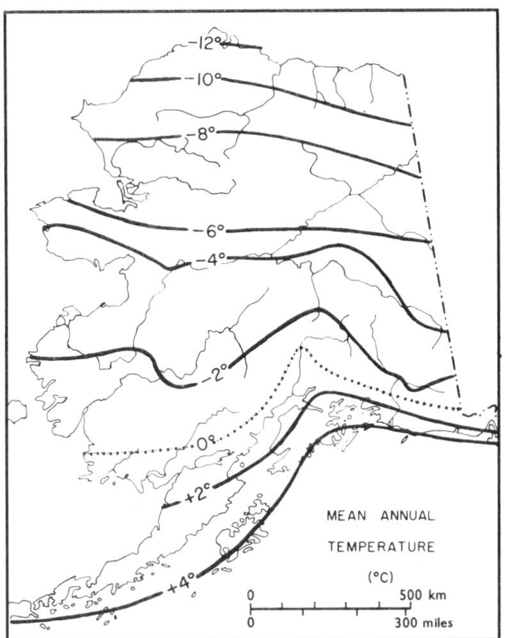

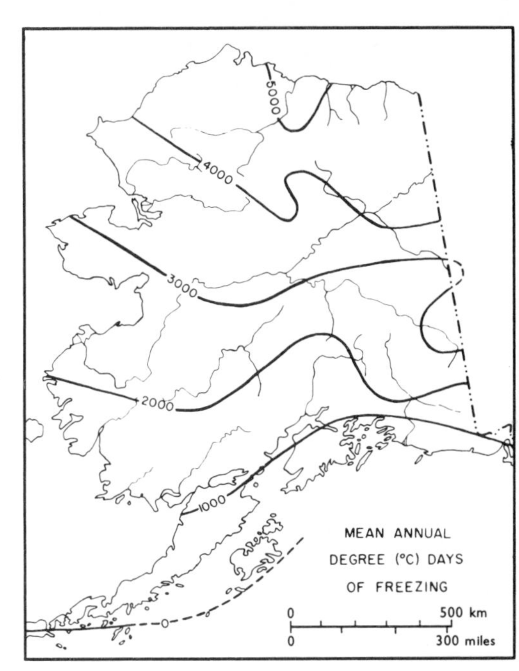

4.43 (*Above left*) Distribution of ice wedges and permafrost in Alaska (*Péwé, 1966a, 78, Figure 3*)

4.44 (*Above centre*) Mean annual air temperature (°C) isotherms in Alaska (*Péwé, 1966a, 79, Figure 4*)

4.45 (*Above right*) Degree (°C) days of freezing in Alaska taken from monthly mean temperature data (*Péwé, 1966a, 79, Figure 5*)

In each case the forms would be incontrovertible evidence of former permafrost if correctly identified. Fossil ice-wedge polygons and sand-wedge polygons are widespread (cf. review by Dylik and Maarleveld, 1967, 14–21) and constitute the basis for many paleoclimatic reconstructions of periglacial environments. The crux of the problem is to distinguish the fossil forms from other features that may closely resemble them, especially in cross-section. Examples are differential solution in calcareous deposits (Yehl, 1954), decay of tree roots (C. S. Denny and Walter Lyford, 1956, personal communication), large-scale desiccation cracks in warm arid climates (Knechtel, 1951; Lang, 1943; Neal, 1965; Neal, Lange, and Kerr, 1968; Willden and Mabey, 1961) and in temperate climates (cf. E. M. White and Bonesteel, 1960; E. M. White and Agnew, 1968; E. M. White, 1972), and perhaps seasonal frost cracks in non-permafrost areas (cf. section on Frost cracking in this chapter). Some of the problems were discussed by Dżułyński (1956), Johnsson (1959), and Wright (1961, 941–2). Among critical criteria are (1) the polygonal character

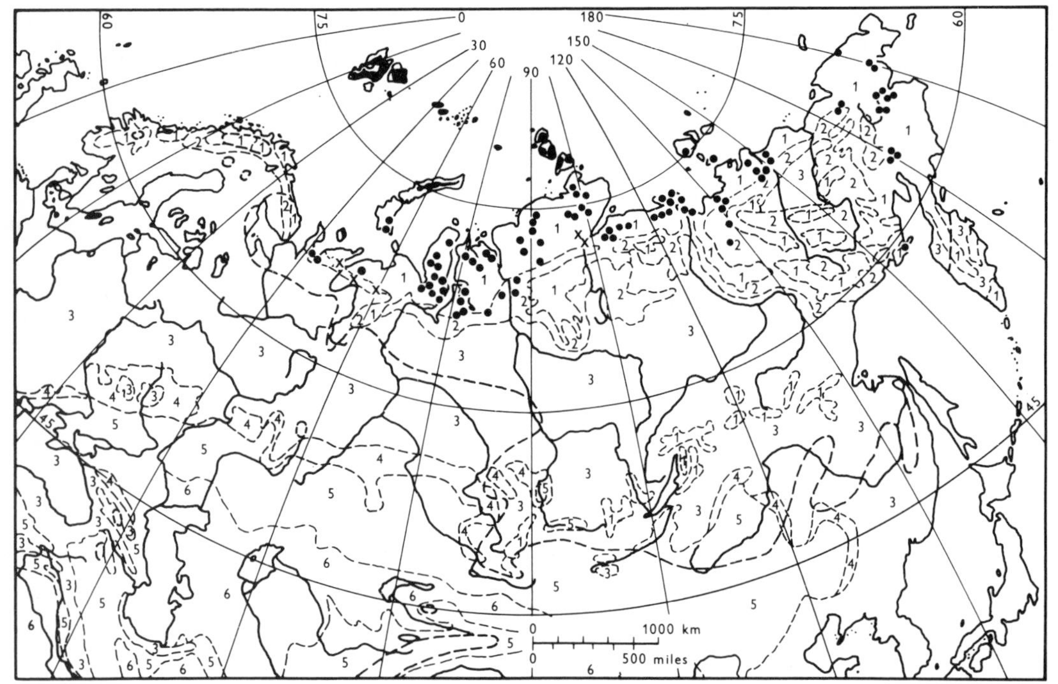

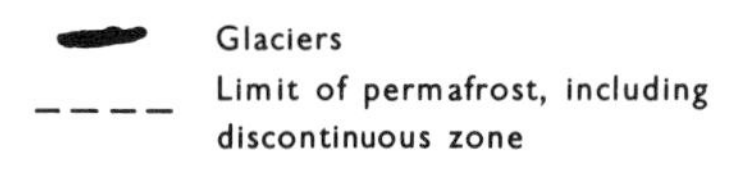

4.46 Distribution of active (•) and subfossil (×) ice-wedge polygons and related forms and oriented lakes in northern Eurasia (*Frenzel, 1960a, 1010, Figure 7*)

of the crack pattern in plan, (2) the depth of the cracks, and (3) the nature and organic content of the surrounding and infilling sediment. The presence of polygonally oriented casts would eliminate non-cracking processes, deep cracks if of frost-crack origin would indicate either permafrost or very deep seasonal frost approaching permafrost conditions, and the nature and organic content of the sediment might differentiate between frost cracking and large-scale desiccation cracking in an arid environment. The problem of differentiating frost-cracking from other cracking processes is discussed further under Origin (p. 137).

Nonsorted polygons (and stripes) that appear to extend beneath glacier ice, and that may be fossil in the sense of being exhumed, are reported from the Antarctic (Stephenson, 1961).

4.47 (*Opposite*) Small sorted polygons, Mesters Vig, Northeast Greenland. Scale given by 16-cm rule (*cf. Washburn, 1969a, 162, Figure 103*)

5 Sorted polygons

'Sorted polygons are patterned ground whose mesh is dominantly polygonal and has a sorted appearance commonly due to a border of stones surrounding finer material' (Washburn, 1956*b*, 831) (Figures 4.47–4.48). As with nonsorted polygons, it is convenient to classify sorted polygons into small forms (diameter < 1 m) and large forms (diameter > 1 m). The minimum mesh size of sorted polygons is probably in the order of 10 cm across and thus slightly larger than that of nonsorted forms, but otherwise there is no size distinction in the small forms. However, the maximum reported mesh size of large sorted forms is some 10 m and thus an order of magnitude smaller than of large nonsorted polygons. Both small and large forms are most frequent on nearly horizontal surfaces and they do not appear to occur on as steep slopes as do some nonsorted polygons. In places small sorted forms occur in the central bare areas of larger sorted forms.

The central areas are similar to those of nonsorted forms in having a concentration of fines, either with or without stones. The bordering stones tend to increase in size with the size of the polygons but to decrease in size with depth whatever the polygon dimensions. Tabular stones tend to be on edge and oriented parallel to the border.[1] The borders themselves narrow downward and terminate in some forms (Figure 4.49), whereas in other forms they broaden with depth and merge into a stony layer (Figure 4.50). In some small forms the downward-narrowing borders are coincident with a crack pattern.

Small sorted polygons occur in many different environments. Cold-environment forms are known from polar regions to temperate highlands. Superficially very similar features also occur in warm-arid regions, including Death Valley, California, where forms up to 2·7 m (9 ft) across are associated with a crack pattern in rock salt and in gypsiferous and stony silt (Hunt and Washburn, 1966, B118). However, large sorted polygons are best developed in permafrost environments. Whether permafrost, or a substitute impermeable horizon, is required for the origin of large periglacial forms is not known for sure. Active sorted polygons 1·8–3·7 m (6–12 ft) in diameter are reported from the Avalon Peninsula of Newfoundland where permafrost is absent but an impermeable till could be regarded as a substitute (Henderson, 1968). Because large-sorted polygons are best developed in permafrost environments, fossil forms (if not in saliferous material and of the warm-arid type) are commonly regarded as reasonable evidence of former permafrost. At the same time, pending further study, they should not be accepted as proof.

6 Nets

Nets are patterned ground whose mesh is neither dominantly circular nor polygonal. In most other

[1] See footnote to section on Sorted circles in this chapter.

4.48 (*Opposite*) Large sorted polygons, Långfjallet, Dalecarlia, Sweden (*G. Lundqvist, 1949, 338, Figure 3*)

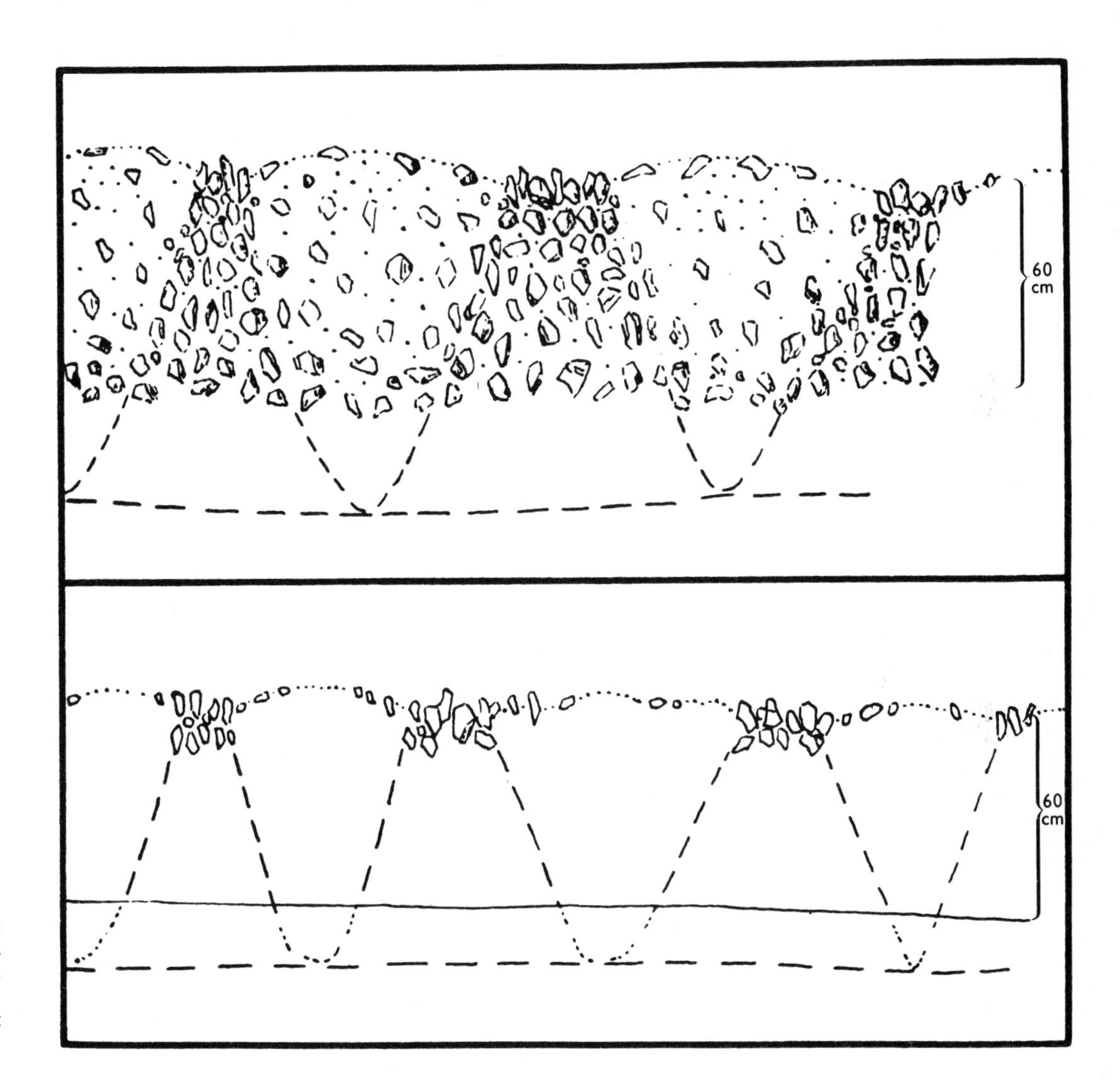

4.49 (*Opposite*) Excavation across stony border of sorted net (*cf. Washburn, 1969a, 167, Figure 107*)

4.50 Stony border of sorted polygons merging with stony layer at depth (Ahlmann, 1936, *10. Figures 4–5*). Horizontal dashed lines show approximate lower limit of active layer; horizontal solid line shows depth of snow at time of observation

respects such intermediate forms are similar to impinging circles and polygons, and the same nonsorted and sorted terminology applies. The size range and slope relation of most nets are similar to those of circles and polygons except that no nets are known to be as large as the largest nonsorted polygons.

The constitution of most nets parallels that of circles and polygons except for special non-sorted forms known as hummocks (thufur in Iceland), which are characterized by a knob-like shape and vegetation (Figures 4.51–4.52). Some forms (earth hummocks) have a core of mineral soil; others (turf hummocks) may or may not have such a core or may have a core stone (Raup, 1965). Well-developed hummocks are up to 50 cm high and 1–2 m in diameter.

Aside from hummocks, nets have no special distribution distinct from that of the circle or polygon type to which a given net is most closely allied. Hummocks are usually considered to be most common in subpolar and alpine environments, but they are also prominent in some polar regions such as Ellesmere Island, Arctic Canada (Habrich, 1938), Northeast Greenland (Raup, 1965), and northern Eurasia (Figure 4.53). They occur with or without permafrost. Because of their organic nature, hummocks do not form in vegetation-free areas. Some small-sorted nets occur in caves (Pulina, 1968).

The recognition of inactive or fossil forms is subject to the same conditions as those cited for analogous types of circles and polygons.

7 Steps

Steps are patterned ground with a step-like form and downslope border of vegetation or stones embanking an area of relatively bare ground upslope. They are restricted to slopes, and their downslope border is a low riser fronting a tread whose gradient is less than that of the general slope. They are thus terracette-like forms but are derived from circles, polygons, or nets rather than developing independently. Like their parent forms, they can be nonsorted or sorted, depending on whether the riser is characterized by vegetation only (Figure 4.54) or by stones (Figure 4.55). Probably nonsorted steps are derived mainly from hummocks – perhaps exclusively so. Presumably, sorted steps can be derived from either sorted circles or sorted polygons. They have been reported on gradients of 5°–15° in Alaska (Sharp, 1942*a*, 278) and on an estimated gradient of 10° in Northeast Greenland (Washburn, 1969*a*, 150). The longest dimension of steps tends to be in the direction of steepest gradient. Some sorted steps are clearly an intermediate stage between sorted polygons and sorted stripes (Sharp, 1942*a*, 277–8). Except for the tendency towards a longer downslope dimension, the size of steps generally conforms to their parent form.

The constitution of steps is similar to that of the upslope forms from which they were derived. Stones in the fronts of sorted steps tend to be imbricated as well as on edge. Where vegetation is

4.51 (*Opposite*) Hummocks, Mesters Vig, Northeast Greenland

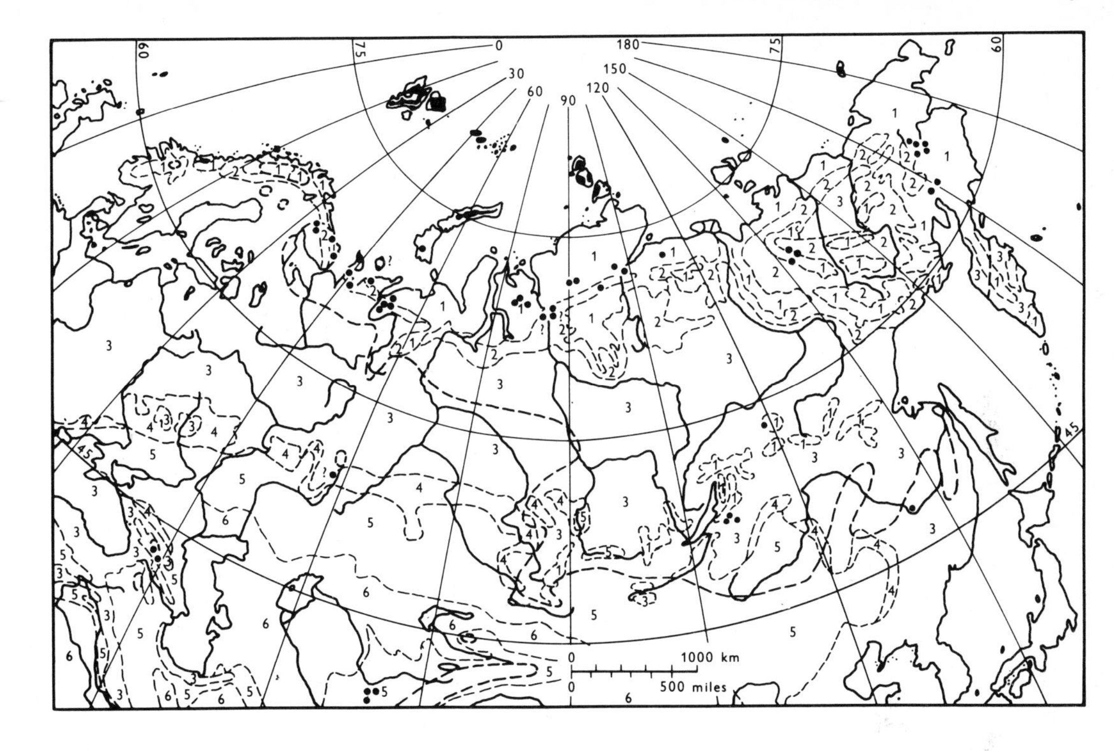

Glaciers

Limit of permafrost, including discontinuous zone

1 Tundra
2 Forest tundra
3 Forest
4 Forest steppe
5 Steppe
6 Desert

4.53 Distribution of active hummocks in northern Eurasia (*Frenzel, 1960a, 1013, Figure 9*)

4.52 (*Opposite*) Cross-section of earth hummock, Mesters Vig, Northeast Greenland. (*cf. Raup, 1965, 35, Figure 2*)

present, it can be overridden and buried by the fronts so that it extends back upslope as an organic layer beneath the tread. The environmental association of steps is similar to that of their parent forms. Inactive forms may be identifiable as such but it is doubtful if fossil forms in section can be distinguished from the parent form unless the presence of overridden vegetation is established.

8 Nonsorted stripes

'Nonsorted stripes are patterned ground with a striped pattern and a nonsorted appearance due to parallel lines of vegetation-covered ground and intervening strips of relatively bare ground oriented down the steepest available slope' (Washburn, 1956*b*, 837). Both small forms (Figure 4.56) and large forms (Figure 4.57) occur. The large forms shown in Figure 4.57 were on a

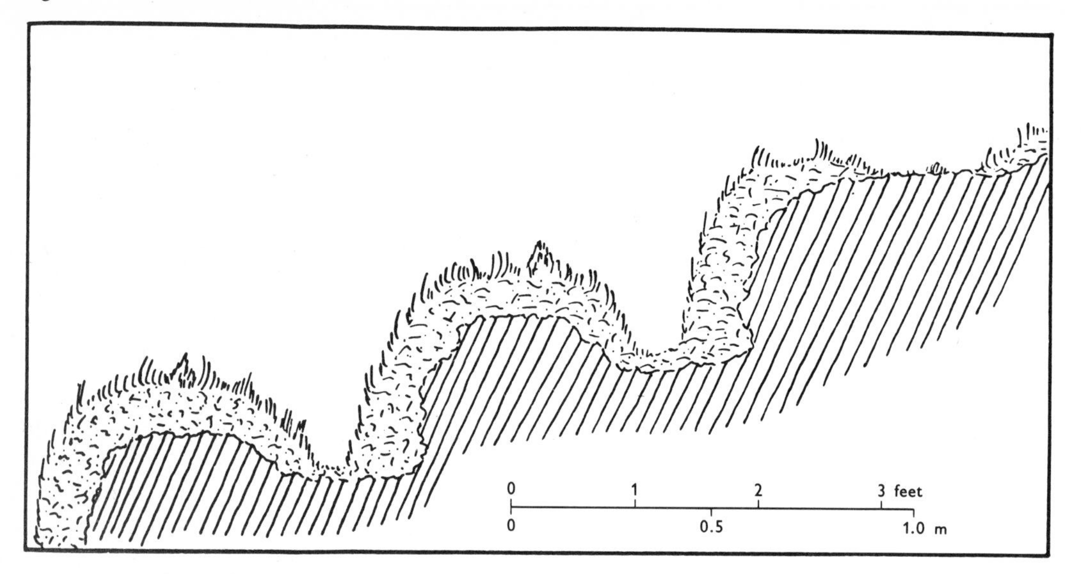

4.54 Nonsorted steps (*Sharp, 1942a, 283, Figure 7*)

gradient of 5–6°, downslope from nonsorted polygons on a more gentle gradient. In places these vegetated stripes and those without vegetation were about equally wide; in other places the vegetated stripes were 0·3 to 0·6 m (1–2 ft) wide and were spaced some 3·0 to 4·6 m (10–15 ft) apart (Washburn, 1947, 94).

The nonsorted stripes described above consisted of a diamicton. The vegetated stripes and associated organic soil narrowed downward and were thus wedge-shaped in cross-section.

Except for being slope features, nonsorted stripes have probably somewhat the same distribution as nonsorted circles and some nonsorted polygons except that they appear to be less common. Warm-climate forms known as 'wavy gilgai' are the slope analogue of 'normal gilgai' (Costin, 1955; Hallsworth, Robertson, and Gibbons, 1955, 2–7, Plate 4).

Fossil nonsorted stripes have been described from eastern England (R. B. G. Williams, 1964), and features that are totally vegetation covered and perhaps inactive forms of nonsorted stripes have been reported from North Wales (Ball and Goodier, 1970, 208–9). Nonsorted stripes that appear to be exhumed from beneath glacier ice occur in the Antarctic (cf. section on Nonsorted polygons). The general scarcity of reports suggests that inactive or fossil nonsorted stripes are rare, difficult to recognize, or both. To the extent that pedogenic horizons might be differentially developed as between vegetated and non-vegetated stripes, the repetitive and more or less

4.55 (*Opposite*) Sorted steps, Mesters Vig, Northeast Greenland. Scale given by 1-m tape (*cf. Washburn, 1969a, 150, Figure 93*)

4.56 (*Opposite*) Small nonsorted stripes, Mesters Vig, Northeast Greenland. Scale given by 16-cm rule at centre (*cf. Washburn, 1969a, 152, Figure 94*)

4.57 (*Above*) Large nonsorted stripes, Mount Pelly, Victoria Island, Northwest Territories, Canada (*cf. Washburn, 1947, Figure 2, Plate 27*)

regular nature of the differentiation may assist recognition. By contrast, nonsorted circles and small polygons may be less persistent features.

9 Sorted stripes

'Sorted stripes are patterned ground with a striped pattern and a sorted appearance due to parallel lines of stones and intervening strips of finer material oriented down the steepest available slope' (Washburn, 1956*b*, 836). Like nonsorted stripes, both small (Figure 4.58) and large (Figure 4.59) sorted stripes occur. They are seldom on gradients of less than 3°, and if derived by downslope extension of sorted polygons or sorted nets the transition gradient ranges from about 3° to 7° and may be characterized by sorted steps. Some sorted stripes occur independently of these other forms. Maximum gradients on which sorted stripes have been recorded rarely exceed 30°; however, they are considerably steeper in places (L. W. Price, personal communication, 1971). The stripes are usually parallel and in places sinuous. The stony stripes may be several centimetres to 1·5 m or more wide, and the intervening finer stripes tend to be several times wider than the coarse. Sorted stripes can be up to 120 m long (Washburn, 1969*a*, 180).

The stones of the coarse stripes can range from pebbles to boulders, depending on the size of the stripes. The intervening finer material can be stone free or contain stones and be a diamicton. The fabric of sorted stripes tends to be characterized by the long axis of stones being primarily in the vertical plane parallel to the stripe. As a result tabular stones are commonly on edge, especially in the coarse stripes.[1] A secondary long-axis maximum, at right angles to the first is reported to be present in the coarse stripes but absent in the intervening finer stripes (Furrer, 1968; Furrer and Bachmann, 1968, 5–8, 11–13). The size of stones generally decreases with depth. Sorting, again depending on size of the forms, extends from a depth of a few centimetres to about a metre, and the stony accumulations normally narrow downward as in the borders of sorted polygons.

Like sorted polygons and nets, sorted stripes occur in many environments, ranging from polar to warm arid. The largest forms are commonly associated with permafrost. Small ones such as those described by Poser (1931, 202–3; 209, Figures 6–7; 211, Figure 9) from the vicinity of Reykjavik, Iceland, can occur where it is absent (Steche, 1933, 221–3). Warm-arid forms are known from Death Valley (Hunt and Washburn, 1966, B125).

Inactive forms can be recognized by the same criteria that apply to sorted polygons and nets. Fossil forms in stratigraphic sections have been reported by Dücker (1933) and Nørvang (1946, 61–2) but are probably rarely recognized.

[1] See footnote to discussion of sorted circles.

4.58 Small sorted stripes, east side Glacial Lake, Seward Peninsula, Alaska. View upslope. Scale given by 17-cm rule

10 Origin

a General The genesis of many forms of patterned ground is problematical, not only quantitatively but even qualitatively. The following discussion, which reviews various possibilities, follows Washburn (1969*a*, 1970) and is based on the following premises: (1) Patterned ground is of polygenetic origin; (2) similar forms of patterned ground can be due to different genetic processes; (3) some genetic processes may produce dissimilar forms; (4) there are more genetic processes than there are presently recognized terms for associated forms, and (5) it is desirable to maintain a simple and self-evident terminology. Table 4.4 serves both as a summary and as an attempt at a genetic classification of patterned ground; it combines, in matrix form, existing terms for geometric patterns with existing terms for genetic processes. The resulting combined terms are readily understandable in light of the matrix approach. However, some combinations would be awkwardly long without being shortened and therefore the terms nonsorted and sorted in the combined forms have been abbreviated to N and S.

Only initial processes ('first causes') with respect to origin of the basic geometric pattern (circles, nets, polygons, or stripes) are reviewed. Once a basic pattern is established, various processes (cf. this chapter under Sorting by frost action) can modify it to produce sorted forms. For instance, a cracking process in itself provides only a nonsorted pattern but a sorting process acting on an initially nonsorted pattern can change it to a sorted pattern. These sorting processes are not specifically included in the table unless they also act as an initial patterning process. Genetic processes in Table 4.4 are divided into (1) processes in which cracking is the essential and immediate cause of the resulting patterned-ground forms, and (2) processes in which cracking is non-essential. Involutions are omitted from the discussion, since they are probably of various origins and associated with many kinds of patterned ground. The table also omits any mention of solifluction, although this process is present in many forms of patterned ground on slopes. Through 'microsolifluction' it may contribute to sorting, and through 'macrosolifluction' to the development of nets, steps, and stripes. However, to what extent it is an initial process in forming patterned ground is uncertain.

In order better to assess the environmental significance of patterned ground, discussion of origin is not limited to frost-action processes.

11 Cracking essential

4.59 (*Opposite*) Large sorted stripes, Hesteskoen, Mesters Vig, Northeast Greenland (*cf. Washburn, 1969a, 179, Figure 113*)

a General Cracking is a very widespread and important process in the initiation of sorted as well as nonsorted patterned ground. Sorted forms initiated by cracking obviously involve the addition of a sorting process. Basic patterns are typically polygonal but solifluction, possibly ice wedging

GEOMETRIC PATTERNS		PROCESSES												
		CRACKING ESSENTIAL						CRACKING NON-ESSENTIAL						
				THERMAL CRACKING										
					FROST CRACKING									
		DESICCATION CRACKING	DILATION CRACKING	SALT CRACKING	SEASONAL FROST CRACKING	PERMAFROST CRACKING	FROST ACTION ALONG BEDROCK JOINTS	PRIMARY FROST SORTING	MASS DISPLACEMENT	DIFFERENTIAL FROST HEAVING	SALT HEAVING	DIFFERENTIAL THAWING AND ELUVIATION	DIFFERENTIAL MASS-WASTING	RILLWORK
CIRCLES	NONSORTED								Mass-displacement N circles	Frost-heave N circles	Salt-heave N circles			
	SORTED						Joint-crack S circles (at crack inter-sections)	Primary frost-sorted circles, incl. ? Debris islands	Mass-displacement S circles, incl. Debris islands	Frost-heave S circles	Salt-heave S circles			
POLYGONS	NONSORTED	Desiccation N polygons	Dilation N polygons	Salt-crack N polygons	Seasonal frost-crack N polygons	Permafrost-crack N polygons, incl. Ice-wedge polygons, Sand-wedge polygons	Joint-crack N polygons ?		Mass-displacement N polygons ?	Frost-heave N polygons ?	Salt-heave N polygons ?			
	SORTED	Desiccation S polygons	Dilation S polygons	Salt-crack S polygons	Seasonal frost-crack S polygons	Permafrost-crack S polygons	Joint-crack S polygons	Primary frost-sorted polygons ?	Mass-displacement S polygons ?	Frost-heave S polygons ?	Salt-heave S polygons ?	Thaw S polygons ?		
NETS	NONSORTED	Desiccation N nets incl ? Earth hummocks	Dilation N nets		Seasonal frost-crack N nets, incl. ? Earth hummocks	Permafrost-crack N nets, incl. Ice-wedge nets and ? Sand-wedge nets			Mass-displacement N nets, incl. ? Earth hummocks	Frost-heave N nets, incl. Earth hummocks	Salt-heave N nets			
	SORTED	Dessication S nets	Dilation S nets		Seasonal frost-crack S nets	Permafrost-crack S nets ?		Primary frost-sorted nets	Mass-displacement S nets	Frost-heave S nets	Salt-heave S nets	Thaw S nets		
STEPS	NONSORTED								Mass-displacement N steps	Frost-heave N steps ?	Salt-heave N steps ?		Mass-wasting N steps	
	SORTED							Primary frost-sorted steps ?	Mass-displacement S steps	Frost-heave S steps	Salt-heave S steps	Thaw S steps ?	Mass-wasting S steps	
STRIPES	NONSORTED	Desiccation N stripes	Dilation N stripes ?		Seasonal frost-crack N stripes ?	Permafrost-crack N stripes ?	Joint-crack N stripes ?		Mass-displacement N stripes	Frost-heave N stripes	Salt-heave N stripes		Mass-wasting N stripes ?	Rillwork N stripes ?
	SORTED	Desiccation S stripes	Dilation S stripes		Seasonal frost-crack S stripes ?	Permafrost-crack S stripes ?	Joint-crack S stripes	Primary frost-sorted stripes ?	Mass-displacement S stripes	Frost-heave S stripes	Salt-heave S stripes	Thaw S stripes ?	Mass-wasting S stripes	Rillwork S stripes

Table 4.4. (*Opposite*) Genetic classification of patterned ground

in cracks, and some volume changes associated with cracking (for instance, in gilgai formation – cf. Hallsworth, Robertson, and Gibbons, 1955) can deform some polygons to nets. Moreover, given favourable slope conditions, it seems probable that small transverse cracks would be narrowed by mass-wasting, whereas cracks paralleling the slope would tend to remain open and determine the location of small stripes. Although small circular desiccation cracks have been observed (Chambers, 1967, 4, 7), there is no evidence that cracking forms well-developed circles or steps. As mentioned below under Permafrost cracking, the present writer has seen polygonal cracks surrounding a circular central area but in this case the circular aspect is believed to be secondary and the pattern still a polygonal form of patterned ground.

Abundant evidence indicates that true polygons, as distinct from nets and other patterned-ground forms that are sometimes referred to as polygons, are primarily contraction phenomena. Possible causes of contraction include thawing, synaeresis, partial wetting, drying, and lowering of temperature.

Thawing as a cause of polygons lacks convincing evidence but has not been sufficiently investigated. Another possibility that should be studied is the role that synaeresis cracks (Jüngst, 1934; W. A. White, 1961, 1964) may have in polygon formation. 'Synaeresis cracks are fissures that develop in a suspension where waters are expelled from the clay-water system by internal forces; they may resemble mud cracks in the sediments' (W. A. White, 1961, 561). They form subaqueously in a saline environment where clays are flocculated and they are apparently due to adjustments accompanying realignment of clay particles under the influence of electrochemical forces between the particles (W. A. White, 1961, 569). According to Kostyaev (1969, 235–6), synaeresis cracks can form polygons with diameters ranging from a few decimetres to tens of metres. Cracks formed by partial wetting in laboratory experiments were reported by Corte and Higashi (1964, 68, 70–1) who suggested surface tension as a possible explanation. These processes remain to be investigated but to the extent the resulting cracks persist, the cracks probably determine the location of later desiccation fissures. In the absence of such desiccation and pending further data, the above processes are here regarded as probably unimportant in causing polygons.

b Desiccation cracking Desiccation cracking is fissuring due to contraction by drying. It is probably one of the most common and important patterned-ground processes, especially for forms having diameters < 1 m but not excluding larger forms. Desiccation can be due not only to drainage and evaporation but also to withdrawal of moisture to loci of ice formation, a process that Taber (1929, 457–8) demonstrated could produce tiny polygonal fissures in laboratory specimens. Subsequently Taber (1943, 1522–7) suggested that the same process could produce ice-wedge polygons and sorted polygons, and Schenk (1955*a*, 64–8, 75–6; 1955*b*, 177–8) argued that it was important in a wide variety of polygonal forms, including small ones. Laboratory experiments by

Pissart (1964) show that the process described by Taber can produce small nonsorted polygons up to about 10 cm in diameter, but the polygons were subsurface rather than surface phenomena and were very irregular. Although such subsurface desiccation – a kind of freeze drying, aided perhaps by smaller-scale moisture transfer during freezing (Hamilton, 1966) – may influence a surface crack pattern, proof is lacking that is is actually responsible for the pattern. To the extent that desiccation is involved, the fact that polygonal cracks confined to the active layer are best developed at the surface and become rapidly less prominent with depth argues strongly for the predominant effect of sub-aerial desiccation; in fact Büdel (1960, 42) regarded such evidence as conclusive for forms he studied. Pissart's preliminary report clearly supports the desirability of further work and experimental data relating to Taber's hypothesis, particularly as applied to small polygons. Without further evidence, however, sub-aerial drying would seem to be the most probable kind of desiccation process responsible for well-defined polygonally patterned ground.

Salt content appreciably influences volume changes associated with wetting and drying as discussed long ago by Wollny (1897, 20–1, 31–2, 51). Sodium chloride, for instance, appears to affect development and doming of desiccation polygons (Dow, 1964; Elton, 1927, 179; Kindle, 1917, 140–4), it is involved in synaeresis cracking (cf. W. A. White, 1961, 1964) as previously described, and it is known to affect frost heaving (Beskow, 1935, 100–1, 232; 1947, 57–8). Also the wide variety of patterned ground in warm deserts, where salt is an important variable in forms related to desiccation as well as in some other forms (Hunt and Washburn, 1966), suggests the desirability of investigating the role of chemical variations in the mineral soil of patterned ground.

A major problem in the origin of some cracks is to distinguish between thermal contraction and desiccation, whatever the latter's cause. This problem as it effects large nonsorted polygons is discussed below under Seasonal frost cracking and Permafrost cracking.

Small nonsorted polygons in West Greenland (Boyé, 1950, 134) and in the arid environment of North Greenland have been ascribed to desiccation (W. E. Davies, 1961*b*, 49, Figure 2; 50; Tedrow, 1970, 81, Figure 65; 82), as have small polygons in Arctic Canada (D. I. Smith, 1961, 74), in Spitsbergen (Büdel, 1961, 347, 353; Elton, 1927, 176–80, 190–91), and elsewhere in the Arctic, and they are demonstrably due to drying in many environments (Longwell, 1928; Segerstrom, 1950, 115; Shrock, 1948, 188–209; and many others). The conclusion is warranted that in general well-developed and regular nonsorted polygons averaging about 1 m or less in diameter are probably due to desiccation cracking, a conclusion also cited by Tricart (1963, 88; 1969, 91). (That desiccation may produce large polygons as well does not affect the argument.) According to Tricart (1967, 196–7, 201), nonsorted polygons with diameters up to 0·5 m are probably due to air drying, whereas those with diameters of 0·6–3·0 m are more likely to be due to freeze drying.

Desiccation cracking as an initiating cause of some nonsorted stripes and sorted stripes has not been investigated in detail. It is supported by a few field observations (Klatka, 1961*a*, 313–14; 1968, 274–6; Washburn, 1969*a*, 153–4) and some experimental evidence suggesting that desiccation cracks tend to be emphasized parallel to the gradient (Furrer, 1954, 237–9, 242; Kindle, 1917, 137–9).

c Dilation cracking Dilation cracking is fissuring due to stretching of surface materials. Probably it is not a widespread cause of patterned ground but it is clearly a genetic process in places where the cracking is consequent on local differential heaving or on sagging of the ground. In this context the terms differential frost heaving and differential thawing as used in Table 4.4 imply processes in which any associated cracking is incidental and non-essential to the origin of patterned ground.

Dilation cracking as opposed to other kinds of cracking may be identified from associated evidence of pronounced heaving or collapse as, for instance, a medial crack along an elongate heaved feature or a radial and concentric crack pattern on a domed one (Benedict, 1970*a*; Washburn, 1969*a*, 109; 110, Figure 71; 125, Figure 78; 136). In addition Benedict (1970*a*) observed that crack widths exceed those to be expected from frost cracking.[1] Perhaps some cracks that would normally be ascribed to frost cracking are actually dilation features as suggested by arching of beds below the cracks but more evidence is needed (Washburn, 1950, 41–4).

Given the presence of dilation cracking, the process might lead to the various patterned-ground forms associated with cracking but patterns other than nonsorted polygons are probably very rare, since even nonsorted polygons of this origin appear to be uncommon.

d Salt cracking Salt cracking describes the fissuring that initiates nonsorted and sorted polygons in hard rock salt and salt crusts of warm deserts. The exact process is not established but may be mainly thermal contraction (Hunt and Washburn, 1966, B120–30). However, desiccation cracking very probably starts some polygonal patterns in salt-rich ground and it may account for some polygons in crusts that formed or hardened subsequent to the cracking (cf. Tricart, 1966, 20–1; 1970, 434–5). There is no evidence that salt cracking forms other than polygonal patterns, some of which occur on slopes as steep as 37° without becoming nets.

e Seasonal frost cracking Seasonal frost cracking is fissuring due to thermal contraction of frozen ground in which the fissuring is confined to a seasonally frozen layer. In a permafrost environment this layer is the active layer. Patterned ground resulting from this process is perhaps

[1] Benedict regarded dilation cracking as a form of frost cracking because of its relation to frost heaving. However, frost cracking as used here and by most authors is restricted to cracks caused by thermal contraction.

widespread and may be difficult to distinguish from forms developed by other processes. Most of the geometric patterns conform to those developed by other cracking processes as modified by slope conditions and presence or absence of superimposed sorting processes.

By definition, frost-crack polygons or, more accurately, seasonal frost-crack polygons, are not necessarily related to permafrost (Hopkins *et al.*, 1955, 139). Such polygons have been described from Alaska (Black, 1952, 131–2; Hopkins *et al.*, 1955, 137–9) and elsewhere (Black *in* Corte, 1969*a*, 140; Pataleev, 1955; J. R. Reid, 1970*b*; Washburn, Smith, and Goddard, 1963) although their occurrence in Alaska has been challenged (Church, Péwé, and Andresen, 1965, 38; Péwé, Church, and Andresen, 1969, 49). In the case described by Reid, which was on the campus of the University of North Dakota in the winter of 1962–3, cracks developed during several days of $-20°$ to $-25°$ temperatures when snow cover was largely lacking. In places the cracks formed tetragonal polygons and were up to 2 m deep. They remained open at temperatures up to $-5°$ and became filled with wind-blown organic silt as the result of two severe dust storms. However, surface indications of the cracks disappeared by spring. The case described by Washburn, Smith, and Goddard, which was at Hanover, New Hampshire, was similar but the pattern was less well defined and no infill was observed. Although such cracks may be locally common where active, it is doubtful they would normally be preserved as fossil forms in the geologic record. Rather, fossil forms are presumptive evidence of permafrost or possibly, in places, of such deep seasonal freezing as to approach permafrost conditions.

Grain size as a criterion in distinguishing between frost cracking and desiccation cracking remains to be investigated in any detail. Frost cracking occurs in many different soils and the influence of grain size on the process, except as it affects distribution of moisture, is not well known. As for desiccation, the process can be eliminated in the cracking of gravels and coarse sands lacking silt and clay. Also, according to E. M. White (1972, 108):

> Narrow surface desiccation cracks will form in sandy soils with some clay, but desiccation-caused micro-relief does not form unless the soil has more than about 40% clay. Depression soils with 20–40% clay occasionally have 60–90-cm-wide polygons formed by 2–5-cm-wide cracks during droughts.

More information is needed on proportions of clay silt as they affect desiccation cracking and formation of nonsorted polygons but grain size restraints will eliminate the process as a factor in some soils and favour alternatives such as frost cracking.

It has been suggested that seasonal frost cracking may initiate some earth hummocks. There is little evidence as to whether or not it causes stripes but, reasoning by analogy with desiccation stripes for which the evidence is better, seasonal frost-crack stripes are plausible.

f Permafrost cracking Permafrost cracking is fissuring due to thermal contraction of perennially frozen ground – i.e. it is frost cracking that extends into the permafrost and is not confined to the active layer. The resulting patterned ground is very widespread in polar environments and is of considerable significance because of its temperature implications and the fact that fossil forms are more subject than most patterned ground to preservation in the geologic record. The literature dealing with active and fossil permafrost cracks and criteria for their recognition is prolific as discussed in this chapter under Frost cracking; further information is provided below.The best-known examples of the process are permafrost-crack nonsorted polygons (following the terminology of Table 4.4) of which two varieties are commonly distinguished—ice-wedge polygons, and sand-wedge or soil-wedge polygons.

Where cracks bordering polygons can be traced into similar cracks in underlying permafrost, permafrost cracking can be assumed to have originated the polygons.

Confinement of bordering cracks to shallow depth and the active layer in a permafrost environment suggests that the polygons formed by processes other than frost cracking. Except for some bog areas in the discontinuous zone, the maintenance of permafrost without a talik between it and the active layer requires deep penetration of freezing temperatures, and therefore it can be logically assumed that frost cracking would have affected the permafrost and not be confined to the active layer or, as in places, to just the upper part of it. On the other hand, surface desiccation cracks in a permafrost environment are necessarily confined to the active layer. The 'filling soil veins' that are described by Katasonov and Solov'ev (1969, 15–16) as being confined to the active layer do not necessarily counter this argument, since cracks below the veins continue into permafrost.

Crack size and spacing are other possible criteria for distinguishing between permafrost-crack and other polygons. The factors controlling depth and spacing are complex but have been quantitatively considered by Lachenbruch (1961; 1962; 1966) and, for desiccation polygons developed in the laboratory, by Corte and Higashi (1964). According to Lachenbruch (1966, 68)

> The use of temperatures measured in Alaskan permafrost, a 'power-law model' to calculate thermal stress, and an estimated value of G_c [critical rate of strain energy release necessary for fast fracture] to calculate crack depth, lead to computed stress relief compatible with observed polygonal dimensions.

These dimensions, in terms of the diameter, vary 'from a few meters to more than 100 meters' (Lachenbruch, 1966, 63). According to Péwé (1966*a*, 77), ice-wedge polygons are 2 (or 3) to 30 m or more in diameter. Although Black (1952, 130) indicated that ice-wedge polygons may be less than 1 m in diameter, he referred to polygons 1–3 m in diameter as resulting from subdivision of larger ice-wedge polygons, and in general it would seem that ice-wedge polygons have diameters exceeding 1 m. Seasonal frost-crack polygons tend to be larger than ice-wedge polygons occurring in the same area according to Hopkins *et al.* (1955, 139), but more information is needed. As noted above under Desiccation cracking, polygons with diameters measuring less than 1 m are probably mainly of desiccation origin.

Permafrost-crack sorted polygons can develop by activation of a sorting process along the crack in the active layer (Péwé, 1964), and perhaps also in fossil forms by melting of ice wedges and concurrent infilling with surface stones. Nets initiated by permafrost cracking may exist but are probably less common than polygons because deformation of an originally polygonal pattern would tend to be inhibited by permafrost. It is even more questionable whether the pattern could be modified to form stripes.

g Frost action along bedrock joints Frost action in the present context comprises frost wedging and frost heaving. Where sufficiently concentrated along bedrock joint patterns, it can initiate some forms of patterned ground but they are uncommon, judging from the few references to them in the literature (cf. W. E. Davies, 1961*a*; Fleisher and Sales, 1971; Kunský and Louček, 1956, 345–7; Washburn, 1956*b*, 846–7; 1969*a*, 185). Control of the patterns by bedrock joints is the essential criterion for these forms. Sorted polygons are the best-known example; sorted stripes are another (Washburn, 1969*a*, 185). Sorted circles in bedrock have also been observed (Washburn, 1950, 49; 52, Figures 1–2, M. Y. Williams, 1936). If the latter are controlled by joint intersections as suggested, they could be termed joint-crack sorted circles. The existence of joint-crack nonsorted polygons and nonsorted stripes remains to be demonstrated but seems probable, considering the known sorted forms and the likelihood that ice wedges would develop in preexisting bedrock joints underlying unconsolidated material. If the bedrock fissuring were due to frost cracking, however, the forms would be frost-crack or permafrost-crack polygons. The fact that joint-crack patterns are confined to bedrock joints minimizes the possibility of net patterns developing by deformation of polygons. Joint-crack polygons are also known that have no periglacial significance (Troll, 1944, 593–4, Figures 22–3; 1958, 38, Figures 22–3).

12 Cracking non-essential

a General A number of hypotheses of patterned-ground origin involve processes that do not depend on initial cracking, and the more probable of these are grouped in Table 4.4. The selection is incomplete and, in view of the uncertainties involved, some important initiating processes may have been omitted. Furthermore, the question exists as to whether or not non-cracking processes can develop strictly polygonal, as opposed to circular, net, or striped patterns. Theoretically, continued mutual impingement of circles would develop polygonal patterns simulating those of crack origin, especially if secondary bordering cracks developed. However, because proof appears to be lacking that this sequence of events does in fact occur in patterned ground, polygonal forms in this section of Table 4.4 carry a question mark. By definition, steps are derived from other patterned ground by mass-wasting; their origin is therefore keyed to that of other forms even though terracettes similar to steps are initiated by mass-wasting.

b Primary frost sorting Primary frost sorting includes upfreezing and the ejection of stones from fines by repeated freezing and thawing, the sorting effects of needle ice, and the gradual particle-by-particle sorting that Corte termed vertical and lateral sorting (cf. discussion in this chapter under Frost sorting). A number of frost-related sorting processes probably modify established patterns to produce sorted forms. The evidence that primary frost sorting can also initiate basic patterns is less clear, and the criteria are correspondingly vague, but the process seems more likely than many to be also a first cause. In listing primary frost-sorted forms in Table 4.4, the usual 'S' designation for sorting is omitted as redundant. Nonsorted forms of patterned ground are excluded by definition. Primary frost sorting would develop sorted circles or, if closely spaced, sorted nets. The development of clearly defined polygons by impingement is questioned for reasons discussed above (12*a*). It is also uncertain whether primary frost sorting would produce debris islands, since normally the coarse rubble in which they occur has a scarcity of fines near the surface and the fines of many debris islands appear to be derived from depth.

c Mass displacement Mass displacement as applied to patterned ground is the *en-masse* local transfer of unfrozen mineral soil from one place to another within the soil as the result of frost action, as discussed in this chapter under Mass displacement. Suggested processes include cryostatic pressure and changes in intergranular pressure. The latter process appears the more probable but both remain to be demonstrated as applied to the genesis of patterned ground. Criteria for recognizing patterned ground due to mass displacement are therefore uncertain, and in practice are mainly negative, involving elimination of other origins. Supposedly mass displacement can form both nonsorted and sorted patterns. Thus movement of a mass of fines through other fines, or through slightly coarser but not obviously stony material, could produce nonsorted varieties, whereas sorted varieties would depend upon movement of fines into distinctly stony material. Here, again, most of the normal geometric patterns are expected but strictly polygonal ones are questioned in Table 4.4 for reasons noted above (12*a*). Earth hummocks and steps are included but questioned because some varieties cannot be excluded from such an origin, despite the lack of convincing evidence for it.

d Differential frost heaving Local differential heaving, discussed in this chapter under Frost heaving and frost thrusting, is frost heaving that is significantly greater in one spot than in the area around it because of less insulation or other factors, so that growth of ice lenses and needle ice is especially favoured. As applied to patterned ground in Table 4.4, 'differential frost heaving' and the adjective 'frost-heave' imply the resulting blister-like expansion of the ground, which may well start many nonsorted and sorted patterns. Frost sorting and the fact that sorting can of itself lead to volume increase of soil (Corte, 1966*b*, 213–16) mean that differential frost heaving can be a self-reinforcing process. Criteria for recognizing patterned ground of this origin are largely negative. For instance, forms consisting of non-heaving mineral soil are excluded. Also strictly polygonal forms are more consistent with other origins (unless the development of well-developed polygons by impingement is accepted) as are forms showing mass displacement of material from depth as opposed to surficial or very shallow forms like the nonsorted circles (stony earth circles) described by P. J. Williams (1959*a*; 1961, 345). Earth hummocks with a core stone whose upfreezing caused the hummocks would be of frost-heave origin. Yet, although hummocks are probably of diverse origins (Beschel, 1966) and core stones are locally common and may be responsible for some hummocks (Dionne, 1966*a*), many hummocks lack them and it is difficult to prove that upfreezing of the stones is really the primary factor even where core stones are present.

e Salt heaving Salt heaving is volume expansion due to growth of salts. The process probably causes several varieties of patterned ground in warm-arid deserts (cf. Hunt and Washburn, 1966). It has many analogies with local differential frost heaving and is probably capable of initiating the same general types of patterns.

f Differential thawing and eluviation Differential thawing and eluviation in the present context constitute a sorting process between fines and stones, which stems from variable thawing rate and from preferential development of rills in coarse material. Such sorting probably initiates some small sorted nets (Washburn, 1956*b*, 855–6), but criteria for recognizing patterns of this origin are unsatisfactory except for shallow forms developed on thawing ice. Nonsorted forms are excluded. Of sorted forms, polygons are again questioned (cf. 12*a*); sorted stripes seem possible but are not known to occur, and other sorted forms seem unlikely.

g Differential mass-wasting Differential mass-wasting is the faster downslope mass movement of some materials than others. Little is known about the process relative to the initiation of patterned ground but it may account for some sorted steps and stripes (R. B. King, 1971, 384–5; Washburn, 1969*a*, 186–8), and perhaps for some nonsorted ones as well.

h Rillwork Rillwork is the process that starts small channels and it may thereby also initiate some striped forms of patterned ground. In this context the process is usually limited to small-sorted stripes (Washburn, 1956*b*, 857–8). Although most sorted stripes normally carry drainage, in many places this is probably a result rather than a cause of the pattern. The possibility that some nonsorted stripes are due to rillwork is not excluded.

i Differential weathering Although not represented in Table 4.4, differential weathering may create local pockets of fines amid coarser debris. This first cause might then be reinforced by sorting processes and heaving. Meinardus (1912*a*, 254–5; 1912*b*, 12–13, 32) suggested that some debris islands result from the weathering of especially susceptible stones in a boulder field. More generally, Nansen (1922, 111–20) thought that a wide variety of sorted circles and sorted polygons (formed from circles by impingement) were started by differential weathering as the result of moisture variations caused by inequalities of the ground surface. More recently the hypothesis has been applied to large sorted polygons on the basis that large stones weather more slowly than smaller stones, which have a larger surface area to volume ratio and are therefore more subject to frost wedging (R. B. King, 1971, 383). In a slightly different form the hypothesis has also been extended to nonsorted polygons and their development from sorted polygons by comminution of stone borders (Elton, 1927, 173; Huxley and Odell, 1924, 219–23, 228). Without denying the importance of differential weathering in producing fines and some sorting effects in patterned ground, the process seems unlikely to be a basic patterning process in the same sense as the others listed in Table 4.4. Only if it can be demonstrated that differential weathering *in situ* has led to the concentrations of fines within each mesh of a sorted pattern would there be a parallel.

X INVOLUTIONS

Involutions (Denny, 1936, 338) were defined by Sharp (1942*b*, 115) as 'aimless deformation, distribution, and interpenetration of beds produced by frost action'. Involutions (other names include Brodelboden, cryoturbations, Taschenboden, and Tropfenboden) may be very complex (Figure 4.60). Even if they are correctly identified as frost features, it is difficult or impossible to determine the exact type of frost action responsible. However, cryostatic pressure and/or changes in density and intergranular pressure may be among the most important. Various investigators have regarded involutions as evidence of permafrost (cf. Poser, 1948*a*, 54, 58) but it is now clear that involutions cannot be accepted as proof of it as pointed out by Wright (1961, 943) and Black (1969*b*, 233–4). Laboratory experiments by Pissart (1970*a*, 40–4; 1972) show that significant pressure effects can be created in locally closed systems during freezing simulating that in non-permafrost environments.

Contortions appearing very similar if not identical to involutions can be formed by other processes, including load casting, slumping and perhaps other types of mass-wasting, and volume changes unrelated to frost action. In places the exact mode of formation may be problematical but the results can be strikingly like involutions (cf. Anketell, Cegła, and Dżułyński, 1970; Butrym *et al.*, 1964; Dionne, 1971a; Dżyułński, 1963; 1966; McArthur and Onesti, 1970; McCallien, Ruxton, and Walton, 1964). Although in some cases the relation of associated structures

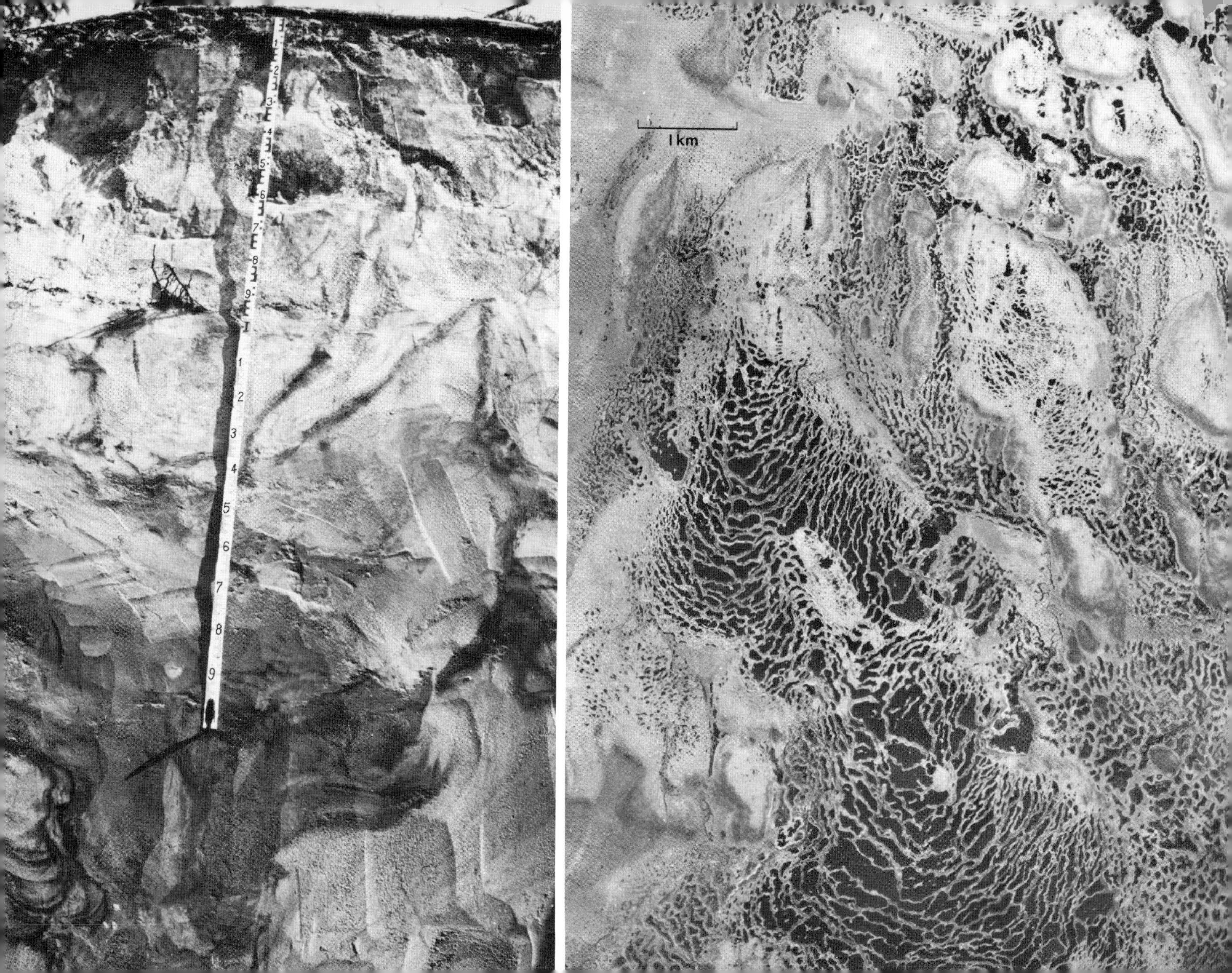
1 km

such as ice-wedge casts may help to identify involutions (Johnsson, 1962, 386–92), the need for supplementary evidence emphasizes the ambiguity of involution-like features. As stressed by Corte (1969*a*, 141), a critical study is needed of the origin of involutions and similar features.

XI STONE PAVEMENTS

1 General

Stone pavements are accumulations of rock fragments, especially cobbles and boulders, in which the surface stones lie with a flat side up and are fitted together like a mosaic. Other names that have been applied to these accumulations include alpine subnival boulder pavements, boulder pavements, dallages de pierres, Pflasterboden, and Steinpflaster.

Well-developed periglacial stone pavements usually occur in valleys in association with lingering snowdrifts or Aufeis but some also occur in snow-filled depressions such as nivation niches.

2 Constitution

Cobbles and boulders are the predominant constituents of periglacial stone pavements but pebbles and smaller grain-size fractions occur locally in the cracks between the larger stones of the mosaic. Any stones beneath the surface tend to decrease in size with depth. The stones may be derived from underlying bedrock or be transported from elsewhere.

3 Origin

The origin of periglacial stone pavements is not well understood but a variety of processes are probably involved. S. E. White (1972, 195) has summarized a widely held view by indicating that they may be the combined result of upfreezing of stones, ground saturation and the removal of fines by meltwater, the rotation and sifting of the stones in the saturated ground under their own weight and the weight of overlying snow, with possibly snow creep and, as suggested by Sørensen (1935, 60) and L.-E. Hamelin (in S. E. White, 1972, 199), gelifluction being additional causes. The role of snow weight has been questioned by Tricart (1967, 239; 1969, 100) but supported by S. E. White (1972, 199) and others. The weight of Aufeis should be added to this list, since well-developed stone pavements beneath Aufeis have been observed by Porter (1966, 82–3) in the Anaktuvuk Pass area, Alaska (Figure 3.10), and by the writer in the Mesters Vig district of Greenland.

4 Environment

Periglacial stone pavements are widely distributed, occurring in the Arctic (Sørensen, 1935, 60–62; Porter, 1966, 82–4; Poser, 1931, 230–31; 1932, 54–5), in the mid-latitude highlands (Troll,

4.60 (*Far opposite*) Involutions in sand and peat, right bank Lena River, 90 km north of Yakutsk, Yakutia, USSR. Scale given by 2-m tape

4.61 (*Near opposite*) String bogs northwest of Moosonee, Ontario, altitude 30–45 m (*Canada Energy, Mines & Resources Photo A 15090–65*)

1944, 549, 654; 1958, 3, 82; S. E. White, 1972), and in the Antarctic where it has been observed at Marble Point, McMurdo Sound, by the writer. However, their use as evidence of a periglacial environment is handicapped by the fact that they have not been extensively studied and their origin and quasi-equilibrium range is uncertain. Furthermore, different processes in warm-arid regions give rise to desert pavements, which are somewhat similar, although generally characterized by smaller stones.

5 Inactive or fossil forms

The value of inactive or fossil stone pavements as a criterion of former periglacial conditions is uncertain, pending further study. S. E. White (1972, 199) was appropriately cautious in drawing paleoclimatic conclusions from features he studied in the Colorado Front Range, some of which may still be active.

XII STRING BOGS

1 General

String bogs (Aapamoore, tourbières réticulées, Strangmoore) are muskeg areas characterized by ridges of peat and vegetation, interspersed with depressions that often contain shallow ponds (Figure 4.61). The ridges are up to about 2 m high and may be tens of metres long, and some are surrounded by palsas (Hamelin, 1957, 91–2). Ridges transverse to the gradient are commonly asymmetrical with the steepest side facing down the regional slope and the gentler, upslope side damming a pond. The regional slope is very gentle, and many string bogs occur on gradients of less than 2°.

Three types of ridge patterns are commonly recognized (cf. Hamelin and Cook, 1967, 164; Tricart, 1967, 229; 1969, 98): (1) Linear transverse ridges, commonly convex downslope, (2) irregular ring-like anastamosing ridges, and (3) ridges forming a net-like pattern. Either vegetation or water may be the major element in a given bog sector. The patterns tend to be much more irregular than patterned ground, and string bogs are usually separately classified.

2 Constitution

The ridges consist of peat containing ice lenses for at least part of the year. Ice lenses as thick as 40 cm have been reported (Dionne, 1968, 3). The ridges tend to be colonized by xerophytic plants and the depressions by hydrophytic varieties.

3 Origin

The origin of string bogs is still not established and a number of hypotheses exist. These include:

(1) Solifluction and wrinkling of the bog surface (Andersson and Hesselman, 1907, 79; cf. Auer, 1920, 29–31). (2) Raising ridges by frost thrusting from intervening ponds combined with ruptures of frozen bog surfaces by water under hydrostatic pressure (Helaakoski, 1912, iii–iv, 69–70, 77–9; cited by Tanttu, 1915, 14). (3) Differential frost heaving combined with detachment and drifting of vegetation into windrows (Sohju theory) (Tanttu, 1915, 16–23). This hypothesis was accepted by Auer (1920, 37–40, 118–26) for some occurrences but not as a general explanation, and was favoured by Drury (1956, 66, 72–5) if combined with a tendency for meltwater flowing over a surface to form initial riffles and ridges. (4) A variety of botanical and physical processes with stress on solifluction and changes in moisture regimen (Auer, 1920, 135–43). (5) Disruption of a bog by growth of ice lenses during a cold period and subsequent solifluction during climatic warming (Hamelin, 1957, 102–5). (6) A 'combination of biological and hydrological factors' with emphasis on the habit of *Scirpus caespitosus* to grow in rows (Jan Lundqvist, 1962, 83). (7) Differential thawing of permafrost and local collapse within bogs (Schenk, 1966, 157–8). (8) Principally differential frost heaving and frost thrusting (Tricart, 1967, 230; 1969, 98).

There is a vagueness about most of these hypotheses, and the origin (or origins) of string bogs still remains to be demonstrated as witness the many unanswered questions and problems raised by the various hypotheses and by some commentaries and reports (cf. Henoch, 1960; Mackay, 1958*a*; P. J. Williams, 1959*b*).

4 Environment

In general, string bogs appear to have a distribution somewhat similar to palsas but to occur slightly farther south in the Northern Hemisphere (Troll, 1944, 639–44; 1958, 72–4). They are commonly regarded as characteristic of the subarctic taiga. Drury (1956, 63) described them as closely associated with tree line. They are widespread in western Siberia and the Hudson Bay region of Canada. In the Quebec–Labrador area they may cover 10 per cent of the terrain (Hamelin, 1957, 95). Although string bogs or closely similar features have also been observed far north of tree line and well within the zone of continuous permafrost (Henoch, 1960, 335; Mackay, 1958*a*), most investigators agree they are not necessarily indicative of permafrost. However, Schenk (1966, 157–8) regarded them as necessarily associated with thawing permafrost and concluded they can be delicate indicators of climatic change. Hamelin's conclusion that net-like string bogs are caused by bog disruption in a cold period and solifluction in a warm period also requires that such forms be a record of climatic change but Hamelin (1957, 105–6) emphasized the hypothetical nature of his conclusion.

5 Inactive and fossil forms

The low relief and probably rapid modification of string bogs by environmental changes suggest that inactive and fossil forms would be difficult to recognize. A frequent lateral alternation between xerophytic and hydrophytic plant remnants in stratigraphic section would be suggestive but no such deposits are known to the writer.

XIII PALSAS

1 General

Palsas are mounds containing ice lenses and occurring in bogs (Jan Lundqvist, 1969). Forms include circular mounds, 'fairly' straight ridges, and winding ridges. Widths are commonly 10–30 m, and lengths 15–150 m. Heights range from less than 1 m to 7 m (Forsgren, 1968, 117), or about 10 m at a maximum (Jan Lundqvist, 1969, 213). Large forms tend to be considerably less conical than small ones (Sten Rudberg and P. Wrammer, personal communication, 1971). In places palsas combine to form complexes several hundred metres in extent. Palsa surfaces are frequently traversed by open cracks.

2 Constitution

The critical constituent of a palsa is a permafrost core characterized by ice lenses that are commonly no thicker than 2–3 cm (Jan Lundqvist, 1969, 208), although locally lenses 10–15 cm thick are common (Forsgren, 1968, 113). The ice lenses generally distinguish palsas from pingos (discussed below), which are characterized by cores of massive clear ice. There are two types of palsas, those with a peat core and those with a core of mineral soil, usually silt. The peat-core type is the most common, the other being regarded as exceptional (Jan Lundqvist, 1969, 208–9) but more common than thought formerly (Sten Rudberg and P. Wrammer, personal communication, 1971). The vegetation of a palsa may comprise low shrubs and lichen in addition to the sedges characterizing the peat.

3 Origin

According to Jan Lundqvist (1969, 209), the growth of palsas is caused entirely by the development of the ice lenses but the mode of water transfer to the lenses is not clear. He indicated that hydrostatic pressure is involved in some cases but that the transfer process accompanying the growth of segregated ice is probably more common. Mackay (1965) described mounds domed by gas pressure in the Mackenzie Delta area of northern Canada, and he subsequently suggested that such doming may initiate some palsas (J. R. Mackay, personal communication, 1972).

4 Environment

Palsas are characteristic of the Subarctic and commonly occur in areas of discontinuous permafrost. In some places they extend into underlying permafrost; in others they rest on an unfrozen substratum (Friedman *et al.*, 1971, 138; Jan Lundqvist, 1969, 208–9). Palsas are almost exclusively associated with bogs, and commonly occur in areas where the winters are long and the snow cover tends to be thin. In Iceland the southern limit of palsas (rústs) is probably close to the 0° mean annual isotherm (Thorarinsson, 1951, 154–5). In Sweden 200–210 days with temperatures below 0° delimits the general distribution of palsas, and their southern limit very roughly correlates with the −2° to −3° mean annual isotherm (Jan Lundqvist, 1962, 93).

5 Inactive and fossil forms

Palsas appear to go through a developmental cycle that eventually leads to thawing and collapse. The fact that palsas in various stages of growth and decay occur together shows that their collapse is not necessarily indicative of climatic change (Jan Lundqvist, 1969, 209). The recognition of fossil forms is problematical. As pointed out by Lundqvist, stratigraphic evidence may be lacking, since the structure of a palsa is such that collapse would merely tend to restore the former surface of a bog.

Because a small palsa tends towards an originally more conical shape, collapse is more likely to leave a raised rim than in the case of a large palsa.

XIV PINGOS

1 General

Pingos, also called hydrolaccoliths and bulgunniakhs, are large ice-cored mounds. The term pingo, suggested by Porsild (1938), is an Eskimo name for hill. Pingos range in height from a few metres to some 70 m (230 ft) (Leffingwell, 1919, 150–1), and in diameter from several metres to some 600 m (2000 ft) (Mackay, 1962, 73).[1] In the absence of a cross-section, the distinction between small pingos and some palsas can be nebulous. Gaping dilation cracks radiate from the apex of many pingos as the result of growth of the ice core, and in many the summit area is collapsed because the ice core has thawed, producing thermokarst (Figures 4.62–4.63).

2 Constitution

In contrast to palsas, typical pingos are larger and contain a massive ice core rather than a series

[1] Small ice-cored mounds less than 50 cm high and ranging in diameter from 2 to 5 m occur within ice-wedge polygons on Banks Island in the Canadian Arctic. Although thought to have originated in essentially the same way as some pingos (closed-system type), these pingo-like forms are not regarded as true pingos because of their small size (French, 1971*a*).

of ice lenses (Jan Lundqvist, 1969, 206–10). However, drilling shows that some pingos have ice-lensed cores, rather than discrete planoconvex cores of clear ice, to depths greater than the pingo heights (Mackay and Stager, 1966, 363–7). Pingo ice is described briefly in the chapter on Frozen ground. The material enveloping the core can range from well-sorted clay to gravel, or be a diamicton or even bedrock. Bedrock occurrences in dolomite are reported from northern Canada (Craig, 1969), and in sandstone from Northeast Greenland (Müller, 1959, 13–55; 1963, 9–34).

3 Origin

Pingos are thought to form in two main ways – by cryostatic pressure and by artesian pressure. There are two varieties of the cryostatic hypothesis. (1) A lake becomes covered or filled by encroachment of vegetation from the margins, and being in a permafrost environment this leads to the water becoming entrapped by progressive freezing from top, bottom, and sides. Eventual freezing and expansion of the entrapped water results in heaving and doming of the ground. (2) Draining or diversion of a water body that had insulated the underlying high-moisture sediments from freezing, leads to progressive all-sided freezing of the sediments. This causes a massive ice body to form, either by (*a*) expelling unfrozen water to where it freezes as injection ice (Figure 4.64), or (*b*) providing a ready supply of water for development of segregation ice.

The pingos in the Mackenzie Delta area of northern Canada, where 98 per cent of the 1380 pingos that have been mapped here (Stager, 1956) are clearly located in (or marginal to) present or former lake basins, support the cryostatic concept. Doming of some of the pingos was accompanied by faulting (Mackay and Stager, 1966, 363–7). The origin of these pingos, known as the closed-system or Mackenzie type, was discussed by Porsild (1938), Müller (1959, 97–103; 1963, 59–64), and Mackay (1962; 1963*b*; 1966*a*; 1972*a*, 17–19) among others.

According to the artesian concept, ground water flowing under artesian pressure below thin permafrost or in taliks within permafrost forces its way to near the surface where it freezes as injection ice, again forming an ice core that heaves the surface (Figure 4.65). Such pingos, known as the open-system or East Greenland type, can develop in either soil or bedrock. Their origin has been discussed in detail by Müller (1959, 60–72; 1963, 37–47). Other Greenland occurrences have been described by Cruickshank and Colhoun (1965), Lasca (1969, 24–5), O'Brien, Allen, and Dodson (1968), O'Brien, 1971, and Washburn (1969*a*, 100–5). The hypothesis as presented by Müller has been questioned by Scheidegger (1970, 375) on the basis that water in an artesian tube in permafrost would freeze, and that drilling has failed to reveal water pockets beneath pingo ice.

The cryostatic, closed-system type and the artesian, open-system type of pingo can differ in

4.62 (*Opposite*) Pingo, Wollaston Peninsula, Victoria Island, Northwest Territories, Canada

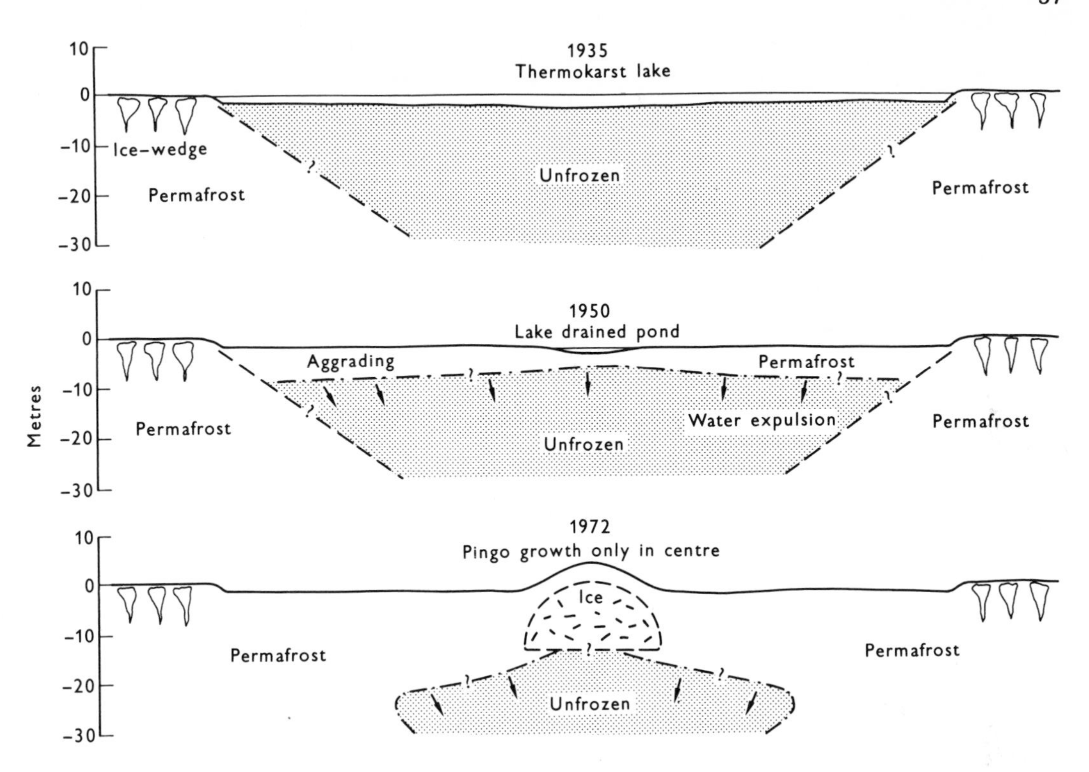

4.64 Growth of closed-system pingo in Mackenzie Delta area, Arctic Canada. Schematic cross-section (*courtesy J. R. Mackay, 1972; cf. Mackay, 1972a, 19, Figure 22*)

4.63 (*Opposite*) Pingo, Wollaston Peninsula, Victoria Island, Northwest Territories, Canada (*cf. Washburn, 1950, 50, Figure 1, Plate 13*)

continuity of water supply and thereby in pingo habit. Whereas the closed-system pingos tend to be isolated and located on near-horizontal surfaces characterized by lakes, the open-system type is probably most common in drainages where the artesian water pressure is sometimes continuous enough to cause a series of pingos. Holmes, Hopkins, and Foster (1968, H15) observed that the open-system pingos they studied in Alaska tended to cluster and mutually interfere with each other. Watson (1971, 381–8) reported fossil forms that he believed reflected an upslope migration of injection centres, and Mückenhausen (1960) and Pissart (personal communication, 1971) explained the elongate shape of some fossil pingos on a slope of $2°–4\frac{1}{2}°$ as reflecting headward growth.

Despite such differences between some closed-system and open-system pingos, these types

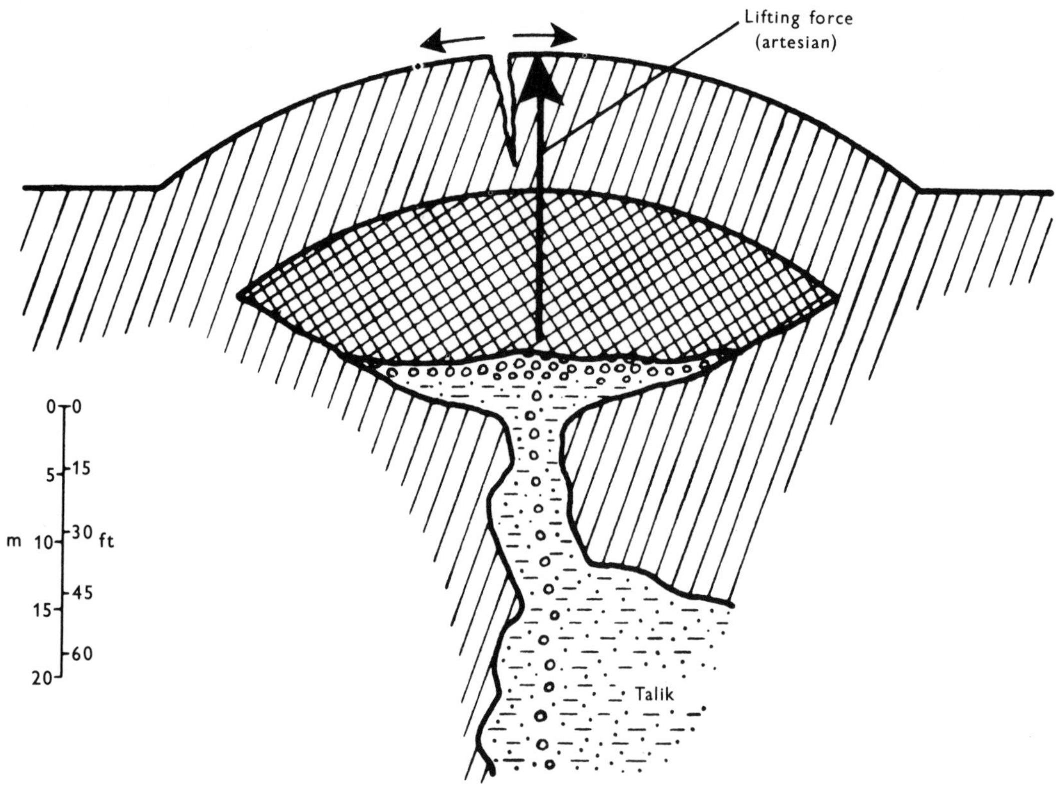

Sand,etc.

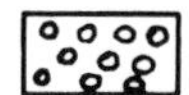

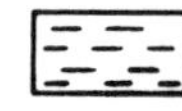

4.65 Growth of open-system pingo. Hypothetical cross-section (*Müller, 1968, 846, Figure 2*)

probably converge. Pissart (1970*a*, 31–3) has suggested that perhaps the open-system type is really formed under temporary closed-system conditions resulting from winter freezing of open taliks. Moreover, Mackay (1971*b*; 1972*a*, 15–19) has advanced the intriguing concept that there are all gradations between pingos and massive, horizontally layered ice masses such as occur at Herschel Island and eastward along the coastal plain of Arctic Canada.

In addition to cryostatic and artesian pressures, several other, much more questionable causes of pingo growth have been suggested. One hypothesis is that density differences between thick ice masses and overlying heavier material causes upward intrusion of ice diapirs (Gussow, 1954), but the close association of pingos with lakes and former lake basins in the Mackenzie Delta and

Anderson River areas, as well as other considerations, cast serious doubt on the hypothesis as applied to these areas (Mackay, 1958*b*, 55; 1962, 32).

Another suggestion is that tectonic subsidence depresses permafrost to sub-permafrost levels at which thawing and resulting compaction of saturated sediments lead to hydrostatic pressures that force water upward through favourable spots in the permafrost to generate pingos (Bostrom, 1967). The hypothesis was proposed for pingos in the Mackenzie Delta area but there appears to be no evidence that pipes of thawed sediments, maintained unfrozen by upward seepages, connect the lakes and active pingos[1] with sub-permafrost levels as required by the hypothesis.

Still another hypothesis postulates that ice segregates during initial freezing in an open system and, as a pingo grows, that more ice is added to the base of the pingo ice and to the permafrost to replenish ice that is lost by thawing at the top (Bobov, 1970; Scheidegger, 1970, 376). Details of the process and the evidence for it appear to be largely lacking in the English literature.

Swinzow (1969, 187) cited reports of a pingo that was under sufficient pressure to rupture explosively and throw blocks of frozen ground a distance of 15–20 m (Plaschev, 1956); the accompanying noise, similar to a gun shot, was heard 7 km away (Strugov, 1955). Small amounts of water and incombustible gas issue from some pingos (Mackay and Stager, 1966, 363–6).

Ages of several pingos in Siberia have been reported to range from 106 to 162 years (Schostakowitsch, 1927, 418, following Sukachëv, 1912, 82–9). In the Mackenzie Delta area, pingos up to about 6 m high are known to have formed after 1935, following coastal recession that led to draining of lakes. Other pingos have formed since 1950. The vertical growth rate of one pingo, whose crest was characterized by an open dilation crack 0·5–1·5 cm wide and 15–18 m deep, was about 15 cm/year from 1969–71 (Mackay, 1972*b*, 146–7, personal communication, 1972). In general the vertical growth rate of pingos ranges from a minute amount to more than 0·5 m/year (Shumskii, 1964*b*, 226). According to Jaromír Demek (personal communication, 1969), a 15-m high pingo he studied had an average growth rate of 1·5 m/year.

4 Environment

Pingos are necessarily associated with permafrost. Like ice-wedge polygons and sand-wedge polygons, they are key indicators of polar or subpolar environments. However, open-system pingos, believed to be in balance with the present climate, occur in the discontinuous permafrost zone of central Alaska where the mean annual temperature ranges from −2·2° to −5·6° (22°–28°F) (Holmes, Hopkins, and Foster, 1968, H7), or −1° to −2° in places (Péwé, 1969, 3). Thus pingos do not necessarily imply temperatures as low as that under which ice wedges originate in Alaska

[1] As discussed below in connection with growth rates, a number of pingos in the Mackenzie Delta area are known to be developing at the present time.

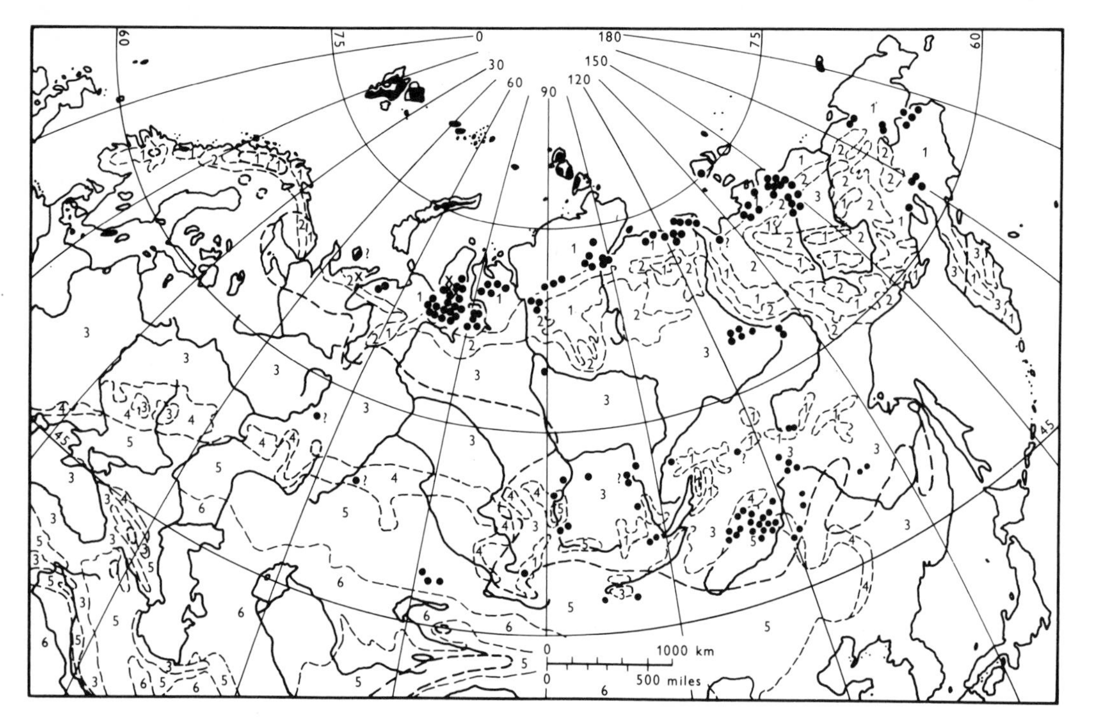

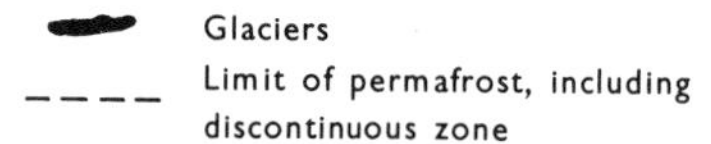

1 Tundra
2 Forest tundra
3 Forest
4 Forest steppe
5 Steppe
6 Desert

4.66 Distribution of active (•) and subfossil (×) pingos in northern Eurasia (*Frenzel, 1960a, 1021, Figure 13*)

(−6° to −8°) (Péwé, 1966*a*, 78; 1969, 4). In addition to the Alaska, Greenland, and Mackenzie Delta occurrences, pingos are reported from the Yukon (O. L. Hughes, 1969), Canadian Arctic Islands (Fyles, 1963; Pissart, 1967; 1970*b*; Washburn, 1950, 41, 43), USSR (Frenzel, 1960*a*, 1021, Figure 13; 1967, 29, Figure 17; Grave, 1956), and elsewhere in the Arctic. Their distribution in northern Eurasia is shown in Figure 4.66.

Numerous features averaging 30 m high and 400 m in diameter and believed to be pingos occur at depths of some 30–70 m on the floor of the Beaufort Sea in the vicinity of the Mackenzie Delta (Shearer *et al.*, 1971). If they are pingos, they may have originated on land and been submerged by post-glacial rise of sea level, or they may have formed beneath sea level as favoured by Shearer *et al.* In either case the observed sea-bottom temperatures, which range from below −1° to −1·8° (−1·8° being the lowest possible without freezing of the salt water here), are indicative of permafrost. Freezing of fresh water in the sediments would be expected, and in fact fresh-water ice was found in a sea-bottom core taken from a depth of about 35 m (20 fathoms) (Yorath, Shearer, and Havard, 1971, 244).

5 Inactive and fossil forms

Like palsas, inactive pingos, i.e. partially collapsed pingos in permafrost environments, are not necessarily indicators of environmental change. This is because pingo growth can lead to its own destruction by the dilation cracking of the material capping the ice core and exposure of the core to thawing. Fossil pingos in contemporary non-permafrost environments are proof of former permafrost conditions. Fossil forms are recognized in places by their pattern, ring-like rim, and internal collapse structure. Where characterized by a thermokarst depression only, they might be very difficult to distinguish from other thermokarst depressions or from kettles of glacial origin (Black, 1969*b*, 232–3). They are being increasingly widely reported from former periglacial environments in Britain (Figure 4.67), Europe, and Asia (Frenzel, 1960*a*, 1021, Figure 13; 1967, 29, Figure 17; Maarleveld, 1965; Mitchell, 1971; Mullenders and Gullentops, 1969; Pissart, 1956; 1963; 1965; Svensson, 1969; Watson, 1971; 1972; Watson and Watson, 1972; Wiegand, 1965). Possible fossil pingos have also been reported from North America (Flemal, Hinkley, and Hessler, 1969; Flemal, 1972; Wayne, 1967, 402). According to Mullenders and Gullentops, (1969, 321–8), the forms described by Pissart from Belgium probably developed during the Late Dryas (*c.* 10500–11000 C^{14} years BP).[1] However, recent pollen data indicate these pingos originated during the more rigorous climate of the Würm maximum (*c.* 17000–20000 C^{14} years BP), probably in an area of discontinuous permafrost as suggested by their distribution field relations (A. Pissart, personal communication, 1971).

[1] Before Present (taken to mean before 1950).

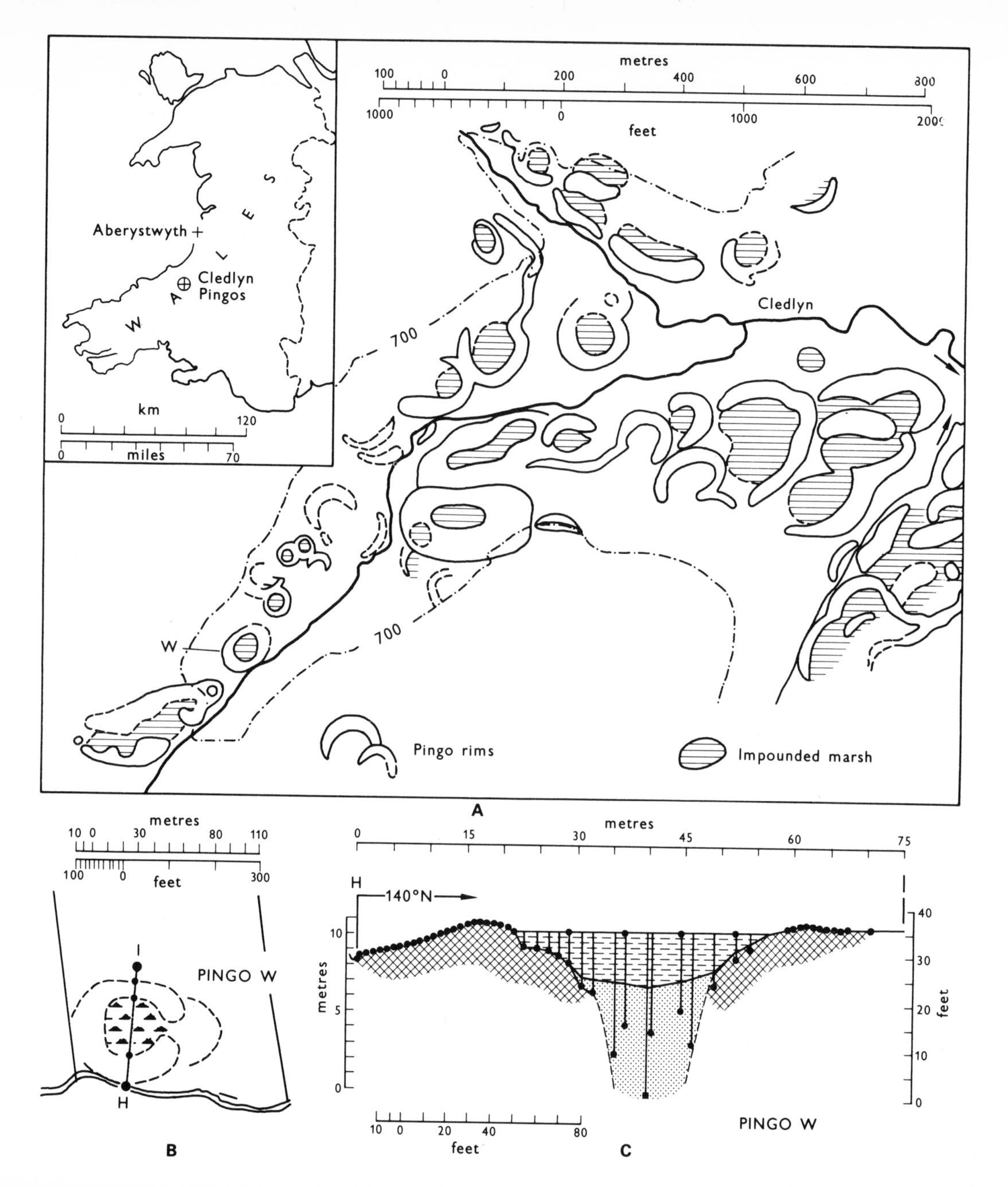

4.67 Fossil pingos, Cledlyn basin, central Cardiganshire, Great Britain. **A.** Map of pingos. **B.** Pingo W. **C.** Cross-section of pingo W (*after Watson and Watson, 1972, 213–14, Figure 1, and Figure 2 – in part*)

5 Mass-wasting

I INTRODUCTION

Mass-wasting is 'The gravitative movement of rock debris downslope, without the aid of the flowing medium of transport such as air at ordinary pressure, water, or glacier ice' (Longwell, Flint, and Sanders, 1969, 162, 654). The term was first used by Longwell, Knopf, and Flint (1939, 41).

Although many types of mass-wasting occur in periglacial environments, some types are particularly widespread, including avalanching, slumping, frost creep, and gelifluction. Slushflow is common locally.

II AVALANCHING

1 General

Avalanching as the term is used here is the sudden and very rapid movement of snow and/or rock debris down a slope. The term has been applied to free fall, sliding, and flow but some investigators restrict it to movements where flow is dominant (Longwell, Flint, and Sanders, 1969, 166–7). Regardless of the exact mechanism, rapidity of movement is a key characteristic, with calculated rates for rock avalanching ranging from 1 to 100 m/sec (Longwell, Flint, and Sanders, 1969, 166, Figure 8–3). Some glossaries define avalanches in terms of snow and ice only (Howell, 1960, 20), but avalanches of rock debris (including rockfall and debris avalanches) are also commonly recognized (Varnes, 1958, Plate 1 opposite 40), so that three broad categories can be identified – snow avalanches, rock avalanches, and mixed or snow-rock avalanches.

On a small scale, avalanching can occur on high banks and elsewhere in lowlands but it is essentially a highland phenomenon, characteristic of steep slopes, which if of bedrock can be steeper than the angle of repose. Frost creep and gelifluction on the other hand are confined to debris slopes whose gradient usually does not exceed the angle of repose. The minimum gradient on which avalanches can start depends on conditions and varies widely. Some avalanches attain sufficient momentum to carry them several hundred metres up opposing slopes. Studies by Shreve (1968*a*; 1968*b*) and the speed of avalanches make it probable that many avalanches are air lubricated in that they ride on a cushion of entrapped air.

Although rock avalanches as well as snow and mixed avalanches are common in periglacial highlands (cf. Rapp, 1960*a*, 97–137), rock avalanches are also common in other highlands and need have no special periglacial significance.

2 Snow avalanches and mixed avalanches

Snow avalanches are of many kinds, ranging from dry powder-snow varieties to wet-snow avalanches. The kind is dependent on climate and particularly on day-to-day weather conditions before, during, and after a snowfall. Because snow-avalanche danger is of critical concern in mountaineering, skiing, and highway maintenance through alpine passes, the prediction and control of avalanching has been intensively investigated and a vast literature has accumulated (US Forest Service, 1968).

Many avalanches that start as snow avalanches pick up rock debris *en route* and become mixed avalanches. Both kinds can drastically alter the landscape by ploughing through a forest and leaving a swath of broken and overturned trees or smashed buildings in their wake, but only the mixed avalanche and the rock or debris avalanche can leave a deposit or landform that may become a more enduring record. A characteristic feature of such a deposit is its nonsorted nature and hummocky topography. In places such a deposit is difficult to distinguish from a moraine, and usually there would be still greater difficulty in establishing that an ancient deposit of this kind had been made by a mixed avalanche rather than by a rock or debris avalanche having no periglacial implication.

In distribution and frequency, snow and mixed avalanches are probably most common in temperate alpine regions and progressively less so in subpolar and polar alpine environments, since temperate alpine highlands would generally have more frequent snow storms and thaw periods than polar highlands.

III SLUSHFLOW

Slushflow has been defined as the 'mudflowlike flowage of water-saturated snow along stream courses' (Washburn and Goldthwait, 1958). However, the results of essentially the same process

5.1 (*Opposite*) Beginning of slushflow, Mesters Vig, Northeast Greenland. Note saturated snow (dark)

have been described as slushers (Koerner, 1961, 1068; Ward and Orvig, 1953, 161–2) and as slush avalanches (Caine, 1969; Nobles, 1966; Rapp, 1960*a*, 138–47). They may occur on glaciers (cf. 'slushers' above; also Nobles, 1961, 758–9) where a relationship to stream courses is less well defined or lacking as compared with glacier-free areas, although natural levees of snow are formed in both places. A more comprehensive definition would therefore be simply 'slushflow is the predominantly linear flow of water-saturated snow'.

Slushflow is transitional to fluvial action on the one hand and to true avalanching on the other. It can be a discontinuous slow process as well as very rapid, and is characteristic of periglacial environments in areas where intense snow thaw in the spring produces more meltwater than can drain through the snow (Figures 5.1–5.2). In places like the Mesters Vig district of Northeast Greenland it is the common method of stream break-up and can be an important agent of erosion and deposition (Raup, 1971*b*, 50–68) (Figures 5.3–5.4). A characteristic depositional feature in mountainous terrain is a whale-backed fan of predominantly nonsorted debris at the mouths of gullies. The fan is formed when a slushflow spews forth from the gully and suddenly drops its debris load. The resulting diamicton can be remarkably till-like, the particle sizes ranging from fine sand to boulders (Figure 5.4). Slushflows are probably more common in arctic and subpolar highlands than in the Antarctic with its infrequent thaw periods.

IV SLUMPING

Slumping as defined by Sharpe (1938, 65) (who used the term slump) is 'the downward slipping of a mass of rock or unconsolidated material of any size, moving as a unit or as several subsidiary units, usually with backward rotation on a more or less horizontal axis parallel to the cliff or slope from which it descends'.

Slumping is a widespread process in many regions but because of ground ice is especially common in periglacial environments. It tends to characterize steep slopes of unconsolidated material wherever permafrost with an appreciable ice content is subject to thawing. Coastal cliffs and river banks are typical sites. Mackay (1966*b*, 68–80), working in northern Canada, stressed that the pattern of slumping in periglacial environments is very different from that elsewhere. Scarp retreat in the Mackenzie Delta cliff sites he studied ranged from about 1·5 to 4·6 m/day (5–15 ft/day) where ground ice was especially prominent. In some cases the ice-soil ratio was so high that scarp retreat resulted in very little export of material beyond the toe of the scarps. Several cycles of slumping, separated by scarps averaging 4·6–6·0 m (15–20 ft) in height, could be recognized in places, and he reported that slumps can remain active for decades.

5.2 (*Opposite*) Slushflow, Mesters Vig, Northeast Greenland

5.3 Lejvelv slushflow, Mesters Vig, Northeast Greenland, 1958. Stones were scattered 20 m above the stream bed and a diamicton and fragmented tundra vegetation were spread over a third of the Lejvelv fan, which was 3/4 km wide at its toe

5.4 (*Opposite*) Slushflow fan deposit, Mesters Vig, Northeast Greenland. Scale given by notebook (*c.* 15 × 20 cm) at centre

V FROST CREEP

1 General

Frost creep 'is the ratchetlike downslope movement of particles as the result of frost heaving of the ground and subsequent settling upon thawing, the heaving being predominantly normal to the slope and the settling more nearly vertical' (Washburn, 1967, 10). The following discussion emphasizes observations in arctic environments and is based largely on Washburn (1967). Detailed studies by Benedict (1970*b*) in an alpine environment provide additional information and support the main points.

2 Characteristics

Upon freezing, the ground tends to heave at right angles to the slope, since this is the predominant cooling surface. As shown by Taber's (1929, 447–50; 1930, 308; 1943, 1456) classic experiments, the direction of ice-crystal growth and heaving is normal to the cooling surface. Figure 5.5 illustrates the results. A particle at P_1 tends to move to P_2; as measured with respect to the vertical plane at P_1, the movement would be the distance A-P_2, which represents the potential frost creep – i.e. the maximum possible movement by frost creep if there were no further freeze-thaw cycles. The potential frost creep would be the true frost creep if the particle at P_2 dropped vertically to P_3 upon thawing of the ground. However, there is a tendency for the movement during thawing to be towards some point P_4 between P_1 and P_3, as the result of a retrograde component that is 'probably due to cohesion and to interference of particles with each other' (Ahnert 1971, 36), or is related to capillary pressures (Washburn, 1967, 109–15). Thus the slope tends to settle back against itself, reducing the amount of frost creep that would otherwise be present.

The retrograde component was observed in laboratory work by Davison (1889) who ascribed it to cohesion in a classic paper that first described the mechanism of frost creep. The observation was overlooked for some 70 years until confirmed by field observations like those illustrated in Figure 5.7, where the retrograde component is prominent in the graph of the 1961 movement, starting in early July following gelifluction in June. The arrow with the number 1·3 at the beginning of the graph indicates that the horizontal component of potential frost creep was 1·3 cm when the first observation was made in 1961; the adjusted slope-parallel measurement is 1·6 cm. In view of the absence of later freeze-thaw cycles in the ground, the subsequent downslope movement was solifluction (or gelifluction as discussed in the next section) rather than frost creep. The retrograde movement starting in July amounted to 0·9 cm (adjusted), so the true frost creep

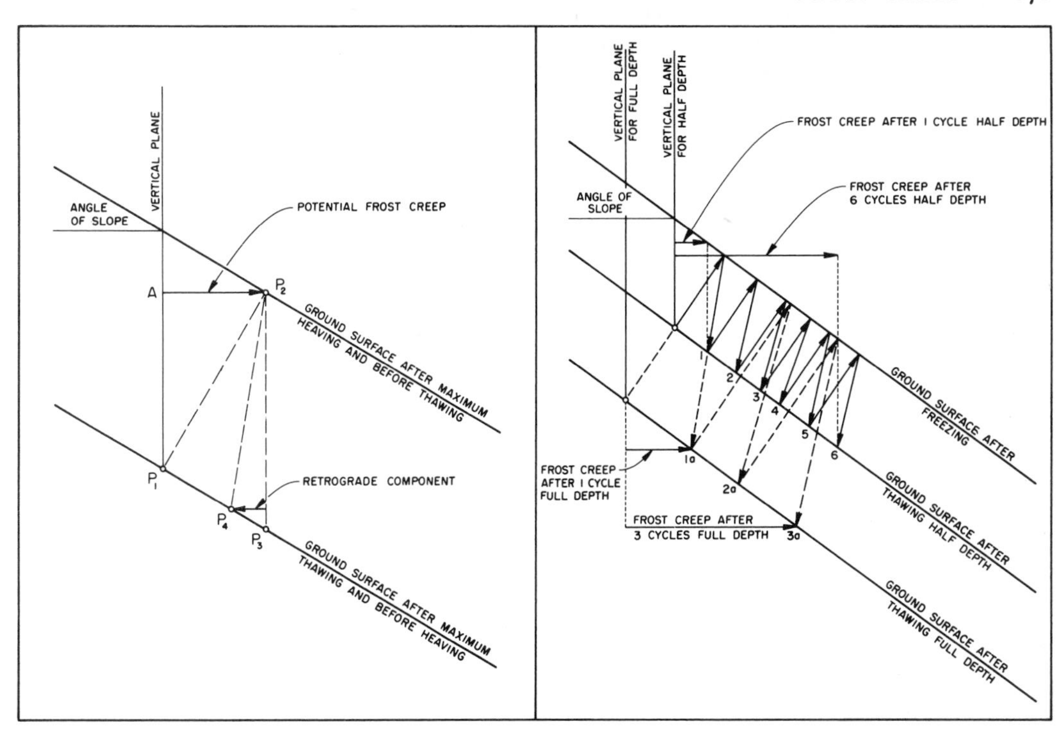

5.5 (*Right*) Diagram of frost creep during one freeze-thaw cycle

5.6 (*Far right*) Diagram of frost creep during several freeze–thaw cycles

for that year was 1·6 cm less 0·9 cm, or 0·7 cm (cf. Washburn, 1967, 46–50). Retrograde movement was also discussed by Benedict (1970*b*).

There are very few measurements of the rate of frost creep alone but it must be highly variable. Figure 5.5 depicts the effect of only a single freeze-thaw cycle. The effect of several cycles is shown in Figure 5.6, which illustrates that the amount of frost creep varies directly with their number and depth. Clearly it also varies directly with the tangent of the gradient, the amount of moisture available for freezing, and the frost susceptibility of the soil. Thus

$$FC = f\,(n, d, s, w_i, f_s)$$

where FC = frost creep, n and d are number and depth of freeze-thaw cycles, s = gradient, wi = amount of moisture transformed to ice, and f_s = frost susceptibility of the soil. In general

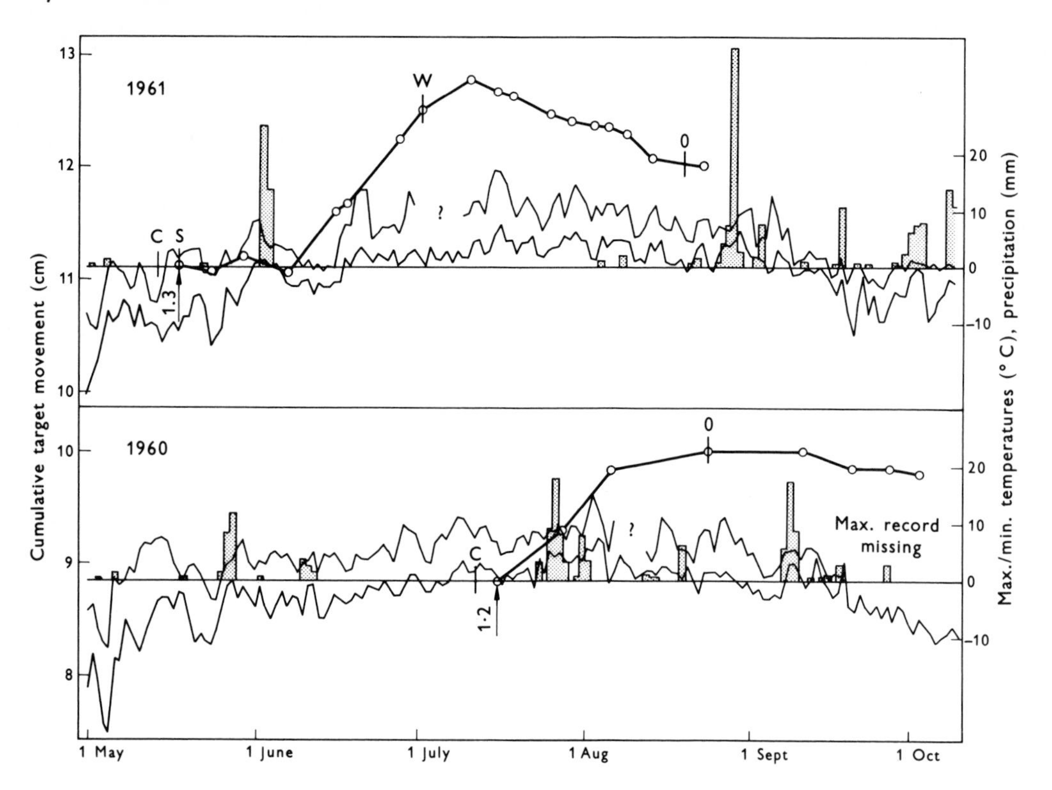

5.7 Movement of target 36, Experimental Site 7, Mesters Vig, Northeast Greenland (*cf.* *Washburn, 1967, 193, Plate C2–31*)

the rate of movement decreases rapidly with depth in both frost creep and solifluction.

Other factors remaining constant, frost creep should be greater in a subpolar or even temperate climate than in a high polar environment where there tend to be fewer freeze-thaw cycles and the depth of thaw is shallow. Measurements of near-surface movement on a silty diamicton slope of 10°–14° at Mesters Vig, Northeast Greenland, indicate that frost creep tended to exceed other forms of mass-wasting (mainly gelifluction) 'but by not more, and probably less, than 3:1 over a period of years, and either process can predominate in a given year. . . . Absolute values of mass-wasting due to frost creep and gelifluction . . . ranged from a mean of 0·9 cm/yr in sectors subject to desiccation during summer, to a mean of 3·7 cm/yr in sectors that remained saturated' (Washburn, 1967, 118). Many more measurements in various regions are needed to assess the significance of frost creep as opposed to other types of mass-wasting but it is clearly one of the most important.[1]

VI GELIFLUCTION

1 General

Andersson (1906, 95–6) defined solifluction as follows: 'This process, the slow flowing from higher to lower ground of masses of waste saturated with water (this may come from snow melting or rain), I propose to name *solifluction* (derived from *solum*, "soil", and *fluere*, "to flow").' Gelifluction was defined by Baulig (1956, 50–1; cf. also Baulig, 1957, 926) as solifluction associated with frozen ground. Thus, gelifluction is one kind of solifluction. This terminological distinction is necessary to avoid ambiguity because solifluction, by definition and observation, is not restricted to cold climates, yet its prominence there has led many writers to imply such an association. As a result the sense in which the term solifluction is being used, whether broad or restricted, is not always clear, whereas gelifluction is unequivocally periglacial. Terminological questions are reviewed by Dylik (1967) and Washburn (1967, 10–14).

The relation of gelifluction to frost creep is illustrated by Figure 5.8. Detailed discussions of gelifluction include those of Zhigarev (1967) and Benedict (1970*b*).

2 Characteristics

Gelifluction like frost creep must decrease with depth but data are lacking as to whether there is any significant difference between them in this respect. Where both are involved and undifferentiated (the usual case), the vertical velocity profile tends to be concave downslope (P. J. Williams, 1966, 196–201).

The predominant characteristic of gelifluction, like solifluction, is its dependency on moisture. The prominence of gelifluction in cold climates is due to (1) the role of the frost table or, especially, permafrost table in preventing downward movement of moisture and thereby promoting saturation of the soil, and (2) the role of thawing snow and ice (including frozen ground) in providing moisture. Under these conditions gelifluction is possible on gradients as low as 1° (St.-Onge, 1965, 40).

[1] Carson and Kirkby (1972, 272–300, 324–37), published a useful review of frost creep and solifluction in the context of soil creep and periglacial landscapes while the present volume was in proof.

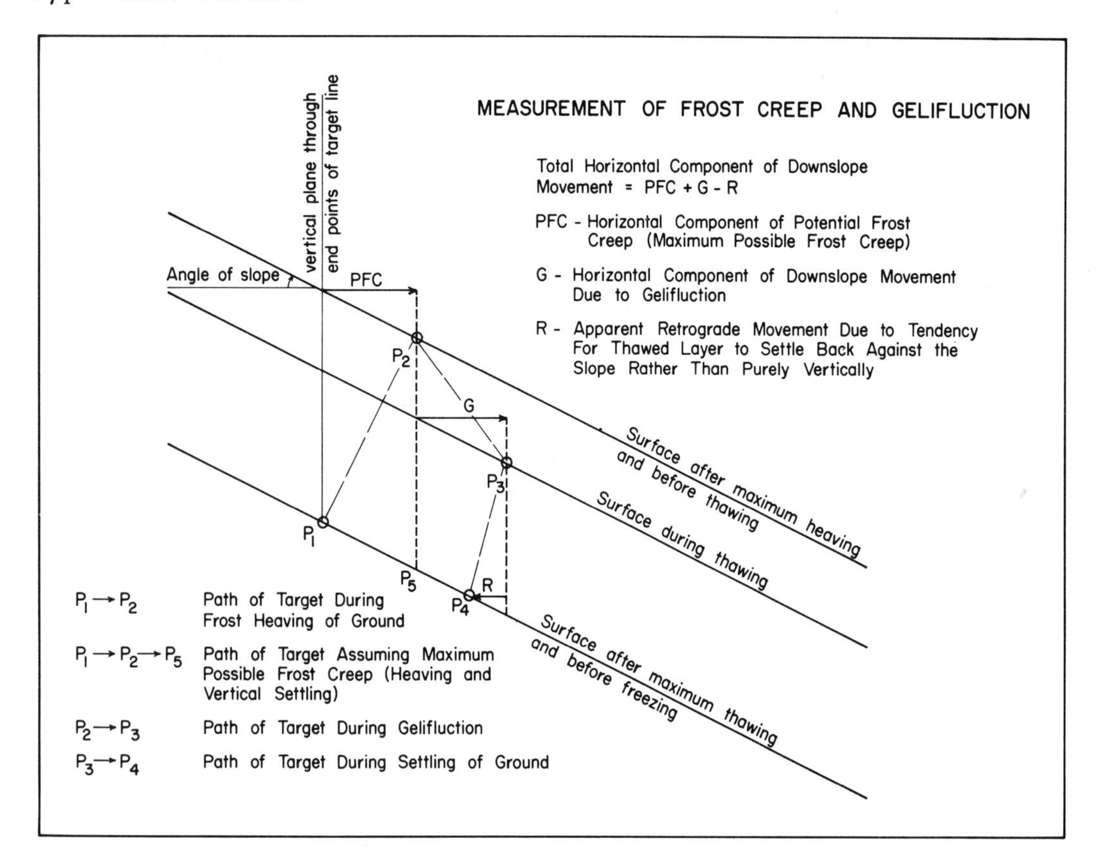

5.8 Relation of gelifluction to frost creep and retrograde movement with reference to a vertical plane through the end points of a line at right angles to movement (*cf. Washburn, 1967, 20, Figure 5*)

The role of moisture is clearly demonstrated by observations in many places, including Signy Island in the South Orkney Islands (Chambers, 1970, 93), and Mesters Vig, Northeast Greenland. At Mesters Vig, averaged over a five-year period, rates of near-surface movement (including both frost creep and gelifluction) on a 10°–14° gradient ranged from a minimum of 0·6 cm/year in relatively dry spots to a maximum of 6·0 cm/year in wet spots (Washburn, 1967, Appendix C, Table CII).

The importance of moisture is also demonstrated by the fact that within limits its influence is more significant than the binding effect of vegetation or the effect of gradient. Thus at Mesters Vig the highest rates of movement at one of the experimental sites was in the best vegetated sector where the gradient was somewhat less than in the drier part of the slope (Figures 5.9–5.10). Furthermore, observations here and at other sites indicate that significant gelifluction probably occurs only at moisture values approximating or exceeding the Atterberg liquid limit – i.e. values at which soils have little if any shear strength (Washburn, 1967, 104–8). However, Fitze (1971) on the basis of observations in Spitsbergen found no consistent correlation in this respect. Although movement was restricted to moisture contents above the plastic limit, movement took place below the liquid limit in some places and not until well above it in others, depending on the grain size of the soil. Fitze also regarded gradient as exerting less influence than grain size.

Gradient is important, since in flow the component of gravity acting parallel to a slope increases as the sine of the gradient. Observations suggest a straight-line relation between the sine (or tangent where low gradients are involved) and the rate of gelifluction and frost creep in continuously wet sectors (Table 5.1, Figure 5.11).[1]

Table 5.1. Mean annual target movements, experimental sites, Mesters Vig, Northeast Greenland (*cf. Washburn, 1967, 94*)

(Targets were wood cones on pegs inserted to depths of 10 and 20 cm)

ES		*Mean gradient (degrees)*		*Movement (cm/yr)*	
		'Dry'	'Wet'	'Dry'	'Wet'
6	1956–1961	—	2·5	—	1·0
7	1956–1961	12·5	10·5	0·9	3·4
8	1956–1961	12·5	11·5	2·9	3·7
15	1957–1960	—	3·5	—	1·1 (transverse)
		—	3·0	—	3·1 (axial)
17	1957–1959	—	12·0	—	7·6 (transverse)
		—	12·0	—	12·4 (axial)

Gelifluction is strongly influenced by grain size (cf. Washburn, 1967, 101–3). The high porosity and permeability of pure gravel and coarse sand promote good drainage and do not favour saturated flow except where porewater pressures reduce intergranular pressures. On the other hand, fines tend to remain wet longer than coarse grain sizes, and silt is particularly subject to flow because it lacks the cohesion of clays and slakes readily. Moreover, the Atterberg liquid limit is lower in silt than clay so that less moisture is required for flow. The fact that silty diamictons

[1] Frost creep increases with the tangent rather than sine of the gradient (cf. section on Frost creep). However, on gentle slopes the difference is negligible, and where gelifluction is also present the sine function is here arbitrarily applied to the total movement.

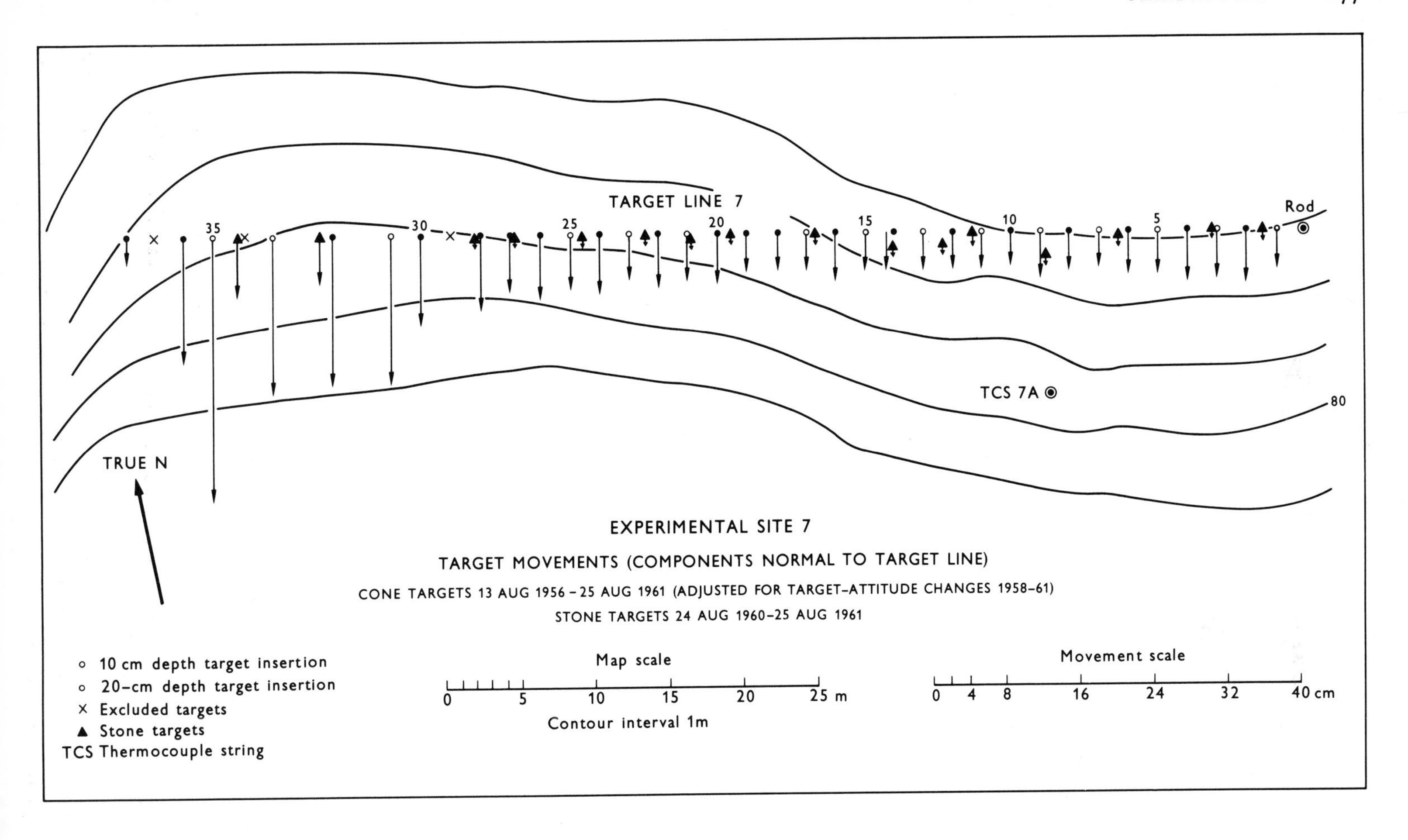

5.9 (*Opposite*) Target line 7, Mesters Vig, Northeast Greenland. View east along target line; target 40 is circled (*cf. Washburn, 1967, 38, Figure 13*)

5.10 (*Above*) Target line 7, Mesters Vig, Northeast Greenland. Target movements (*cf. Washburn, 1967, 36, Figure 11*)

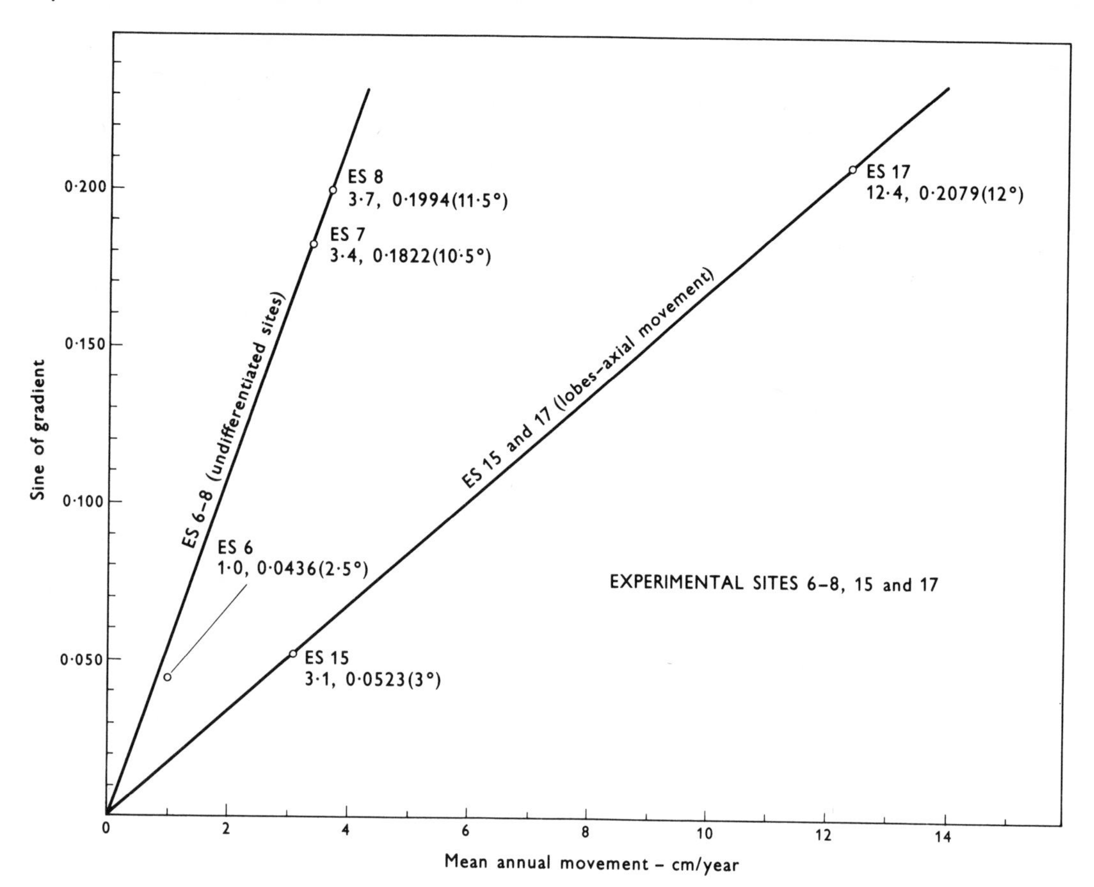

5.11 Mean annual target movements in wet areas at experimental sites, Mesters Vig, Northeast Greenland (*cf. Washburn, 1967, 95, Figure 43*)

are especially prone to flow has been recognized by many investigators (Johansson, 1914, 91–2; Schmid, 1955, 52–3; Sigafoos and Hopkins, 1952, 181; Sørensen, 1935, 21–8; and others). In cold climates mechanical weathering tends to be more important than chemical weathering (Blanck, Rieser, and Mortensen, 1928, 689–98; Meinardus, 1930, 45–72; Washburn, 1969*a*, 43–4), so that silt tends to predominate over clay and clay-size particles. This may be one of the reasons for the prevalence of gelifluction (and frost creep) in such climates.

Considering the above influences, gelifluction should be a function of depth, sine of the gradient (force parallel to the slope), amount of liquid moisture, and grain size of the soil. The function is similar to that for frost creep except for substituting the sine for the tangent of the gradient, liquid moisture for ice, and omitting number and depth of freeze-thaw cycles as factors.

As discussed below under Frost creep and gelifluction deposits, gelifluction tends to orient the long axes of stones up and down the slope and to produce several characteristic landforms, including (1) gelifluction sheets (Figure 5.12), (2) gelifluction benches (Figure 5.13), (3) gelifluction lobes (Figures 5.14–5.15), and (4) gelifluction streams (Figure 5.16) (Washburn, 1947, 88–96). They are designated as gelifluction features because of their flow-like nature and because the slopes on which they occur are known as gelifluction slopes, although frost creep, regarded here as a separate process, is an important contributor to movement. For instance, observations at Mesters Vig, Northeast Greenland, show that 'Either frost creep or gelifluction can predominate in different places on the same slope, depending on variations in local conditions' (Washburn, 1967, 118).

Data regarding the relative importance of gelifluction and frost creep are very limited. Examples of Mesters Vig observations are illustrated in Figures 5.7 and 5.18. In Figure 5.7, the true frost creep for target 36 in 1961 was 0·7 cm. The gelifluction indicated by the rising graph in June and early July amounted to 2·0 cm so that gelifluction in this year was some three times greater than frost creep. In the preceding year gelifluction and frost creep were nearly equal (1·4 cm and 1·2 cm, respectively). At target 35 (Figure 5.18), 2 m away, they were also nearly equal in 1959–60 (2·3 cm and 2·2 cm, respectively) but in 1960–61 there was about three times as much gelifluction as frost creep (*c.* 5 cm and 1·8 cm, respectively). Thus the ratio of gelifluction to frost creep differed in successive years for both targets but each target showed a similar ratio in any one

Table 5.2. Some rates of frost creep and gelifluction in Greenland, Lapland, Norway, and Spitsbergen

Location	*Gradient (degrees)*	*Rate (cm/yr)*	*Reference*
Greenland			
Mesters Vig	10–14	0·9–3·7 (mean)	Washburn (1967, 118)
Lapland			
Kärkevagge	15 (mean)	2 (mean to 25-cm depth)	Rapp (1960*a*, 182)
Norra Storfjäll area	5	0·9–3·8 (mean)	Rudberg (1964, 199, Table 2, item 1)
Tärna area	5	0·9–1·8 (mean)	Rudberg (1962, 317, Table II)
Norway			
Okstindon	5·17	1·0–6·0	Harris (1972, 169, Table V)
Spitsbergen			
Barentsoya	11	1·5–3 (mean)	Büdel (1961, 365; cf. 1963, 277)
N side, Hornsund	3–4	1–3	Jahn (1960, 56; 1961, 12–13)
N side, Hornsund	7–15	5–12	Jahn (1960, 56)

5.12 (*Opposite*) Gelifluction sheet encroaching on emerged strand lines, Mount Pelly, Victoria Island, Northwest Territories, Canada (*cf. Washburn, 1947, Figure 2, Plate 33*)

5.13 (*Above*) Gelifluction bench, Mount Pelly, Victoria Island, Northwest Territories, Canada (*cf. Washburn, 1947, Figure 1, Plate 25*)

5.14 (*Opposite*) Gelifluction lobe, Mount Pelly, Victoria Island, Northwest Territories, Canada. Stakes were used to measure movement (*cf. Washburn, 1947, Figure 1, Plate 26*)

5.15 Gelifluction lobe, Hesteskoen, Mesters Vig, Northeast Greenland (*cf. Washburn, 1967, 90, Figure 40*)

5.16 (*Opposite*) Gelifluction streams, Mount Pelly, Victoria Island, Northwest Territories, Canada (*cf. Washburn, 1947, Figure 1, Plate 27*)

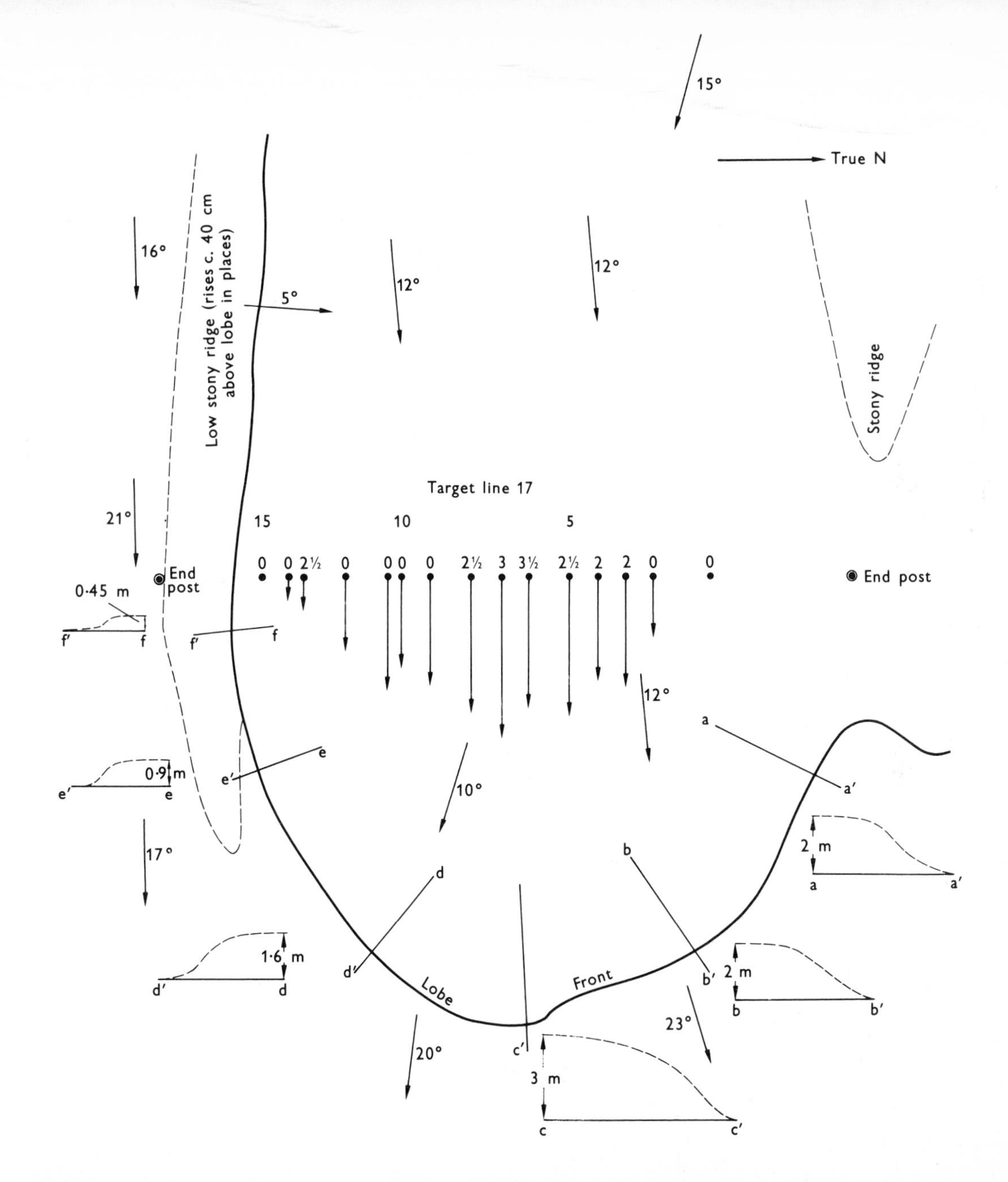

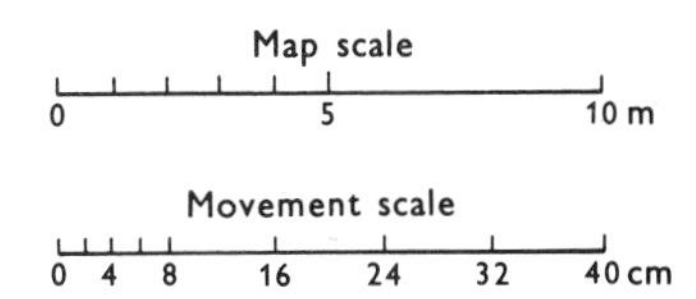

5.17 Target movements, Experimental Site 17, Hesteskoen, Mesters Vig, Northeast Greenland (*cf. Washburn, 1967, 88, Figure 38*)

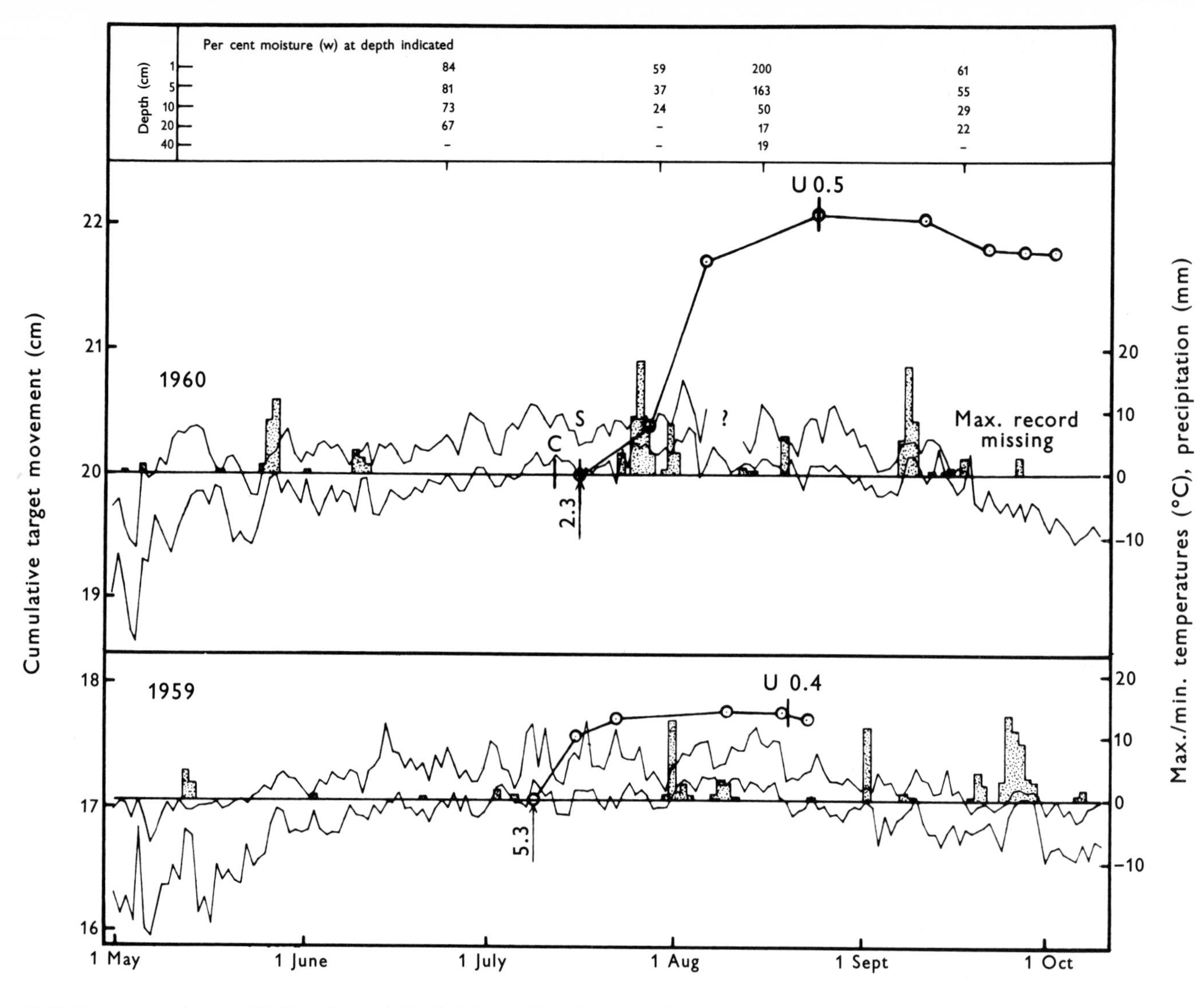

5.18 Movement of target 35, Experimental Site 7, Mesters Vig, Northeast Greenland (*cf. Washburn, 1967, 192, Plate C2–30*)

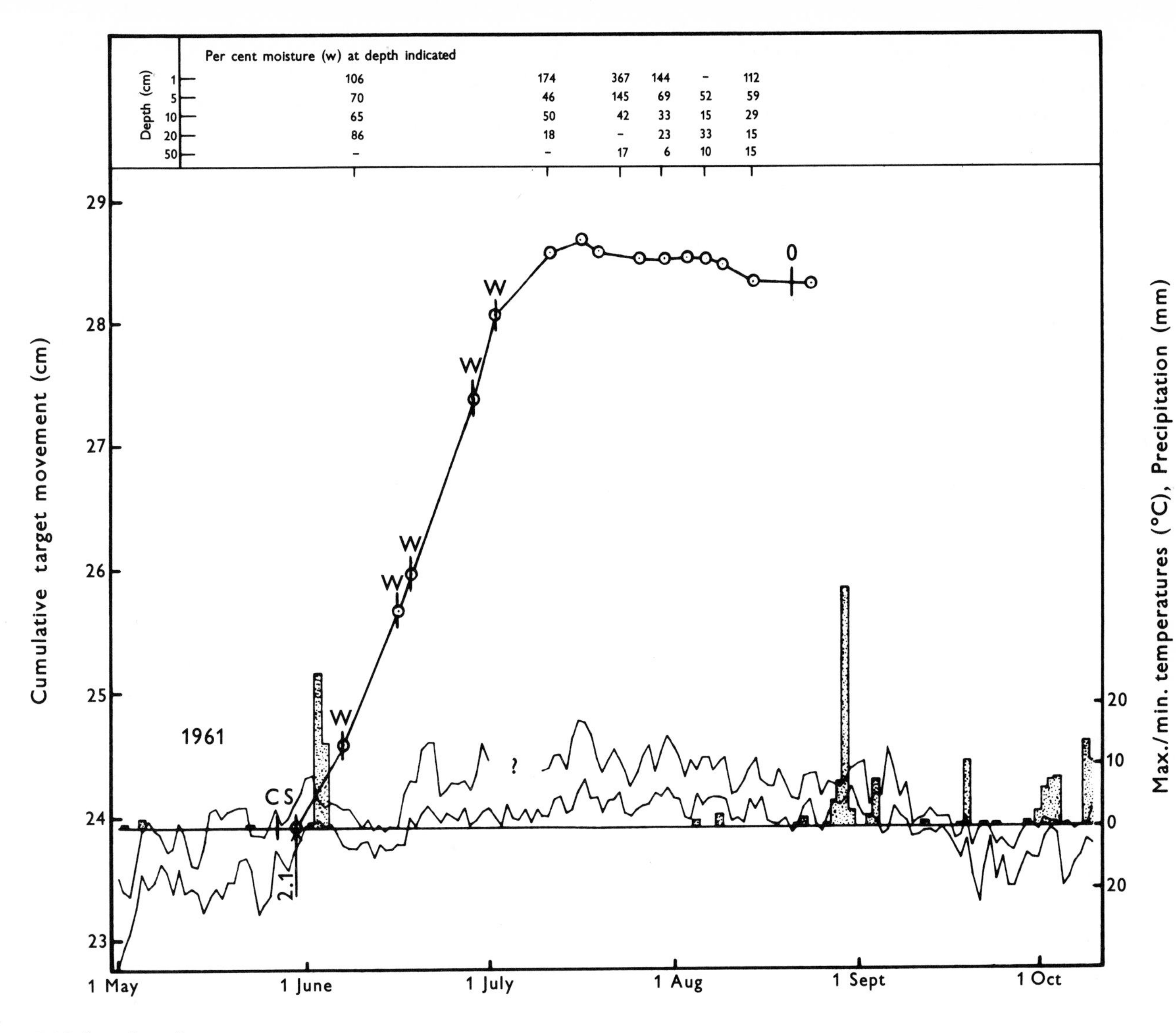

5.18 (*continued*)

year. The greater amount of gelifluction (and frost creep) at target 35 was due to wetter conditions there.

Gelifluction rates quoted in the literature (Table 5.2) generally include frost creep and are of the same order of magnitude as mean annual movements for frost creep and gelifluction at Mesters Vig. Considering the variety of factors involved, the consistent order of magnitude is remarkable. Rates on gelifluction lobes rather than on uniform slopes are much more variable and are strongly dependent on where they are measured, since the maximum movement occurs along the lobe axis and decreases uniformly towards the margins as illustrated by Figure 5.17.

Vegetation has been regarded as impeding gelifluction sufficiently to warrant distinguishing between 'free' (vegetation-free) and 'bound' (vegetation-bound) movement, and recognizing the domain of the former as a frost-debris zone and of the latter as a milder-climate tundra zone. Free gelifluction was thought to be associated with sorted stripes, and bound movement with benches (on gradients of 2° to 15°–20°) and with lobes (on gradients of 15°–20° to 30°) (Büdel, 1948, 30–3, 41–2; 1950, 12–13). According to J. W. Wilson (1952, 249), 'the development of more than a meagre vegetation, however, generally prevents soil movement (and so patterning) by the stabilizing effect of the roots'. On the other hand, although Bertil Högbom (1910, 46; 1914, 331–2, 360–3) recognized an impeding effect of vegetation, he regarded vegetation as being more likely to be affected by gelifluction than vice versa. There is an intimate interaction between gelifluction and vegetation, with some types of growth being more compatible with gelifluction than others (Frödin, 1918, 1–14; Raup, 1969, 1–21, 26–84, 93–143, 146–159, 199–202; Seidenfaden, 1931; Sørensen, 1935, 56). Sørensen (1935, 63–4) argued that, even so, plants were generally so strongly influenced by gelifluction that they were a function of it. Others have stated that vegetation, by retaining water and retarding run-off, actually favours slope movements (Sigafoos and Hopkins, 1952, 182). Furthermore, as already noted (p. 175), measurements at Mesters Vig on vegetation-covered and vegetation-free sectors of the same slope show that the most rapid movement was in the best vegetated sector and that control by moisture was more important there than any binding effect of vegetation (Washburn, 1967, 43–4, 104–5). The role of bound gelifluction in determining the development of benches and lobes is also questionable. The present writer has seen prominent benches and lobes in vegetation-free areas in both the Arctic and Antarctic. Moreover, the location of lobes is controlled not so much by gradient as by concentrations of moisture (Rudberg, 1962, 316; Raup, 1969, 137–9; Washburn, 1967, 96–8).[1]

Future progress in the study of frost creep and gelifluction will probably make a quantum jump if emphasis is put on detailed laboratory investigations (Higashi and Corte, 1971) and on automatic recording of data in the field as well as in the laboratory. The same is true of patterned-

[1] A discussion of frost creep, gelifluction, and their controlling factors by Harris (1972), which appeared while the present volume was in proof, provides additional supporting data.

ground research. Traditional baseline studies and periodic comparisons have great value and will still be essential but are unlikely to bring the insights that modern technology now makes possible.

VII FROST-CREEP AND GELIFLUCTION DEPOSITS

1 Sheets, benches, lobes, and streams

Frost creep and gelifluction are commonly associated, and the resulting deposit is therefore of joint origin. However, it is common practice to assume that gelifluction is the dominant process or includes frost creep, and that the deposit is therefore a gelifluction deposit. In the following, it is assumed that both processes have contributed to such deposits regardless of their designation.

Deposits can occur on gradients of as little as 2° (Louis, 1930, 98). According to topographic form they can be: (1) Gelifluction sheets, characterized by a smooth surface and in places by a bench-like or lobate lower margin (Figure 5.12). (2) Gelifluction benches, characterized by their pronounced terrace form (Figure 5.13). The longest dimension of the benches tends to be parallel to the contour but in places is at angles up to 45° (Frödin, 1918, 29, Tafel IV – also cited by Büdel, 1948, 42). (3) Gelifluction lobes, characterized by their tongue-like appearance (Figures 5.14–5.15). Their longest dimension tends to be at right angles to the contour. (4) Gelifluction streams, characterized (as opposed to lobes) by pronounced linear form at right angles to the contour (Figure 5.16).

Gelifluction deposits tend to be diamictons but some deposits show crude stratification parallel to the slope (Figure 5.19). The material is commonly angular and its arrangement or fabric is characterized by a tendency for the long axis of stones to be oriented in the direction of movement (cf. French, 1971, 725–6; Furrer and Bachman, 1968, 5–8; Galloway, 1961, 349–50; G. Lundqvist, 1949; K. Richter, 1951, cf. Büdel, 1959, 298; Rudberg, 1958). In association with other evidence, this fabric has been used to support the solifluction origin of some deposits (Watson, 1969). In the absence of evidence as to topographic form, such deposits can be easily confused with similar diamictons of other origin (Flint, 1961, 148–9, Table 1; 152; 1971, 152–3; Rapp *in* Watson, 1969, 113). Moreover, as stressed by Black (1969*b*, 230), in the absence of independent evidence gelifluction deposits may be especially difficult to distinguish from solifluction deposits made in nonfrost climates. However, the former would tend to have constituents that were more angular and less chemically weathered.

The suggestion has been made that it is possible to distinguish between periodic (annual) and episodic (occasional) gelifluction deposits (Büdel, 1959, 301–10). According to Büdel, the periodic type is characterized by deposits commonly less than 1 m thick (maximum 2 m) and by laminar structure; the episodic type is characterized by deposits commonly 2 m or so thick (maximum

4 m) and by fold structure including downslope-tipped ice-wedge casts. Büdel associated the periodic type with gradients as low as 1·7° in the present polar environment and with gradients of 4°–6° to 17°–27° in middle-latitude environments during the Pleistocene, the difference being based on an assumed quicker snowmelt and therefore greater desiccation in middle latitudes. However, snowmelt in arctic environments, as at Mesters Vig, Northeast Greenland, can be so rapid that it leads to slushflows and earthflows. Pleistocene episodic gelifluction deposits Büdel associated with gradients of less than 4°–6° in middle latitudes but he found no parallel in present polar environments. The postulated distinction between periodic and episodic deposits, which is not so much one of process as frequency of process, would be a valuable climatic criterion if valid and practical. However, it is largely theoretical and the postulated association is based more on assumed gradient relationships than on the study of deposits forming today.

2 Block fields, block slopes, and block streams

a General Block fields are considerable areas, broad and usually level or of only gentle gradient, covered with moderate-sized or large angular blocks of rock (after Sharpe, 1938, 40). Block slopes are similar areas on slopes (Washburn, 1969*a*, 36) (Figure 4.6). Block streams, made famous by the Falkland Island occurrences described by Andersson (1906, 97–104) and others, are extensive accumulations confined to valleys or forming narrow linear deposits extending down the steepest available slope.[1] As used by Richmond (1962, 19) the term rubble sheet covers both block fields and slopes, and the term rubble stream is synonymous with block stream. The German terms Blockmeer and Felsenmeer include accumulations of rounded boulders as well as of blocks, and the term stone fields, stone slopes, etc., might be used in this sense in English, thereby reserving the terms block for angular, and boulder for rounded, material.

Such accumulations of blocks and boulders have diverse origins but many are so closely related to frost-creep and gelifluction deposits that they are included in this section.

b Constitution Caine (1968, 33–42) found a three-layered structure in the non-vegetated block fields and block streams of Tasmania: (1) blocks with open-work texture at the surface and to

[1] Caine (1968, 6–7) indicated that a gradient of 15° was a reasonable upper limit for block fields in Tasmania but he was not contrasting block fields with steeper slopes as would seem desirable in view of Sharpe's original definition. The present writer suggests that 5° would be an appropriate boundary between block fields and block slopes. In his excellent study Caine recognized block streams and block glacis (aprons) as types of block fields with block glacis having their greatest extent parallel rather than transverse to the contour. In view of the proposed distinction between block fields and block slopes, it seems best to follow Caine in using the terms block glacis and block streams in the pattern sense but not to define them as types of block fields, since otherwise terminological problems arise for similar features on block slopes.

5.19 (*Opposite*) Crude stratification in diamicton of gelifluction slope, Schuchert Dal, Northeast Greenland. Scale given by 60-cm rule

depths of over 3 m in places; (2) an intermediate layer, 10–30 cm thick, characterized by interstitial humic mud between the blocks; and (3) a basal layer lying on bedrock and consisting either of blocks with an interstitial filling of silty sand or, in a few places, of silty sand without blocks. Vegetation-covered block fields had the interstitial filling extending to the surface.

In general the texture of block slopes and block streams is characterized by a fabric in which the long axis of individual blocks tends to be aligned in the same direction as the local gradient but less steeply inclined (Cailleux, 1947; Caine, 1968, 96; Klatka, 1961*b*, 9–13; 1962, 69–83, 120). At the Łysa Gora locality in Poland, the classic site of Łoziński's pioneering periglacial investigations, the fabric is increasingly well developed downslope and the stones smaller, but on a talus-like block slope nearer the source of the blocks the fabric is less pronounced and the imbrication tends to be directed downslope rather than upslope (Klatka, 1961*b*, 6, 9–13; 1962, 69–83, 120). A change from an oriented fabric on a slope to an irregular fabric on a lower gradient where mass-wasting was less prominent was reported by Dahl (1966*a*, 65–8).

Caine's (1968, 43–96, 106) observations in Tasmania confirm these textural trends for the upper layer of the block areas he studied. He also noted that at the downslope toe of those areas the orientation of the long axes changed and became transverse to the gradient.

c Origin Block fields and slopes of the La Sal Mountains, Utah, were studied by Richmond (1962, 62–5, 80–1) who recognized four genetic types: (1) frost-wedged, formed *in situ*; (2) frost-creep, formed with slow downslope movement over bedrock; (3) frost-sorted, formed *in situ* from coarse and fine rubble; and (4) frost-lag, formed with slow downslope movement from coarse and fine rubble.

Weathering *in situ* of massively jointed bedrock can produce both rounded boulders and blocks but only the latter are generally indicative of a predominance of mechanical weathering, usually frost wedging, over chemical weathering, abrasion, and other processes (cf. King and Hirst, 1964). However, as stressed by Caine (1967, 427; 1968, 101–3), angularity in dolerite can be associated with subsurface decomposition in places, and, on the other hand, rounding does not disprove periglacial weathering if the rounding occurred later. Both angular and rounded stones can probably also form as a lag concentrate from erosion of a stony deposit such as till. Eluviation of fines by piping (H.T.U. Smith, 1968) and other processes may help to explain the blocky appearance of a deposit, but time is also a factor in that an accumulation of blocks may reflect insufficient time for frost wedging to comminute material to smaller sizes as noted by Guillien and Lautridou (1970, 45). They also reported that the maximum size of frost-wedged fragments in their experiments was not related to intensity of freezing (Guillien and Lautridou, 1970, 36, 38). However, the production of large blocks would seem to be favoured by deep penetration of freezing into jointed bedrock on a scale difficult to reproduce in the laboratory. Caine (1968, 97–105, 116) concluded that the material of the block areas in Tasmania might have originated by weathering under both periglacial and non-periglacial conditions. Whether block fields and slopes can form by frost sorting as indicated by Richmond (1962, 63), Corte (1966*b*, 234) and others remains to be proved, but the demonstrable upfreezing of large stones supports the possibility.

Although stones can undergo significant transport in various ways, it is commonly assumed that any movement of the whole accumulation is due to mass-wasting with emphasis on solifluction, usually gelifluction. However, the role of creep (frost creep and other forms) is not established and may be significant as suggested by Richmond (1962, 63). Caine (1968, p. 106–11) interpreted the fabric of the deposits he studied as showing that movement occurred while the blocks were in a matrix of finer material similar to the basal layer he described (cf. *b Constitution*). Lack of mixing of these layers and evidence provided by lichens and weathering features indicate no movement at present, as the former movement probably occurred by gelifluction and frost creep under periglacial conditions. He concluded that the matrix had been subsequently eluviated from the upper openwork layer and that some blocks had been sorted out of the basal layer by upfreezing as shown by its locally block-free aspect. The lack of thermokarst features and general dissimilarity to rock glaciers were taken as evidence that interstitial ice was not an essential factor in movement.

d Environment Active block fields, slopes, and streams are widely distributed in the polar zone, especially but not exclusively in highlands. Antarctic occurrences are summarized by Nichols (1966, 26), and Eurasian localities are shown in Figure 5.20. Activity is commonly indicated by lack of lichens on the stones as opposed to their presence on stones in more stable locations in the same region. Active block accumulations probably also occur in subpolar and middle-latitude highlands but activity or stability is more difficult to prove.

e Inactive and fossil forms Block accumulations in middle-latitude highlands have been widely interpreted as recording former periglacial conditions but this conclusion requires proof of present inactivity and elimination of alternative possibilities. Schott (1931, 58–71) doubted the periglacial significance of block accumulations, although granting the influence of increased frost wedging during glacial ages. Büdel (1937) regarded many block accumulations below a minimum gradient as stable today, based on various types of evidence but not on instrumental observations, whereas Schmid (1955, 106–8) questioned the fossil nature of many such block accumulations in the German Mittelgebirge. Widespread accumulations of truly angular blocks are certainly reasonable evidence of former frost wedging if located in an environment where such blocks are not accumulating today, but further interpretations are fraught with difficulty in the absence of additional evidence. Some of the problems in interpreting block fields and slopes are illustrated

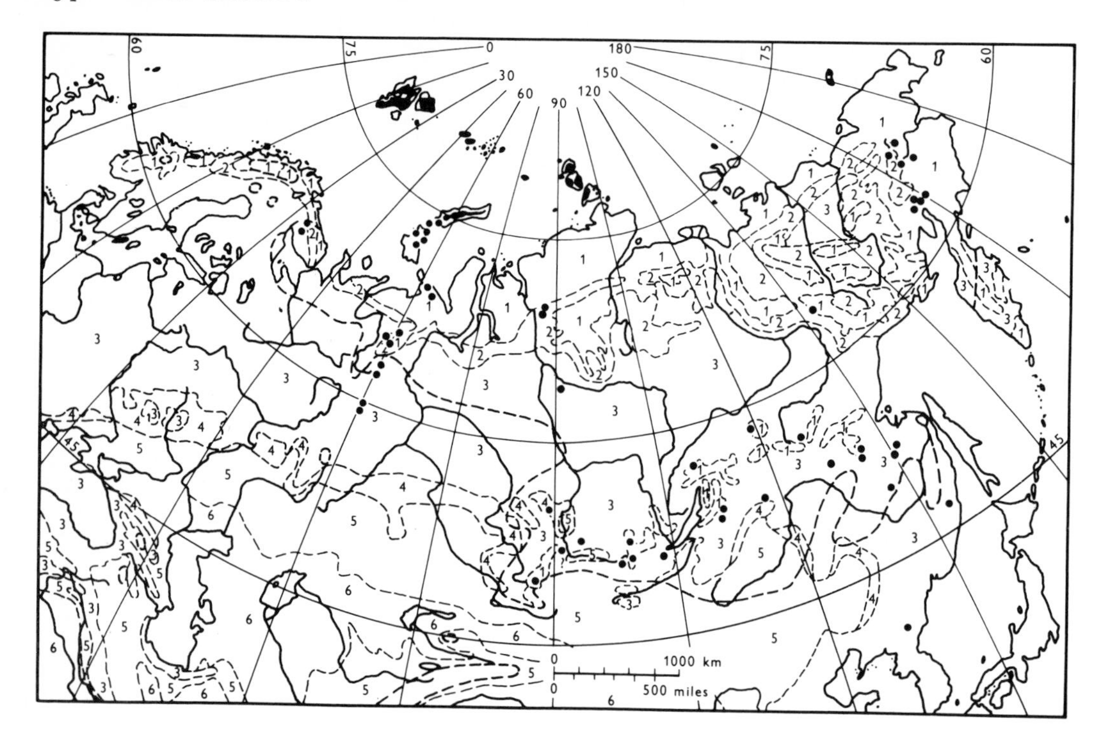

5.20 Distribution of block fields, block slopes, and block streams (•) in northern Eurasia (omitting Scandinavian occurrences) (*Frenzel, 1960a, 1020, Figure 12*)

by deposits in Åland, Finland (C. A. M. King and Hirst, 1964) and a discussion between Dahl (1966*a*; 1966*b*) and J. D. Ives (1966).

3 Ploughing blocks

Ploughing blocks are a special kind of frost-creep and/or gelifluction deposit consisting of isolated, commonly boulder-size stones that leave a linear depression upslope and form a low ridge downslope. The depression and ridge are formed as the stone moves downslope, and their size tends to vary directly with the size of the stone. In general the long axis of the stone lies in the direction of gradient. Particularly studied by Tufnell (1972), ploughing blocks, known as Wanderblöcke in German, have been described from alpine environments by Louis (1930, 97–8),

Mohaupt (1952, 10–11, 34), Poser (1954, 150–2), Schmid (1958, 258–60), and Ball and Goodier (1970, 212–13) among others. They have also been observed in polar regions by the present writer. Poser considered ploughing blocks to be gelifluction (flow) phenomena, whereas Schmid explained their movement by frost creep. Mohaupt concluded that sudden, fast movement was involved in places but did not exclude slow movement. An observation in Northeast Greenland by the present writer also suggests sudden, fast movement in places, since the upslope depression was equally fresh and smooth throughout its length (Figure 5.21) as if it had been formed by a single slip, possibly the same spring. That ploughing blocks may be recognized long after movement has ceased is demonstrated by occurrences in the Harvard Forest at Petersham, Massachusetts, which show post-movement soil profiles (Lyford, Goodlett, and Coates, 1963, 28–30).

VIII ROCK GLACIERS

1 General

Rock glaciers have been defined as 'glacierlike tongues of angular rock waste usually heading in cirques or other steep-walled amphitheaters and in many cases grading into true glaciers' (Sharpe, 1938, 43). They are characterized by steep fronts at or near the angle of repose, and commonly by concentric ridges and finer debris at depth than at the surface (Figure 5.22). Apparently all active rock glaciers contain ice, and if the necessary information as to its amount and distribution is available, active rock glaciers can be divided into ice-cemented and ice-cored types (Potter, 1969, 13).[1]

In some places rock glaciers appear to be transitional to debris-covered glaciers or ice-cored moraines, as illustrated by the exchange between Barsch (1971) and Østrem (1971)

2 Constitution

The surface of rock glaciers tends to be characterized by blocks and smaller stones but all sizes and shapes are possible. The surface appearance is misleading, however, in that the interior of rock glaciers, where known, usually consists of a diamicton in which fines may be plentiful. As noted, all active rock glaciers appear to contain ice.

3 Origin

Rock glaciers probably originate in several ways. Hypotheses include (1) landsliding (Howe,

[1] See also Potter (1972) published while the present volume was in proof.

1909, 52), (2) glaciation (several varieties of hypotheses – Capps, 1910, 364; Cross, Howe, and Ransome, 1905, 25; Kesseli, 1941; Richmond, 1952), (3) flow of interstitial ice (Blagbrough and Farkas, 1968, 821–3; Wahrhaftig and Cox, 1959), (4) a combination of (2) and (3) (Flint and Denny, 1958, 131–3). Landsliding in Howe's sense may contribute material to rock glaciers but does not seem to be responsible for them, since rock glaciers move at slow rates only, commonly

5.21 Ploughing block, Mesters Vig, Northeast Greenland

5.22 (*Opposite*) Rock glacier, northeastern Selwyn Mountains, Yukon Territory, Canada. The bedrock is sandstone and shale. Photo by Clyde Wahrhaftig

less than 1 m/year. All the other hypotheses require ice at some stage, and in some areas, such as the unglaciated San Mateo Mountains of New Mexico, glacier ice can be excluded (Blagbrough and Farkas, 1968, 822). Although it has been suggested that 'rock-glacier creep' (Sharpe, 1938, 42–8) is mainly the result of frost creep and solifluction (Chaix, 1923, 32; Kesseli, 1941, 226–7), movement appears to be dependent on the presence of permafrost, which occurs in many places as indicated by interstitial ice or an ice core. The basal shear stress calculated from the thickness and surface slope of active rock glaciers considerably exceeds that required for frost creep and solifluction alone (Wahrhaftig and Cox, 1959, 403–4), but the exact mechanism of movement still remains to be established.

4 Environment

Active rock glaciers are reported from polar, subpolar, and middle-latitude highlands. That active forms occur in middle-latitude highlands has been demonstrated by movement observations such as those indicated below (cf. also R. L. Ives, 1940, 1277–8).

Table 5.3. Some rates of rock-glacier movements

Location	*Rate (cm/yr)*	*Observation period (yr)*	*Reference*
Macun, Lower Engadin, Swiss Alps	25–30	2	Barsch (1969, 11, 27)
Val Sassa and Val 'dell Acqua, Swiss Alps	136–158	21	Chaix (1943, 122)
Galena Creek, Wyoming	4–83	3	Potter (1969, 52, Table 7; 130, Figure 21)
Colorado Front Range	2–15	5	S. E. White (1971, 55–7, Tables 2–4)
Clear Creek, Alaska Range	36–69 (1·2–2·3 ft)	8	Wahrhatfig and Cox (1959, 395, Figure 6)

As discussed above, movement is apparently dependent on the presence of permafrost. However, the perennial ice in some rock glaciers is buried glacier ice, and the resulting deposits would be transitional in origin.

5 Inactive and fossil forms

In general inactive or fossil rock glaciers appear to have about the same paleoclimatic significance as moraines of former valley glaciers—i.e. former lower temperature or greater precipitation or both. As stressed by Potter (1969, 76) correlating the time significance of rock glaciers with that of moraines left by valley glaciers is dangerous because rock glaciers may remain active much longer than glaciers that wasted rapidly because of being relatively free of debris. Although fossil rock glaciers would normally lack the prominent outwash trains expectable from the rapid melting of such comparatively 'clean' glaciers (Potter, 1969, 77), this contrast might not hold in some high-latitude arid environments where runoff from glaciers is restricted.

Judging from the distribution of active rock glaciers today and their nature, inactive and fossil forms indicate a former more rigorous climate consistent with at least discontinuous permafrost. Some of these forms apparently developed in the absence of former glaciers as on Kendrick Peak in northern Arizona (Barsch and Updike, 1971, 105–6, 110–12) and in the San Mateo Mountains of New Mexico (Blagbrough and Farkas, 1968), so that the permafrost here would not be related to buried glacier ice.

IX TALUSES

Taluses are apron-like accumulations of sliderock. The debris is typically angular and usually grades from coarse at the top of a talus to finer near the toe, except where the momentum of a large rock carries it unusually far. Stratification is poor to absent.

Taluses are due to mass-wasting by definition. The breaking away of rock from the cliff face may result from various wedging effects, such as hydration and root growth, in addition to ice wedging. Therefore, although taluses are most common in periglacial environments they are not limited to them. Taluses have been reported as forming in Virginia today (Hack, 1960), and having no special climatic significance (Wilhelmy, 1958, 17–18). However, stabilized taluses may indicate former greater frost wedging as suggested for the Baraboo, Wisconsin, area by H. T. U. Smith (1949*b*, 199–203). Clearly their climatic significance must be interpreted with caution.

X PROTALUS RAMPARTS

A protalus rampart (Bryan, 1934, 656) is 'A ridge of rubble or debris that has accumulated piecemeal by rock-fall or debris-fall across a perennial snowbank, commonly at the foot of a talus' (Richmond, 1962, 20) (Figure 5.23). Sliding is usually also involved (Richmond, 1962, 61). Protalus ramparts occur in many cirques. The material is dominantly angular, coarse, and non-oriented, and much of it is produced by frost wedging. Some sorting may be present in view of the greater momentum of larger blocks. Fines tend to be absent because they lack the momentum to bounce, slide, or roll across the snow as pointed out by S. E. White (*in* Blagbrough and Breed, 1967, 768). Protalus ramparts up to 9 m (30 ft) high have been repeatedly reported (Behre, 1933, 630; Daly, 1912, 593), and heights up to 24 m (80 ft) have been observed by Blagbrough and

Breed (1967, 766) who reported basal widths up to 14 m (45 ft) and lengths up to 91 m (1300 ft). Probably most protalus ramparts are arcuate and concave upslope but some form linear benches.

In addition to the influence of climate, including its indirect effects such as vegetation cover, the location of protalus ramparts is strongly influenced by topography and by the susceptibility of the bedrock or unconsolidated material on the slope above to disintegration and mass-wasting. In location, material, and form, protalus ramparts are probably similar to many small end moraines, especially in cirques. Except in some cirques, a distinguishing characteristic would be the orientation of the ridge, which in a protalus rampart would be oriented parallel rather than transverse to a valley. Confusion might arise in the case of a lateral moraine fragment but the lithology of ridge stones as between local material and that from up the valley might be diagnostic. Protalus ramparts can be readily confused with the toes of some taluses that have moved by creep or solifluction but some ramparts can be distinguished by their comparative lack of fines (Blagbrough and Breed, 1967, 769).

If located where they are no longer forming, protalus ramparts indicate a former cooler climate (H. T. U. Smith, 1949*a*, 1503), greater snowfall, or both. They do not necessarily prove former greater frost action, since frost wedging activity might remain unchanged while locus of deposition varied with change in size and/or position of the snowbank. Protalus ramparts indicate a snow environment and the high probability of frost wedging, and they may be a guide to a former approximate orographic snowline, since they are clearly related to it. Multiple protalus ramparts in southern Utah appear to record two Wisconsin glaciations (Blagbrough and Breed, 1967, 771–2).

Protalus ramparts have been observed in the Antarctic by the present writer and are probably reasonably common in polar, subpolar, and middle-latitude highlands (cf. Blagbrough and Breed, 1967; Bryan, 1934; Bryan and Ray, 1940, 35; Richmond, 1964, D28, D30).

XI GRÈZES LITÉES

Grèzes litées are bedded slope deposits of angular, usually pebble-size rock chips and int rstitial finer material (Dylik, 1960; Guillien, 1951). Grèzes litées (Figure 5.24) are described by Guillien (1951) as occurring on gradients of 7°–45° and consisting of fragments up to 2·5 cm in diameter, and are reported by Malaurie (*in* Malaurie and Guillien, 1953, 712) as forming alternating beds of grus and larger fragments up to 3·5 cm in diameter. The origin of grèzes litées has been ascribed to frost wedging, sheetflow, and rillwork, in association with perennial snowdrifts (perhaps beneath them) on vegetation-free slopes (Guillien, 1951, 159–61). It has also been suggested that an alternation of processes may be involved – creep, sheetflow, rillwork, and gelifluction (after Bout, 1953) and/or differential creep (after Baulig) (Malaurie, *in* Malaurie and Guillien, 1953,

5.23 (*Opposite*) Protalus rampart, Mountain Lakes Wild Area, Oregon. A remnant snow patch lies immediately upslope on concave inner side of the protalus rampart. Photo by G. A. Carver

5.24 Grèzes litées, Klemencice, Poland. Scale given by 15-cm rule (*cf. Dylik, 1960*)

712–13). Contemporary deposits have been recorded from West Greenland by Malaurie, and grèzes litées have been described as constituting a sub-arctic cryonival phenomenon by Guillien. Laboratory experiments by Guillien and Lautridou (1970) indicate that frost wedging of certain limestones can produce the same kind of material that constitutes the Pleistocene grèzes litées of the Charente region of France. These investigators reaffirmed the belief that the sorting and bedding show that meltwater from snowdrifts was the essential agent in depositing the coarser material and eluviating the finer, and they regarded grèzes litées as indicative of nivation. However, the exact nature and origin of grèzes litées do not yet appear to be firmly established, and more observations are needed of such deposits in course of formation. Although occurring in periglacial environments, grèzes litées are still too poorly known to be regarded as diagnostic of them.

6 Nivation

I Introduction: II Nivation benches and hollows: III Altiplanation terraces

I INTRODUCTION

Nivation is the localized erosion of a hillside as the result of frost action, mass-wasting, and the sheetflow or rillwork of meltwater at the edges of, and beneath, lingering snowdrifts. The term was introduced by Matthes (1900, 183).

As discussed by Embleton and King (1968, 544–58) in an excellent summary of nivation, the effectiveness of nivation is strongly influenced by the thickness of the snowdrift and the presence or absence of permafrost beneath it. To the extent that the ground surface remains frozen beneath the snow, as is usual in permafrost regions except at the feather edge of a snowdrift, frost wedging and gelifluction are inhibited and nivation is limited to the feather edge and the immediately adjacent snow-free area. As a snowdrift becomes thinner and contracts, the site of maximum effectiveness of nivation follows and eventually traverses the area freed from snow. Where permafrost is lacking and extensive thaw areas occur beneath the snow, nivation can affect the snow-free and snow-covered areas simultaneously. There is some difference of opinion as to the role of snow creep but that it can displace unfrozen particles is evident (Costin *et al.*, 1964; Haefeli, 1953, 248; 1954, 62; Mathews and Mackay, 1963; P. J. Williams, 1962, 358).

The effect of nivation is to countersink lingering snowdrifts into hillsides and thereby produce nivation hollows. Thus the distribution of such features is strongly influenced by topography and is probably closely related to the position of the orographic snowline. Nivation has been reported from many periglacial environments and has a wide quasi-equilibrium range. Reports include those of Matthes (1900, 179–85), Ekblaw (1918), Lewis (1939), McCabe (1939), Nichols (1966, 29–30), Berger (1967), Henderson (1959*b*, 70–81), Ohlson (1964, 85–95), Gardner (1969), St.-Onge (1969), and many others.

As noted in the chapter on Mass-wasting, it is widely believed that frost wedging and flow of meltwater from lingering snowdrifts are the key processes in the production and deposition of grèzes litées. They have therefore been ascribed to nivation (Guillien and Lautridou, 1970, 42). Some grèzes litées are probably of this origin but it remains to be demonstrated that all deposits with the characteristics of grèzes litées are necessarily nivation phenomena.

II NIVATION BENCHES AND HOLLOWS

From observations in the Canadian Arctic, St.-Onge (1969) concluded that nivation is strongly influenced by lithology, as reflected in various nivation landforms. Thus nivation terraces, the largest features, were in gabbro that produced coarse debris; nivation benches were in sandstone that disintegrated into somewhat finer material; and other forms such as semicircles, hollows, and ledges were in shale that broke down into fine sand and silt. Although the choice of terms is open to criticism, since one man's terrace may be another man's bench, and hollow is often used for a variety of landforms, the general concept that lithology plays a critical role is sound. St.-Onge (1969, 5–6) summarized the evolution of a nivation terrace in gabbro or bedrock of similar hardness as follows

> Snow accumulates in an area oriented to the wind or in an initial irregularity in the slope. Conditions favourable to frost shattering [frost wedging] are created. This process yields large quantities of coarse fragments and few fines. The fines are removed by water percolating through the boulders. The backwall recedes leaving an apron of coarse fragments. These boulders act as the base level of the terrace since they can no longer be saturated. As the terrace widens, the effects of percolating water and sheet wash are greatly reduced. Away from the outer edge, the open-work boulder field is gradually clogged with a matrix of fine material. The moisture retention capacity increases with a corresponding increase in frost shattering. Eventually the percentage of fine material is sufficiently high to render geliturbation [frost heaving and frost thrusting] possible.

'Fossil' nivation depressions may be related to former colder climate, greater snowfall, or both. Their identification is complicated by similar-appearing features such as earthflow scars. Nivation depressions constitute a useful periglacial criterion if their significance relative to present and former snowlines can be established for a region.

III ALTIPLANATION TERRACES

Altiplanation terraces (Eakin, 1916, 77–82), also known as cryoplanation terraces or goletz (golec) terraces in Europe and Asia, are hillside or summit benches that are cut in bedrock, lack predominant structural control, and are confined to cold climates (Figure 6.1). They have a veneer of solifluction debris, which may be imprinted with patterned ground. Structure, such as jointing and lithology, may help to determine location and degree of development of terraces but their origin is largely independent of such controls. In this respect altiplanation terraces are a periglacial analogue of warm-desert pediments (cf. Demek, 1969*a*, 66), although the origin and implications of these two land forms are very different. Altiplanation terraces appear to range in size from about 10 m across to widths of 2–3 km and lengths exceeding 10 km; the gradient of the terrace tread is commonly 1°–12° (Demek, 1969*a*, 42, 44, 64).

6.1 Altiplanation terrace and tor near Eagle Summit, Alaska. Photo by Troy L. Péwé

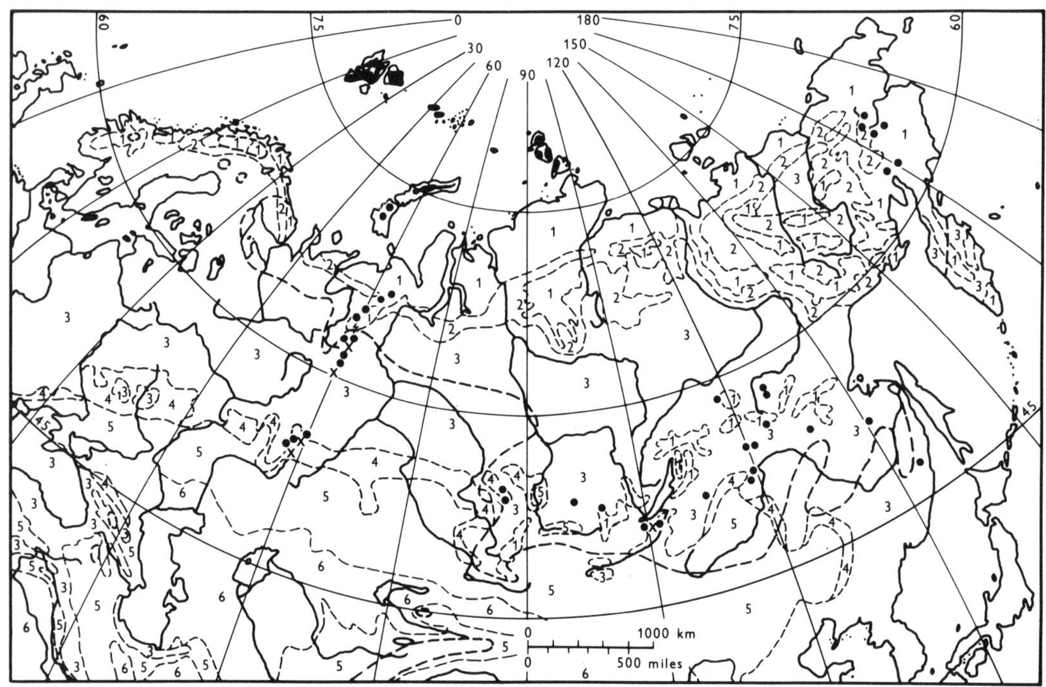

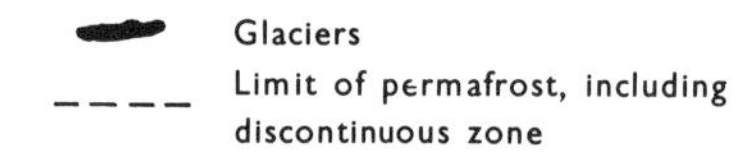

6.2 Distribution of active (•) and fossil (×) altiplanation terraces in northern Eurasia (*Frenzel, 1960a, 1019, Figure 11*)

6.3 Distribution of altiplanation terraces in the Soviet Union and Mongolia. No distinction is made between active and fossil forms (*Demek, 1969a, 11, Figure 3*)

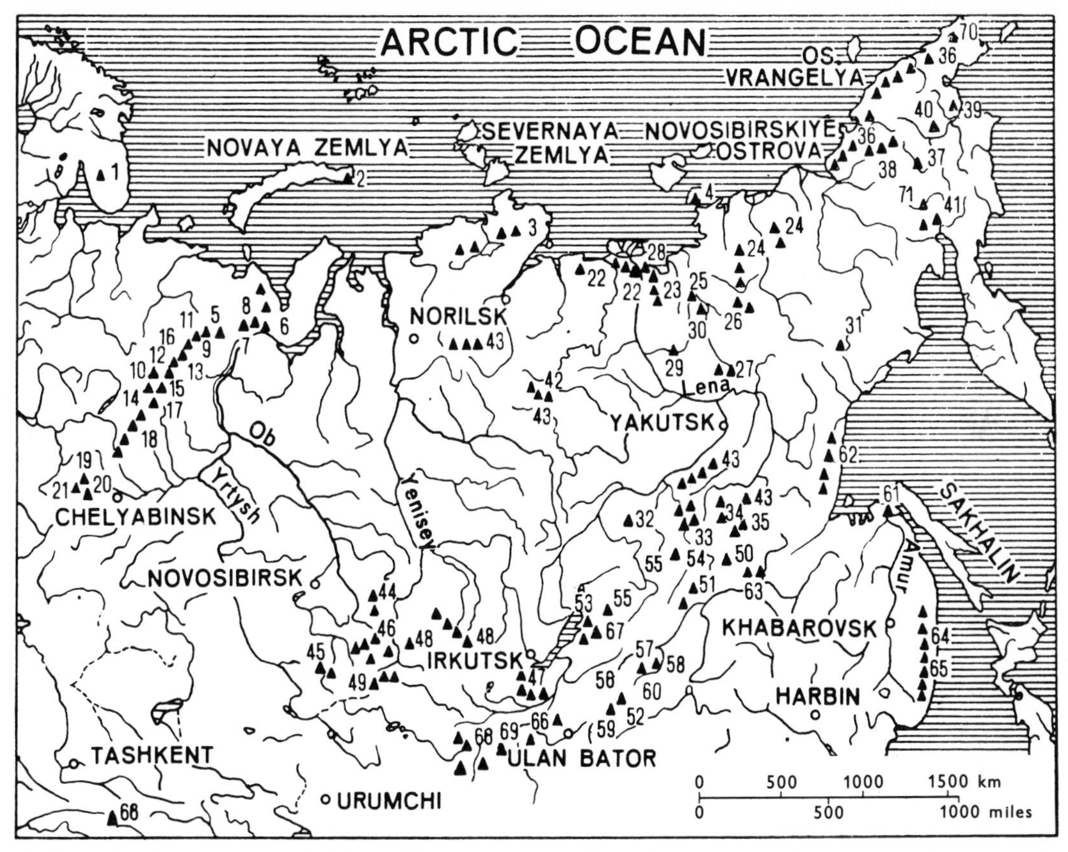

USSR – 1 Koliskiy Peninsula (*Perov, 1959; Mozlov, 1966*); 2 Novaya Zemlya (*Miloradovich, 1936*); 3 Taymyr Peninsula–Gory Byrranga (*Kaplina, 1965*); 4 Novosibirskye Islands–Bolshoy Lyakhovskiy Islands (*Bielorusova, 1963*); 5 Polyarniy Ural Mts. (*Baklund, 1911; Panov, 1937*); 6 Polyarniy Ural Mts.–Ray-Iz Massif (*Zavaritskiy, 1932*); 7 Polyarniy Ural Mts. (*Sofronov, 1945*); 8 Polyarniy Ural Mts.–western slope of the Pechora River Basin (*Kaletzkaya and Miklukho-Maklay, 1958*); 9 Pripolyarniy Ural Mts. (*Aleshkov, 1936; Dolgushin, 1961; Lyubimova, 1955; Chernov and Chernov, 1940; Dobrolyubova and Sochknai, 1935*); 10 Severniy Utal Mts.–Upper Vishera River Basin (*Duparc and Pearce, 1905; Duparc, Pearce, and Tikanowich,* 1909); 11 Severniy Ural Mts. (*Alekshkov, 1936; Edelshteyn, 1936; Panov, 1937*); 12 Severniy Ural Mts. (*Varsanofyeva, 1932*); 13 Severniy Ural Mts. (*Boch and Krasnov, 1943, 1946; Tarnogradskiy, 1963*); 14 Severniy Ural Mts.–Visherskiy Ural (*Suzdalskiy, 1952*); 15 Severniy Ural Mts.–Khrebet Chistop (*Gorchakovskiy, 1954*); 16 Severniy Ural Kamen Mt. (*Strigin, 1960*); 18 Sredniy Ural Mts. (*Aleksandrov, 1948*); 19 Yuzhniy Ural Mts. (*Bashenina, 1948*); 20 Yuzhniy Ural Mts.–Iremel Mt. (*L. O. Tyulina, 1931; Tsvetayev, 1960*); 21 Yuzhniy Ural Mts.–Yaman Tau (*Gorchakovskiy, 1954*); 22 Yakutia–Kryazh Chekanovskogo (*Sochava* in *Obruchev, 1937; Gakkelya and Korotkevich, 1962*), Kryazh Prontchistcheva (*Rusanov* et al. *1967*); 23 Yakutia–Gory Kharaulakh (*Sochava and Gusev* in *Obruchev, 1937*); 24 Yakutia–Alazeyskoye Ploskogorye, Polousniy Khrebet (*Gakkelya and Korotkevitch, 1962; Rusanov* et al. *1967*); 25 Yakutia–khrebet Kular (*Rusanov* et al., *1967; Demek, 1967*); 26 Yakutia–khrebet Cherskogo (*Kropachev and Kropacheva, 1956*); 27 Yakutia–Verkhoyanskiy khrebet (*Korofieyev, 1939; Lazarev, 1961; Yegorova, 1962*); 28 Yakutia–Tuora-Sis (*Lazarev, 1961*); 29 Yakutia–Verkhoyanskiy khrebet (*Yegorova, 1962*); 30 Yakutia–surroundings of the town of Batagay (*Demek 1967*); 31 Nyerskoye ploskogorye Mts. (*Tskhurbayev, 1966*); 32 Yakutia–Olekma–Vitim Region (*Obruchev, 1937*); 33 Yakutia–Olekma River Basin (*Timofieyev, 1965*); 34 Yakutia–Aldanskoye Nagorye Mts. (*Rabotnov, 1937; Dolgushin, 1961*); 35 Yakutia–Aldanskoye Nagorye Mts. (*Kornilov, 1962*); 36 Arctic Ocean Coast (*Obruchev, 1937*); 37 Upper Anadyr River Basin (*L. N. Tyulina, 1936*); 38 Severnyi Anyuyskiy khrebet (*Obruchev, 1937*); 39 Zolotiy khrebet (*Obruchev, 1937*); 40 Ust-Bielskiye Gory (*Kaplina, 1965*); 41 Kolymskiy khrebet (*Sochava* in *Obruchev, 1937*); 42 Anabarskiy Massif (*Yermolov, 1953*); 43 Gory Putorana, Anabarskiy Massif, Aldanskoye ploskorgorye Mts., Prilenskoye Plateau (*Krasnov and Kozlovskaya, 1966*); 44 Kuznetskiy Alatau (*Tolmachev, 1903; Ilyin,1934*); 45 Altai-Terektinskiy khrebet (*Keller, 1910*); 46 Altai (*Makerov, 1913; Ilyin, 1934*); 47 Sayanskoye-Dzidinskoye nargorye Mts. (*Lamakini and Lamakini 1930*); 48 Vostochniy Sayan–Zapadnyi Sayan (*Kushev, 1957*); 49 Zapadniya Tuva (*Kozlov, 1966*); 50 Olekma region (*Kozmin, 1890; Sukachëv, 1910*); 51 Olekminskiy Stanovik, Cheromnagovyiy khrebet, Mogochinskiy khrebet, Tungirskiy khrebet (*Makerov, 1913; Korzhuyev, 1959*); 52 Patomskoye nagorye (*Bashenina, 1948*); 53 Barguzinskiy khrebet–Severo-Muyskiy khrebet–khrebet Verchnie–angarskiy (*Dumitrashko, 1938, 1948; L. N. Tyulina, 1948*); 54 Stanovoye nagorye-khrebet Udokan (*Preobrazhenskiy, 1959, 1962*); 55 Khrebet Udokan, khrebet Ulan-Burgasy, Itatskiy khrebet, Tsipinskiye gory (*Gravis, 1964*); 56 Yablonoviy khrebet (*Dengin, 1930*); 57 Borschchovochniy khrebet (*Ryzov, 1961*); 58 Upper Amur River Basin (*Kaplina, 1965*); 59 Khentey, Yablonoviy khrebet, Borshchovochniy khrebet, Badzhalskiy khrebet (*Nikolskaya and Chichagov, 1962*); 60 Khentey–Borshchovochniy khrebet (*Bashenina, 1948; Nikolskaya, Timofieyev and Chichagov, 1964*); 61 Lower Amur River Basin–gora Praul (*Ganeyshin, 1949*); 62 Khrebet Kzugdzhur (*Kaplina, 1965*); 63 Tukuringra–Dzagdy (*Nikolskaya and Shcherbakov, 1956*); 64 Sikhote Alin (*Solovyev, 1961*); 65 Sikhote Alin (*Pryalukhina, 1958*); 66 Daurskiy khrebet (*Zamoryuev, 1967*); 67 Juzhno-Muyskiy khrebet (*Rudavin, 1967*); 68 Pamir (*Sekyra, 1964*). Mongolia – 69 Khangai (*Richter, Haase, and Barthel 1963*); 70 USSR – Chukotka (*Zigarev and Kaplina, 1960*); 71 USSR – Penzhina River Bassein (*Sochava, 1930*).

Nivation, combined with gelifluction, is believed to be an essential process in the origin of altiplanation terraces in Alaska (Péwé, 1970, 360), Siberia, and elsewhere (Demek, 1968*a*; 1969*a*; 1969*b*). Their distribution in Eurasia, Europe, and North America is shown in Figures 6·2–6·5.

Three stages of terrace development were described by Demek (1969*b*): (1) The first stage is characterized by nivation, which produces a nivation hollow or bench. Elongate snow patches are particularly favourable sites for terrace development. The developmental sequence of a nivation terrace as described above by St.-Onge (1969, 5–6) probably corresponds to this stage. (2) The second stage is characterized by nivation operating on a frost-wedged cliff and by a complex of processes that transport the resulting debris across the tread at the base of the cliff. These processes vary, depending on lithology and other factors, but include gelifluction, slope-wash, eluviation, and piping. The tread commonly has a gradient of about 7°. If the cliff rises above the snow patch, frost-wedged debris falls on the snow patch and tends to accumulate at its base, forming protalus ramparts. Eventually the cliff is worn back. Demek (1969*a*, 64) split this stage into the initial-terrace and the mature-terrace stages. (3) The third stage is characterized by a summit flat whose gradient may be less than 2°. Most processes become less active because of the low gradient; however, those responsible for patterned ground continue unabated, and deflation may be locally significant.

Although the development of altiplanation terraces is confined to cold climates and is facilitated by the presence of permafrost, terraces also form without permafrost (Demek, 1969*a*, 56–7). Given adequate time the essential requirements appear to be climatic and topographic conditions favouring nivation and gelifluction. Altiplanation terraces in central Alaska are best developed on well-jointed resistant rocks and are thought to form perhaps slightly below the snowline (Péwé, 1970, 360). The rate of terrace development is not well known but is probably on the order of tens of thousands of years for large well-developed terraces (cf. Demek, 1969*a*, 66–7). Fossil forms may be widespread but their distribution as distinct from active forms (cf. Figure 6.2) remains to be determined in most regions. Much remains to be learned about the origin and distribution of altiplanation terraces but they constitute an important criterion of periglacial conditions.

6.4 Distribution of altiplanation terraces in Europe. No distinction is made between active and fossil forms (*Demek, 1969a, 10, Figure* 2)

1 Northern shore of southwestern England (*Guilcher, 1950*); 2 Southwest England – Dartmoor (*Te Punga, 1956; Waters, 1962; Demek, 1965*); 3 Southwest England – Mendip Hills (*Waters, 1962*); 4 England – Northeast Yorkshire (*Gregory, 1966*); 5 France – Brittany (*Guilcher, 1950*); 6 France – Vosges; 7 West Germany – Harz (*Hövermann, 1953*); 8 West Germany – Niedersächsisches Bergland (Lower Saxonian Highland) (*Spönemann, 1966*); 9 West Germany – Odenwald; 10 Belgium – Ardennes (*Gullentops* et al., *1966*); 11 Poland – Karkonosze Mts. (*Jahn, 1965*); Czechoslovakia – Krkonose (*Sekyra, 1964*); 12 Czechoslovakia – Rychlebské hory Mts. (*Panos, 1960; Ivan, 1965*); 13 Czechoslovakia – Hruby Jesenik Mts. (*Czudek and Demek, 1961; Czudek, 1964; Demek, 1964b, 1964c, Hrádek, 1967*); 14 Czechoslovakia – Ceskomoravská vrchovina (Bohemian – Moravian Highlands) (*Demek, 1964a, 1964b*) 15 Czechoslovakia – Novohradské hory Mts. (*Demek, 1964d*); 16 Czechoslovakia – Cesky les (Bohemian Forest) (*Balatka* in *Demek, 1965*); 17 Czechoslovakia – Slavkovsky les, Tepelská vrchovina (Slavkov Forest, Tepelská vrchovina Highlands) (*Czudek* in *Demek, 1965*); 18 German Democratic Republic – Erzgebirge, Czechoslovakia – Krusné hory (*Czudek* in *Demek, 1965; Král, 1968; Richter 1965*); 19 Czechoslovakia – Nizky Jesenik (Low Jesenik Mts.) (*Czudek* in *Demek, 1965*); 20 Czechoslovakia – Bobravská vrchovina (Bobrava Highlands) (*Demek, 1965*); 21 Czechoslovakia – Chriby Hills (*Czudek, Demek, and Stehlik, 1961*); 22 Czechoslovakia – Hostynské vrchy (Hostyn Hills) (*Czudek, Demek, and Stehlik, 1961*); 23 Czechoslovakia – Moravskoslezské Beskydy (Moravian-Silesian Beskydy Mts.) (*Stehlik, 1960*); 24 Hungary (*Pécsi, 1963, 1964, 1965*); 25 Bulgaria – Vitosha (*Marruszczak, 1961; Demek and Smarda, 1964*); 26 Bulgaria – Rila (*Demek and Smarda, 1964*); 27 Faeroe Islands (*C. A. Lewis, 1966*); 28 Sweden – surroundings of Kiruna (*Bashenina, 1967*); 29 German Democratic Republic – western vicinity of Dresden (*Richter and Haase, personal communication*)

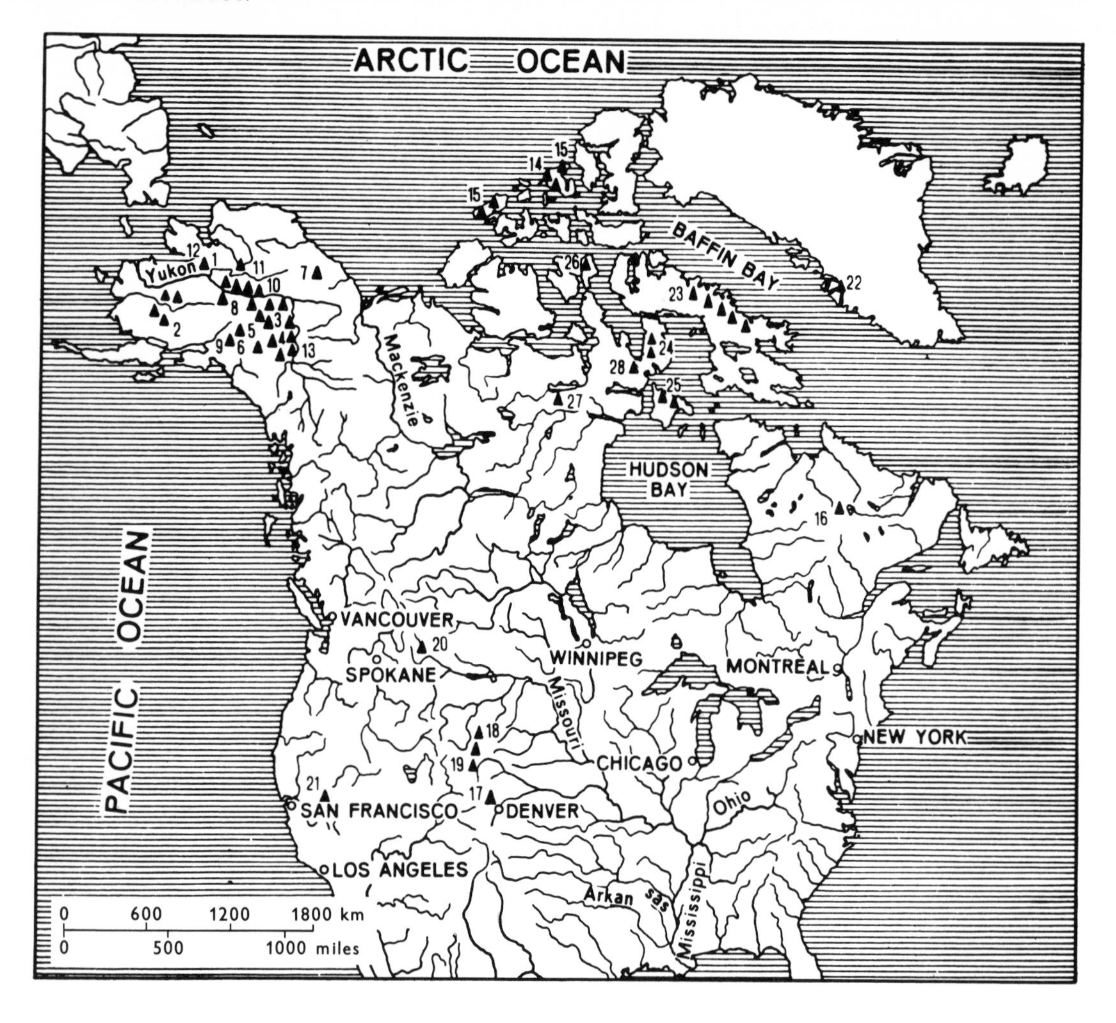

6.5 Distribution of altiplanation terraces in North America. No distinction is made between active and fossil forms (*Demek, 1969a, 13, Figure 4*)

1 Alaska–Yukon–Koyukuk region (*Eakin, 1916*); 2 Alaska–Central Kuskokwim region (*Cady et al., 1955*); 3 Alaska – Yukon – Tanana region (*Prindle, 1905; Mertie, 1937*); 4 Alaska – Yukon – Tanana region (*Jahn, 1961b, 1966*); 5 Alaska – northeast of Fairbanks (*Taber, 1943; Hanson, 1950*); 6 Alaska – Alaska Range (*Jahn, 1961*); 7 Alaska – Brooks Range (*Jahn, 1961*); 8 Alaska – Cosna Nowitna region (*Eakin, 1918*); 9 Alaska – Talkeetna Mts. (*Wahrhaftig, 1965*); 10 Alaska – Indian River Upland (*Wahrhaftig, 1965*); 11 Alaska – Lockwood Hills, Zane Hills (*Wahrhaftig, 1965*); 12 Alaska – Selawik Hills (*Wahrhaftig, 1965*); 13 Alaska – Gulkana Upland (*Wahrhaftig, 1965*); 14 Canada – Ellef Ringnes Island (*St. Onge, 1965*); 15 Canada – Meighen Island, Ellef Ringnes, Prince Patrick Islands (*Heywood, 1957; Robitaille, 1960; Thorsteinsson, 1961; Pissart, 1966c*); 16 Canada – Labrador (*Derruau, 1956*); 17 USA – Colorado (*Russell, 1933*); 18 USA – Bighorn Mts. (*Mackin, 1947*); 19 USA – Colorado Front Range (*B. W. Scott, 1965; Bradley, 1965*); 20 USA – Glacier National Park, Montana; 21 USA – Monitor Pass, California; 22 Greenland (*Ekblaw, 1918*); 23 Canada – Baffin Island (*Bird, 1967*); 24 Canada – Melville Peninsula (*Bird, 1967*); 25 Canada – Southampton Island (*Bird, 1967*); 26 Canada – Somerset Island (*Bird, 1967*); 27 Canada – Marjorie Hills (*Bird, 1967*); 28 Canada – Rae Isthmus (*Bird, 1967*)

7 Fluvial action

I Introduction: II Aufeis: III Break-up phenomena: IV Asymmetric valleys: *1 General; 2 Origin.* **V Dells:** *1 General; 2 Origin*

I INTRODUCTION

Fluvial action in periglacial environments differs from that in other environments mainly in the degree to which it is influenced by frost action. Permafrost, a thick ice cover in winter, adjacent gelifluction slopes, and frost-wedged debris – all can contribute to certain features typical of fluvial action in a periglacial environment.

II AUFEIS

Aufeis may have several origins as discussed in the chapter on Frozen ground. One is by a stream continuing to flow under its frozen surface and spilling through cracks in the ice because of deep freezing, partial blocking of the underflow, and build up of hydrostatic head. As a result, a sheet of ice several metres thick may form over a much wider area than the stream channel itself (Figure 3.10). Such ice may become buried and preserved.

III BREAK-UP PHENOMENA

Ice break-up along rivers in the spring is usually accompanied by floods, especially as the result of downstream ice jams in rivers flowing from warmer towards colder environments. However, the downstream progress of break-up, although common, does not always pertain because of the influence of tributaries and other factors (Mackay, 1963*c*). This downstream flow of ice under flood conditions can cause both erosion and deposition along banks at levels well above the normal summer stage. The strip affected, called a Bečevnik in Siberia – a term adopted by Hamelin (1969) – tends to be bench-like and may show a number of micro features that are formed under such conditions. These features include ice-rafted stones, striated stones, ice-shove indentations and ridges, stone pavements, and disrupted vegetation, among others. A detailed list, based on observations in Siberia, has been presented by Hamelin (1969).

Ice rafting is the transport of rock debris or other material by floating ice, whether in streams, lakes, or the sea. Ice in streams can pick up bottom material (cf. Washburn, 1947, 83–4), or carry material slumped on to its surface from the banks (Figure 7.1). Following break-up of the ice, the debris may be transported far downstream before being deposited. If such erratic debris can be established as ice-rafted, it is clear evidence of a sufficiently cold climate to form ice farther upstream. Similar conclusions can be drawn from the other features.

However, rafting can occur in other ways, as in tree roots (Figure 8.4), and even if ice rafting is established, river ice occurs so far south in temperate regions, and the distance involved in fluvial ice rafting may be so great, that the significance of the evidence is minimized. Moreover, many of the features listed by Hamelin (1969), such as ice-shove indentations and ridges, are subject to destruction by any subsequent ice-free floods and are unlikely to be preserved as a record of former periglacial conditions.

Many other features are subject to alternative explanations. Striated stones, for instance, might be derived from a local deposit of glacial or landslide origin, or might have been rafted far downstream from another environment. It must be concluded that however striking the results of the break-up of rivers, little diagnostic evidence is likely to remain long enough to be of value in environmental reconstructions. The whale-backed slushflow fans of mountain streams, discussed in the chapter on Mass-wasting, may be an exception.

IV ASYMMETRIC VALLEYS

1 General

Asymmetric valleys are valleys that have one side steeper than the other (Figure 7.2). In places the asymmetry can be ascribed to the influence of the periglacial environment on fluvial action.

Asymmetric valleys in the Northern Hemisphere have been variously described as having the steepest slopes facing north or east, or west or southwest (Poser, 1947*a*, 12–13), north or north east (Malaurie, 1952; Malaurie and Guillien, 1953), south or southwest (Troll, 1947, 171), and northwest among other directions (Büdel, 1944, 503; 1953, 255). According to Poser and Müller (1951, 26–8), climatically determined asymmetric valleys are exceptional in polar latitudes because temperature differences due to exposure are less marked and less heat is received than farther south.

2 Origin

Periglacial explanations of asymmetry take several forms. For places where the steepest slopes face south or southwest in a permafrost environment, it has been argued that maximum insolation

7.1 (*Opposite*) Fluvial ice rafting. Mass-wasting of rock debris on to river ice, Mesters Vig, Northeast Greenland

and thawing are concentrated on these slopes, thereby allowing maximum stream erosion at the slope base (Poser, 1947*a*, 13; cf. H. T. U. Smith, 1949*a*, 1503). Greater gelifluction on one slope may force a stream over to the opposite slope, which is then undercut (Büdel, 1953, 255). This last explanation involving greater gelifluction on south-facing slopes has been invoked to explain steep north-facing slopes of Ogotoruk Creek in Alaska (Currey, 1964, 95).[1] Greater gelifluction on one slope and consequent shifting of a stream and undercutting of the opposite slope has also been invoked to explain steep southwest-facing slopes on Banks Island in the Canadian Arctic (French, 1971*b*) and steep north-facing slopes in the central United States (Wayne, 1967, 401). For valleys lacking an eroding stream and having the steepest slopes facing north or east, it has been suggested that greater gelifluction on the opposite slopes has reduced their gradient (Malaurie, 1952; Poser, 1947*a*, 13).

Currey (1964, 92–4) discussed eight ways in which periglacial processes could produce asymmetric valleys, and there are probably others. A number of variables exist, some working in opposition to each other. For instance, it can not always be assumed that the depth of thawing on north-facing slopes is less than on south-facing slopes receiving more insolation, since the vegetation on the latter may produce an overriding insulating effect (L. W. Price, 1971, 645–7). With respect to gelifluction, the distribution of snowdrifts and hence wind direction and topography are important variables influencing its location and effectiveness (Büdel, 1944, 503–4; 1953, 255; French, 1971*b*, 727–9; 1972; Troll, 1947, 171; Washburn, 1967, 93–8). In places gelifluction may be of minor significance. Kennedy and Melton (1967)[2] reported that moisture availability, vegetation, soil permeability, and (at higher altitude) nivation were more important than gelifluction in accounting for asymmetric valleys with steep north-facing slopes near Inuvik in the Canadian Arctic. Moreover, there are additional causes of asymmetry, especially lithology and structure, that must be eliminated before a periglacial origin can be accepted. Considering the number of variables and different explanations, and the possibility of complex origin, the interpretation of asymmetric valleys invites extreme caution. The difficulty is compounded with fossil forms. As noted by Black (1969*b*, 231), their usefulness as indicators of a former periglacial environment (to say nothing of permafrost) is very limited at present.

V DELLS

1 General

Dells are small shallow valleys now devoid of streams (Figure 7.2). They are also known as dry

[1] K. R. Everett (1966, 212–13), referring to steep southeast-facing slopes in the same area, suggested that structure and lithology were responsible.

[2] A shorter, revised version (Kennedy and Melton, 1972) appeared while this volume was in proof.

7.2 (*Opposite*) Asymmetric dell, Grocholice, Poland. Note steeper slopes at left than at right of older valley, incised by younger, dry, valley

valleys but not all dry valleys are dells, the distinction being that dells are characteristically small and shallow. They have been frequently cited as evidence of former periglacial conditions and many discussions of asymmetric valleys include asymmetric dells.

Dells tend to have a gentle, generally U-shaped cross profile. The alluvium is predominantly angular rubble (Büdel, 1953, 255). Valleys believed to be zonal prototypes have been cited by Büdel (1948, 39, 43–4) as particularly common in the 'tundra zone'. However, true dells have been primarily described from non-periglacial regions today and have long been recognized as disequilibrium features in Britain and Europe (Bull, 1940; Geike, 1894, 395–6; Kessler, 1925, 186–91; Clement Reid, 1887).

2 Origin

Dells have been regarded as due to linear erosion by streams and/or mass-wasting, accompanied by an abundant supply of debris contributed by the mass-wasting, in areas where subsurface drainage was inhibited by permafrost or seasonally frozen ground (Büdel, 1953, 255; Clement Reid, 1887, 369–71). The present lack of streams is attributed to thawing of the frozen ground and consequent infiltration of the surface water to depth.

As in the case of asymmetric valleys, alternative possibilities must be eliminated before dells can be accepted as indicators of a former periglacial environment. Other possible explanations for dry valleys or underfit streams that are applicable to dells include solution (in limestone areas), beheading of streams, and, especially, former greater precipitation, which Dury (1965, C15–C40) regarded as the primary cause. Some dry valleys occur where the underlying beds are highly impermeable today and would promote surface stream flow given adequate precipitation, an illustration being the Warwickshire Itchen in England (Dury, 1970, 266).

8 Lacustrine and marine action

I Introduction: II Lacustrine action: *1 General; 2 Varves: a* General, *b* Origin, *c* Significance; *3 Ice rafting; 4 Ice shove and ice-shove ridges.* **III Marine action:** *1 General; 2 Ice rafting; 3 Ice shove and ice-shove ridges; 4 Striated bedrock*

I INTRODUCTION

Periglacial lacustrine and marine action are similar in some ways with respect to the role of water bodies. Ice rafting and ice shove, for instance, are analogous processes differing mainly in scale.

II LACUSTRINE ACTION

1 General

There are only a few geomorphic processes and features that characterize some lakes as being periglacial. They include thawing and development of thaw lakes as described in the chapter on Thermokarst, the formation of ice-shove ridges, and the deposition of varved sediments. Each of these characteristics may be difficult to distinguish from similar features of non-periglacial origin.

2 Varves

a General 'A varve is a pair of laminae deposited during the cycle of the year' (Longwell, Flint, and Sanders, 1969, 380, 661).

There are various kinds of varves. The cold-climate variety, known as glacial varves, consists of a sandy to silty lamina overlain by a finer-grained, silty to clayey lamina; the transition to the next varve pair is very abrupt compared to the usual graded bedding within a varve. Because of the difference in grain size, the fine-grained lamina tends to be much darker than the coarser lamina (Figure 8.1).

b Origin Cold-climate varves commonly reflect the annual freeze-thaw cycle of lakes receiving

abundant sediment from glacial streams. After the ice of streams and lakes breaks up, sediment is poured into the lakes. The coarser material settles first to form the coarse 'summer' lamina, but the finest is kept in suspension until freezing stops the influx of sediment and creates the quiet-water conditions that permit the suspended material to settle out and form the fine-grained 'winter' lamina.

Unfortunately the recognition of such varves is not simple, especially with respect to proof of annual character. The term varve has been loosely and improperly used to designate laminated silts for which the annual origin is not proved. Even for laminae pairs whose glacial origin is demonstrable, some pairs may record variations in meltwater flow and episodes of storm and quiet during the summer rather than reflect the full annual cycle (Hansen, 1940, 418, 422, 468–75).

c Significance Because varves vary considerably in thickness from year to year and form distinctive sequences, varves lend themselves to correlation between a series of former lake basins in a given region. Such correlations are the basis of a Late-Glacial and post-Glacial chronology in Sweden. Established by DeGeer (1912; 1940) and subsequent workers, and checked with radiocarbon dates in places, this varve chronology as proposed spans almost 17000 years dating back from A.D. 1900, but is commonly accepted for only about the last 12000 years (Flint, 1971, 403–5). A varve chronology has also been attempted for parts of North America (New England and eastern Canada) but with less success because of gaps in the record (Antevs, 1925).

Regardless of their chronologic significance, varves of the kind described are cold-climate features and, with the exception of varves deposited in supraglacial lakes, they are deposited in ice-free sites and are therefore, strictly speaking, periglacial phenomena. Moreover, some varves in cold climates may well develop without glaciers being in the immediate vicinity.

Combined with their chronologic implications, cold-climate varves provide information as to how long these climatic conditions persisted in a given locality, and by correlation with deposits in adjacent basins, the spatial as well as temporal march of the conditions may be ascertained. Unfortunately, aside from proving appreciable periods of freezing and thawing, cold-climate varves do not in themselves provide much quantitative information on temperature ranges or on precipitation.

8.1 (*Opposite*) Cold-climate varves near Uppsala, Sweden. Scale given by 15-cm pencil. The varves (at least 16 visible) were deposited in the Baltic Ice Lake. Photo by Richard F. Flint

3 Ice rafting

Ice rafting is most common in lakes that abut against glaciers. Icebergs calving from a glacier carry rock debris of various sizes to places where otherwise only fine sediments are deposited. Such ice-rafted material, often consisting of isolated stones as well as of pockets of nonsorted debris, is a useful criterion in identifying glacial varves. If ice rafting is clearly established as opposed to other processes, the resulting deposit is in itself good evidence of a periglacial environment.

4 Ice shove and ice-shove ridges

Ice-shove ridges are embankments of rock debris pushed up on the shore of a lake or sea by ice. Such ridges are commonly concentrated on points and form distinctive features (Figure 8.2) that differ from massive beach ridges in being more irregular in plan and profile. Where associated with beach ridges they consist of the same material but, as a result of its rearrangement, are usually less well sorted. A detailed descripton of lacustrine ice-shove ridges and related features has been presented by Dionne and Larerdière (1972).

Ice-shove ridges are caused by the grinding action of ice floes driven against the shore by wind-generated waves and currents (cf. Washburn, 1947, 76–7). In places the ridges may be several metres high and extend several metres above water level. In addition to forming by shoreward thrust of ice floes, ice-shove ridges in some lakes develop by thermal expansion of the ice cover with rise of temperature, following any cracking of the ice and freezing of water in the cracks. Pessl (1969) reported that at Gardner Lake, Connecticut, a rise of ice temperature of about 1°/hr for six hours was sufficient to cause ice thrusting and that the shoreward movement of ice during a thirty-day period amounted to 1 m. Ridges produced by ice expansion are probably most prominent in regions lacking permafrost, since the mobility of shore ice is reduced if frozen to permafrost (cf. Washburn, 1947, 76–7).

As with ice-rafted debris in lakes, lacustrine ice-shove ridges are good evidence of a cold climate but, unlike debris that is rafted by icebergs, ice-shove ridges may have no relation to glaciation. In fact most icebergs calved from glaciers ground too far from shore to produce sub-aerial ice-shove ridges.

III MARINE ACTION

1 General

Marine action that is characteristic of periglacial environments and is likely to leave an identifiable record includes formation of ice-shove ridges, the irregular striation of coastal bedrock, and deposition of glaciomarine drift. Evidence of other characteristic aspects, such as the development of an ice foot (shore-fast ice), is less likely to be recognized.

8.2 (*Opposite*) Lacustrine ice-shove ridge at base of Mount Pelly, Victoria Island, Northwest Territories, Canada (*cf. Washburn, 1947, Figure 2, Plate 20*)

2 Ice rafting

Ice-rafted sediments may be deposited in the sea by floating shelf ice and/or drifting icebergs, or even by tree roots (Figures 8.3–8.4). If of glacial origin, such ice-rafted deposits are glaciomarine drift. However, identification of these deposits as glaciomarine drift may be difficult to establish unless they are associated with marine shells *in situ*. Although extensive deposits of glaciomarine drift are periglacial in the sense of being cold-climate phenomena, they are much more closely identified with glacial deposits, and further discussion is omitted.

3 Ice shove and ice-shove ridges

Ice-shove ridges on polar coasts are similar to those on lake shores but the marine environment can produce larger ridges. They tend to be located on unprotected low-lying capes exposed to violent currents and drifting sea ice during the majority of the year (Figure 8.5). Ridge heights as great as 9 m are reliably reported (Bretz, 1935, 219) but reports of 18 m (60 ft) and more are suspect as noted by Nichols (1953). Deep water close to shore favours the shoreward thrust of ice on a gentle beach. Icebergs generally ground before reaching the shore but sea ice may be pushed tens or perhaps even hundreds of metres inland to form massive ridges (cf. Washburn, 1947, 78–80). Leffingwell (1919, 173) reported that ice at Point Barrow, Alaska, reached buildings nearly 60 m (200 ft) inland, and according to James Ross, ice on the north coast of King William Island in Arctic Canada 'travelled as much as half a mile beyond the limits of the highest tide-mark' (Ross, 1835, 416). The ice can striate, gouge, or plane beach deposits, and the ridges or mounds that are formed may be either ice cored or ice free (Hume and Schalk, 1964).

The presence of permafrost extending into the intertidal zone tends to limit the amount of beach debris that can be ploughed up to form ice-shove ridges (Owens and McCann, 1970, 412–13).

Unlike lake ice, thermal contraction and expansion of sea ice is probably a minor factor in producing ice-shove ridges. Sea ice is fragmented into floes and is much more mobile and subject to movement by winds and currents throughout the year than lake ice that remains solidly frozen during the winter.

An observation that ice-shove ridges are either more common or less common along a modern beach than along an emerged beach in the same area needs to be interpreted with great care. Possible causes, in addition to climate change, include differences in topography at different levels, exceptional storms that in the absence of ice could wipe out ridges built over a period of many years, and mass-wasting and other processes that tend to lower and level old ridges. However, some ice-shove ridges, especially those made of gravel, can persist for a long time. Well-

8.3 (*Opposite*) Marine ice rafting. Iceberg with dirt layers, Wolstenholme Fjord, Northwest Greenland

8.4 (*Opposite*) Rafting by tree roots, Mesters Vig, Northeast Greenland. The stone around which the tree root grew must have been carried from a great distance, possibly Siberia

8.5 (*Above*) Ice-shove ridges, Point Barrow, Alaska. The ridges shown were not ice-cored but somewhat similar ridges containing ice cores occurred near by

preserved ice-shove ridges up to 8 m high, constructed of emerged beach deposits and located at an altitude between 110 and 120 m at Skeldal, Northeast Greenland, have a radiocarbon age exceeding 8000 years, judging from the age of near-by shells (Lasca, 1969, 27, 45–8).

4 Striated bedrock

Sea ice grinding against a bedrock coast produces striae somewhat similar to glacial striae but commonly much more discontinuous and irregular. Under certain circumstances they may record the former presence of sea ice in places where it is now absent.

9 Wind action

I Introduction: II Loess: III Dunes: IV Ventifacts

I INTRODUCTION

Wind action is common in periglacial environments but is equally prominent in many others. However, it has certain aspects that are of great value in reconstructing past periglacial environments, and wind action is therefore commonly considered along with other more distinctive periglacial processes.

The prominence of wind action in some periglacial environments is attributed in part to the sparseness of vegetation, especially in broad valleys with braided streams from glaciers, in part to the influence of glaciers in generating katabatic winds. On the other hand there are numerous places where abrasion and deflation by wind are of minor importance, as in some of the western islands of the Canadian Arctic (Pissart, 1966*b*; Washburn, 1947, 74–6), locally at least in West Greenland (Boyé, 1950, 112–17), and in the Mesters Vig district of Northeast Greenland. Descriptions of contemporary periglacial wind action in addition to those noted above include observations by Samuelsson (1927), Hobbs (1931), Teichert (1935), Belknap (1941, 235–8), Troelsen (1952), Bout (1953, 35–52), Fristrup (1953), Mackay (1963*a*, 43–6), Bird (1967, 237–41), McCraw (1967, 399, 404), Nichols (1971, 314–15), French (1972), and many others.

Applying the term periglacial in the restricted sense to ice-marginal phenomona, H. T. U. Smith (1964, 178) listed the following criteria for the periglacial origin of eolian sands

> (1) interfingering or conformable relations with glacial deposits; (2) derivation from source areas which can be best attributed to ice-border conditions or to eustatically lowered sea level; (3) derivation from beaches of ice-dammed lakes; (4) paleontological or paleobotanical data; (5) associated indications of periglacial frost action; or (6) dune form and orientation recording wind direction divergent from that of the present and consistent with that which might occur in a periglacial environment.

Comprehensive studies devoted to former periglacial wind action include Ivar Högbom's (1923) discussion of European dunes and Cailleux's (1942) study of wind effects on Quaternary sand grains. There are a number of regional surveys (cf. Dylik, 1969, for Poland) and some symposia (Schultz and Frye, 1968), but much of the literature on periglacial wind action is scattered in various studies of Quaternary stratigraphy. The status of research on periglacial eolian phenomena in the United States has been reviewed by H. T. U. Smith (1964) who emphasized its elementary character, especially with respect to eolian sands and ventifacts.

II LOESS

Loess is 'Wind-deposited silt, usually accompanied by some clay and some fine sand' (Longwell, Flint, and Sanders, 1969, 309, 653). It tends to be highly cohesive and to form nearly vertical exposures. Stratification is apparently lacking or poorly developed in many exposures but very careful cleaning of an exposure often shows faint, well-developed beds, and stratification is probably more common than is generally supposed. Particles are angular, and the porosity of loess usually exceeds 50 per cent. Loess deposits characteristically blanket the landscape, covering both hills and valleys.

Loess originates in two principal environments – warm deserts and cold deserts. In both places the absence of vegetation is the main factor allowing the wind to pick up and export the silt, and in both the silt tends to be but little weathered chemically. However, associated fossils and the geographic location and distribution of loess deposits generally serve to distinguish which type of loess is represented by a given deposit.

Cold-climate loess is commonly derived from broad floodplains, especially the braided outwash plains of glacial streams. It is clearly a periglacial deposit but much of it is blown far beyond its source and can be deposited in a quite different environment, usually where there is sufficient grass or tree growth to anchor it. Thus while cold-climate loess is derived from a periglacial environment, it is not necessarily indicative of such an environment in the area of deposition.

III DUNES

Like loess, sand dunes form in various environments characterized by little or no vegetation. They develop along sandy shores and in warm deserts and cold deserts. Some dunes in the Antarctic dry valleys are 9–15 m (30–50 ft) high and 61–91 m (200–300 ft) long (Webb and McKelvey, 1959, 127). Although mobile, dunes are unlike loess in that they come to rest in essentially the same kind of environment in which they originated or in one located very close to it. In this respect they are a more useful environmental indicator than loess.

To establish the periglacial nature of dunes requires the elimination of other possibilities, and as in the case of loess this can usually be done by regional considerations. Periglacial dunes occur in the same kind of areas where periglacial loess originates – broad, vegetation-free floodplains and, especially, glacial outwash plains. They also occur along some cold-climate lake shores and sea coasts but such dunes are probably in a minority.

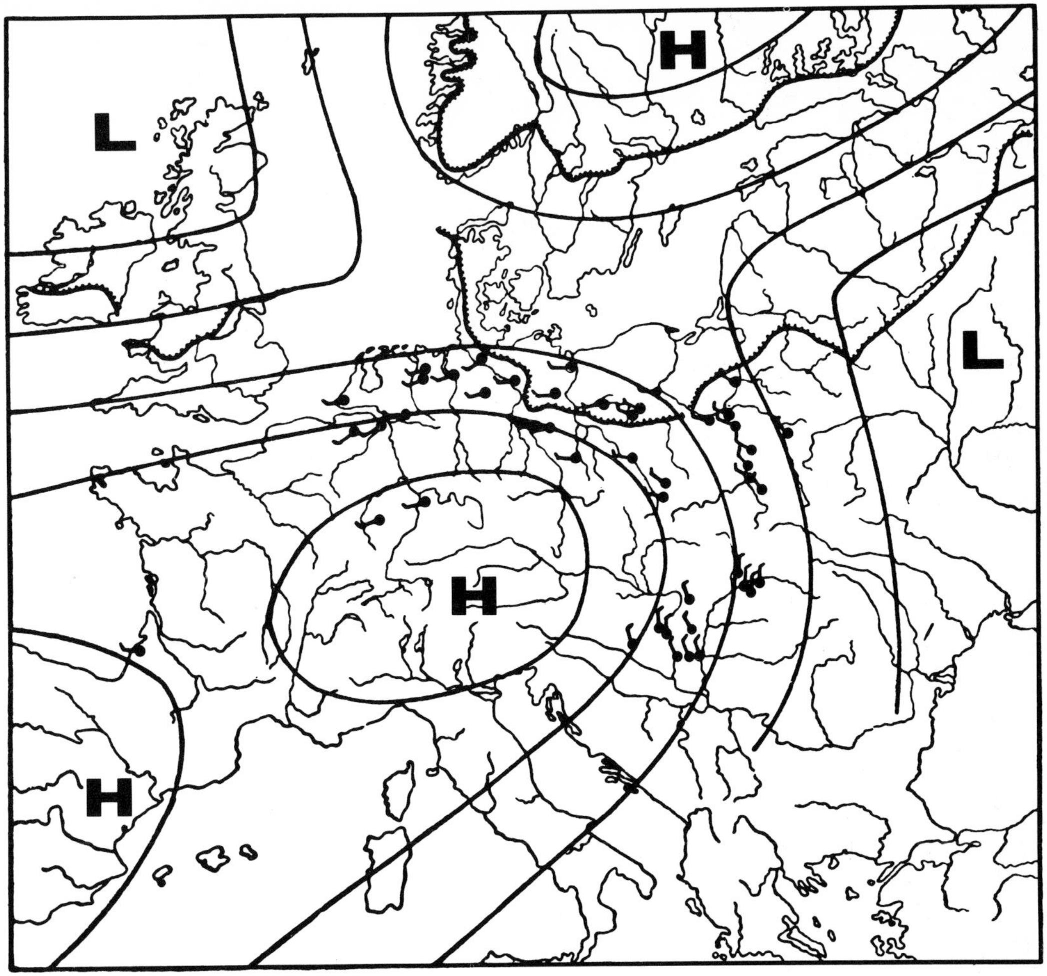

H = highs, L = lows

9.1 Summer atmospheric pressure and wind directions in Europe during Late-Glacial time. (*Poser, 1950, 81, Figure 1*)

Dunes are particularly significant because of the information that their morphology, orientation, bedding, and sand-grain characteristics may provide as to periglacial wind strength and direction. With information from enough regions at hand it may be possible to reconstruct wind conditions marginal to the Pleistocene ice sheets as attempted by Poser (1950) for Europe

9.2 (*Opposite*) Pleistocene ventifact, Warthe moraine, Daszyna, Poland

(Figure 9.1), and thereby help establish the character of the periglacial climates then prevailing. However, to date most studies of stabilized dunes in North America deal with more restricted areas. Investigation of dunes covering an area of 57000 km^2 (22000 mi^2) in Nebrasks led H. T. U. Smith (1965) to infer that some of them were built under a former periglacial wind regime, and a similar conclusion was reached by Henderson (1959*a*, 45–7) for dunes in Alberta, Canada.

IV VENTIFACTS

Ventifacts are stones faceted and polished by wind. Like dunes they can provide much information about periglacial wind conditions. In places like the Antarctic dry valleys, ventifacts present a startling array of sculptured forms rivalling or exceeding those of warm deserts in their degree of development (Nichols, 1966, 35–6). In this connection it may be significant that snow at a very low temperature has sufficient hardness to abrade soft rocks (Fristrup, 1953; Teichert, 1939; Troelsen, 1952). However, the prominent cavernously weathered stones of the Antarctic dry valleys probably owe their origin less to wind than to various weathering processes (Cailleux and Calkin, 1963).

Pleistocene ventifacts of periglacial origin (Figure 9.2) are locally prominent (Powers, 1936). They can show that former wind conditions were quite different from present ones and in places have proved to be very useful in reconstructing periglacial environments (Mather, Goldthwait, and Thiesmeyer, 1942, 1163–73). However, ventifacts are usually studied and interpreted in connection with other eolian phenomena.

10 Thermokarst

I Introduction: II Linear and polygonal troughs: III Collapsed pingos: IV Beaded drainage: V Thaw lakes: *1 General. 2 Oriented lakes.* **VI Alases**

I INTRODUCTION

As noted in the chapter on Frozen ground, the term thermokarst designates topographic depressions resulting from the thawing of ground ice. There are many kinds of thermokarst including linear and polygonal troughs, collapsed pingos, thaw lakes, and alases. Some thermokarst features are due to climatic change but many are caused by such changes of surface conditions as disturbance of tundra vegetation, shift of stream channels, fire, or other non-climatic equilibrium changes that promote thawing. The thickness of the active layer in one burned-over area was shown to be 136–393 per cent greater than before burning (Mackay, 1970, 426–8), and removal of vegetation for placer mining can lower the permafrost table by 4·6 m (15 ft) or more in 1–2 years (O. L. Hughes, 1969, 6).

Mackay (1970) made a case for differentiating between thermokarst subsidence and thermal erosion, the former being without the intervention of flowing water, the latter always with it. Although no such distinction in use of the term thermokarst is made here, it is important to recognize that the evolution of a thaw depression may differ depending on whether or not flowing water is present. The importance of flowing water in thermal erosion along river banks, especially where ice-wedge polygons are present, was emphasized by Walker and Arnborg (1966) and the role of thawing in retreat of lake and sea cliffs has been briefly reviewed by Are (1972).

Given a disturbance that causes removal of a surface insulating layer such as peat, and knowing the original thickness of the active layer and the nature of the terrain and underlying soil, the new depth of thawing in the absence of flowing water can be roughly predicted by use of a diagram (Figure 10.1) developed by Mackay (1970, 430, Figure 13).

Much thermokarst research has been carried out in the Soviet Union, and Kachurin's (1961) comprehensive review of the phenomena there is one of the few monographs on thermokarst (cf. Kachurin, 1962, for brief summary).

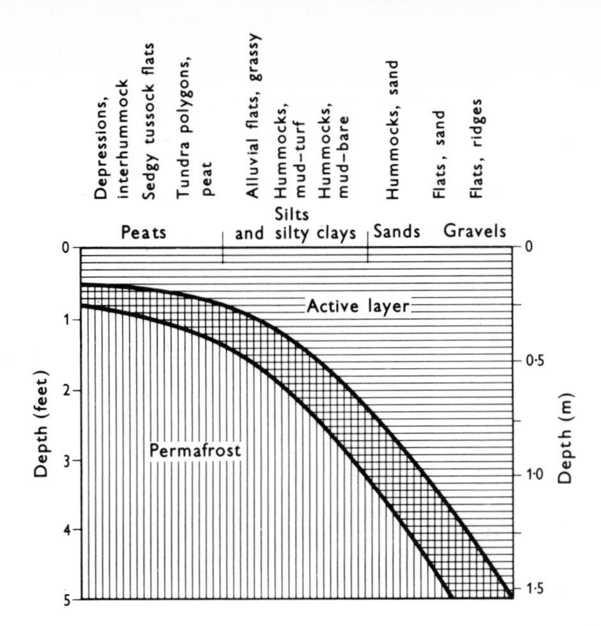

10.1 Approximate range in depth of thaw in common types of periglacial terrain. The diagram can be used to estimate new thaw depth consequent on disturbance of the ground surface (see text) (*Mackay, 1970, 430, Figure 13*)

II LINEAR AND POLYGONAL TROUGHS

Where ice wedges thaw they commonly leave troughs, which in the case of degrading ice-wedge polygons may surround and be responsible for pronounced inter-trough mounds. Some high-centre ice-wedge polygons are of this origin. In Siberia where ice wedges have melted out along river banks, the inter-trough mounds become accentuated, forming conical hills known as baydjarakhs (Czudek and Demek, 1970, 106–7) whose cross-sections parallel to the river bank exhibit striking triangular facets (Figure 10.2).

III COLLAPSED PINGOS

Collapsed pingos may form shallow, usually round or oval, depressions, many of which have a slightly raised rim. However, whether such a rim is an essential characteristic whose absence militates against a pingo origin (Pissart, 1958, 80) is not clear. As described in the chapter on Frost action, partial collapse of a pingo does not necessarily imply environmental change, since pingo growth frequently causes dilation cracking of the material above the ice core and, consequently, exposure of the core to thawing.

More or less completely collapsed pingos are widely reported from Pleistocene periglacial regions in Europe and Asia (Frenzel, 1960*a*, 1021, Figure 13; 1967, 29, Figure 17; Maarleveld,

1965; Mullenders and Gullentops, 1969; Pissart, 1956, 1965, 1965; Svensson, 1969; Wiegand, 1965). Among other places they probably also occur in North America but the reports are less definitive (Flemal, Hinkley, and Hessler, 1969; Wayne, 1967, 402). Depressions caused by collapse of pingos may be very difficult to distinguish from other thermokarst features or from some glacial kettles.

IV BEADED DRAINAGE

Beaded drainage 'consists of series of small pools connected by short watercourses' (Hopkins *et al.*, 1955, 141). The pools result from the thawing of ice masses, which in many places are ice-wedge polygon intersections, and the connecting drainage is commonly along thawing ice wedges and therefore tends to comprise short straight sections separated by angular bends. The pools are some 1–3 m deep (2–10 ft) and up to 10 m (100 ft) in diameter. They have been described from Seward Peninsula, Alaska, and elsewhere in permafrost regions. A somewhat similar drainage pattern occurs in some swampy areas lacking permafrost, and the following criteria are useful in distinguishing beaded drainage from other forms (Hopkins *et al.*, 1955, 141)

(1) Beaded-drainage pools and channels are sharply defined; swamp-drainage courses unrelated to permafrost have indistinct, gradational borders. (2) Beaded-drainage channels generally are straight or consist of series of straight segments separated by angular bends; swamp-drainage courses are straight or smoothly curved. (3) Beaded-drainage courses generally are associated with ice-wedge polygons and locally with pingos and thaw lakes.

V THAW LAKES

1 General

Thaw lakes are lakes whose basins have been formed or enlarged by thawing of frozen ground (Black, 1969*c*; Ferrians, Kachadoorian, and Greene, 1969, 12; Hopkins, 1949, 119; Hussey and Michelson, 1966). They are thus a kind of thermokarst feature. Some lake basins are entirely the result of thawing, others are merely modified by it. Because of drainage changes and concurrent erosion and deposition, thaw lakes in a permafrost environment are dynamic features that tend to change their configuration and, in places, to slowly migrate over the tundra (Tedrow, 1969). Lakes occupying collapsed pingos, and some oriented lakes are varieties of thaw lakes.

10.2 (*Opposite*) Baydjarakhs, left bank Aldan River, 244 km above junction with Lena River, Yakutia, USSR

2 Oriented lakes

Oriented lakes are lakes that have a parallel alignment and are commonly elliptical in plan.

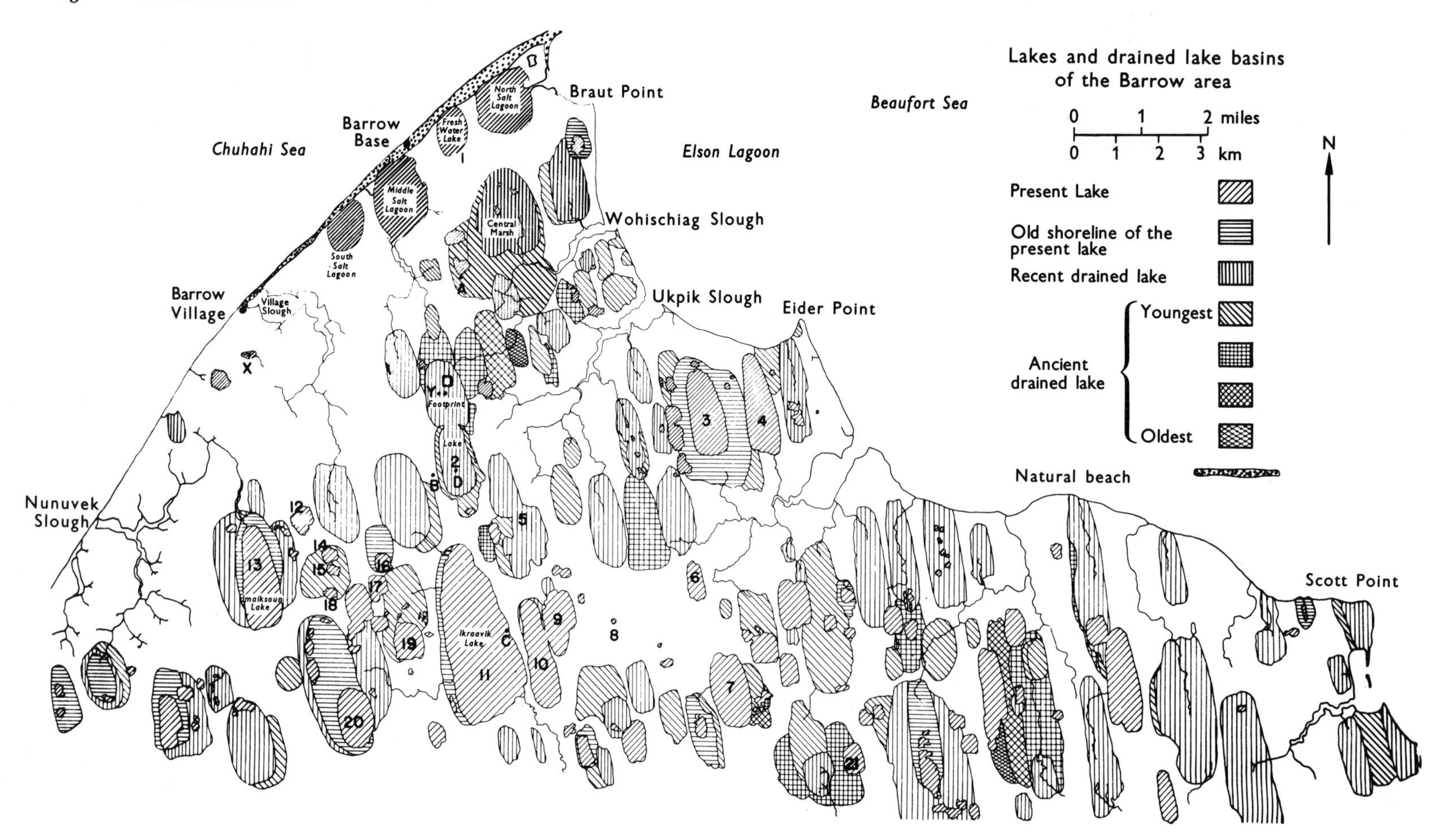

10.3 Oriented lakes, Arctic Coastal Plain, Alaska (*Carson and Hussey, 1962, Figure 2 opposite 420*)

They are described from various environments and as caused by various processes (W. A. Price, 1968) but some are primarily due to differential thawing of permafrost under the influence of predominant winds. An illustration is provided by the oriented lakes of the Arctic Coastal Plain of Alaska (Figure 10.3), described by Black and Barksdale (1949) who suggested that the lakes had formed parallel to a former predominant wind direction by their gradual extension downwind. However, the present view is that they lengthened at right angles to the present predominant (and seasonally opposite) winds as the result of bars forming on the lee shores. As a basin enlarges, mainly by thawing, these bars tend to protect the shores from erosion by waves, currents, and thawing (Carson and Hussey, 1962). Similar oriented lakes occur in the Mackenzie Delta area of Arctic Canada (Mackay 1963*a*, 46–55). Although accepting the cross-wind hypothesis for them, Mackay (1963*a*, 54–5) concluded that

> the precise mechanism of lake orientation remains unexplained. The amount of littoral transport, the aspect of a two-cell circulation, the possibility of thermal effects in a permafrost region, the preference for vegetation growth under favored microclimatic and topographic conditions, and the effect of lake ice on lake orientation need further study.

VI ALASES

An alas is a thermokarst depression with steep sides and a flat grass-covered floor (cf. Czudek and Demek, 1970, 111). The term is of Yakutian origin. Alases (Figure 10.4) are commonly round to

10.4 Alas, Krest-Khal'dzhay, above right bank Aldan River, 360 km above junction with Lena River, Yakutia, USSR. Note small pingos, left centre

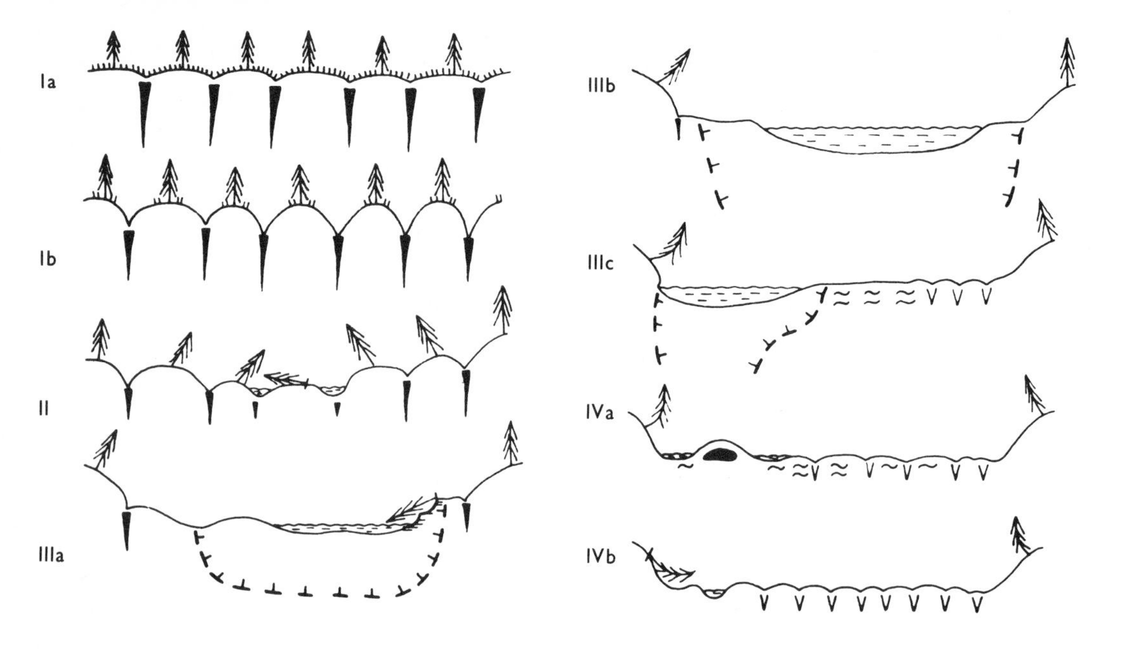

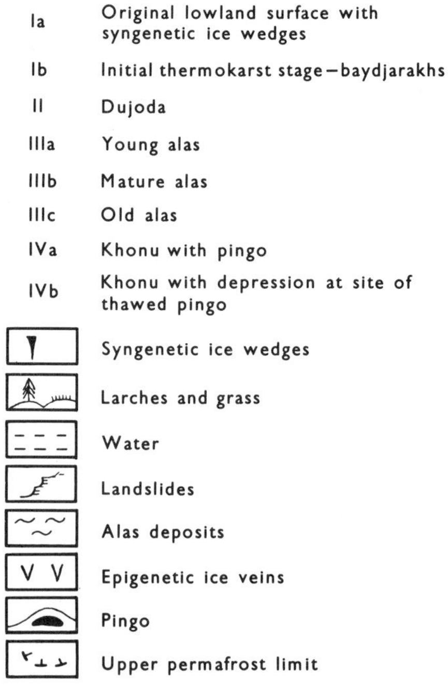

10.5 Sequential alas forms (*Czudek and Demek, 1970, 111, Figure 9*)

oval and many contain shallow lakes. In Central Yakutia, alases range in depth from 3 to 40 m and in diameter from 0·1 to 15 km (Soloviev, 1962, 48). Alas valleys develop as individual forms coalesce. Except for lack of major subsurface valleys, alases resemble true karst topography. They are particularly well developed in Siberia, where they have been extensively studied by Demek (1968*b*, 114–17), Czudek and Demek (1970), Soloviev (1962), and others. Alases may start by a forest fire in the taiga disturbing the thermal equilibrium of permafrost and causing a sequential development of ice-wedge troughs, baydjarakhs, broad collapse areas, lakes, filling in of lakes, growth and collapse of pingos, and joining of adjacent alases into vast alas valley systems tens of kilometres long (Figures 10.5–10.6). In the Central Yakutian lowland, 40 to 50 per cent of the Pleistocene surface has been destroyed by alases, many of which originated during the Hypsithermal interval (*c*. 9000–2500 years C^{14} BP) but some within a human generation (Czudek and Demek, 1970, 113). Although both fluvial and lacustrine action may contribute to alas development, the origin is always as thermokarst.

10.6 (*Opposite*) Alas valley system in the central Yakutian lowland east of Yakutsk, USSR (*Czudek and Demek, 1970, 115, Figure 14*)

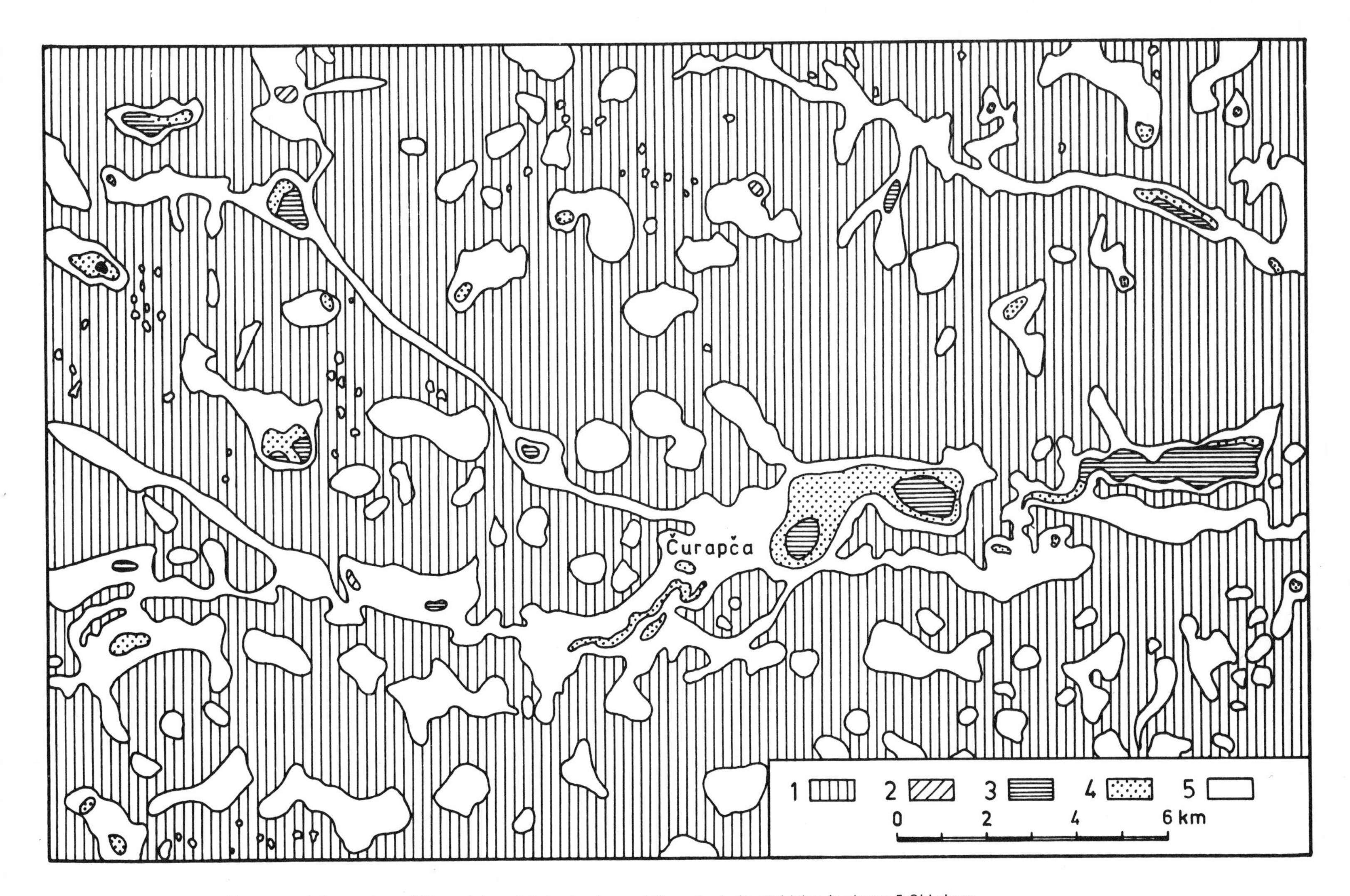

1 Original lowland level with syngenetic ice wedges; 2 Young lakes; 3 Lakes in alases; 4 Recently desiccated lakes in alases; 5 Old alases

11 Environmental overview

Table 11.1 is an attempt to summarize the environmental implications of some of the periglacial processes and features discussed in the preceding chapters. It follows the style of an overview by Lee Wilson (1968*b*, 379, Table 2), which in turn was based on a 1964 Yale Seminar review by the present author. The table is tentative and highly subjective. A generally increasing persistence of snow cover at high altitudes and a decreasing number of freeze-thaw cycles at high latitudes are taken into consideration. Although based on contemporary situations, the table is presumably equally applicable to Pleistocene conditions.

Table 11.1. Quasi-equilibrium range of periglacial processes and features

Predominant ranges are suggested by (1) and lesser ranges in decreasing order by (2), (3), (4). R indicates rare or absent

Processes	*Lowlands*			*Highlands*			
	Polar	*Subpolar*	*Middle latitude*	*Polar*	*Subpolar*	*Middle latitude*	*Low latitude*
Frost action							
Frost wedging	2	1	3	2	1	2	3
Frost heaving and frost thrusting	1	2	3	2	1	3	4
Mass displacement	1	2	R	2	1	3	R
Seasonal frost cracking	2	1	3	2	2	3	R
Permafrost cracking	1	2	R	2	3	R	R
Mass-wasting							
Slushflow	1	2	R	2	3	4	R
Gelifluction	1	2	3	2	1	3	4

Table 11.1. (continued)

Features	*Lowlands*			*Highlands*			
	Polar	*Subpolar*	*Middle latitude*	*Polar*	*Subpolar*	*Middle latitude*	*Low latitude*
Frost creep	2	1	3	2	1	2	3
Rock-glacier creep	R	R	R	2	1	2	4
Nivation	1	2	R	2	1	2	3
Fluvial action							
Ice rafting	1	1	3	4	3	3	R
Lacustrine action							
Ice shove	1	2	3	4	3	3	R
Ice rafting	1	2	3	4	3	3	R
Marine action							
Ice shove	1	2	R	—	—	—	—
Ice rafting	1	2	R	—	—	—	—
Wind action	1	2	3	1	2	3	3
Permafrost	1	2	R	1	2	3	4
Patterned ground							
Nonsorted circles	1	2	R	2	2	3	R
Sorted circles	1	3	R	2	1	R	R
Small nonsorted polygons and nets	1	2	3	2	1	2	2

Table 11.1. (continued)

Features	*Lowlands*			*Highlands*			
	Polar	*Subpolar*	*Middle latitude*	*Polar*	*Subpolar*	*Middle latitude*	*Low latitude*
Large nonsorted (ice-wedge) polygons and nets	1	2	R	2	3	R	R
Small sorted polygons and nets	1	2	4	2	1	3	2
Large sorted polygons and nets	1	3	R	2	3	R	R
Small nonsorted and sorted stripes	1	2	4	2	1	3	2
Large nonsorted and sorted stripes	1	3	R	2	1	4	R
Involutions	1	2	R	2	3	4	R
String bogs	2	1	2	R	2	4	R
Palsas	2	1	R	R	2	3	R
Pingos	1	2	R	R	R	R	R
Slushflow fans	1	2	R	2	3	4	R
Small gelifluction lobes, sheets, streams	1	2	3	2	1	3	3
Large gelifluction lobes, sheets, streams	1	2	R	2	1	3	4
Block fields, block slopes, and block streams	2	2	R	1	1	3	4
Rock glaciers	R	R	R	2	1	3	4

Table 11.1. (continued)

Features	*Lowlands*			*Highlands*			
	Polar	*Subpolar*	*Middle latitude*	*Polar*	*Subpolar*	*Middle latitude*	*Low latitude*
Taluses	3	3	4	2	1	3	3
Protalus ramparts	R	R	R	2	1	1	3
Grèzes litées	2	3	R	2	1	3	4
Nivation benches and hollows	1	2	R	2	1	3	4
Altiplanation terraces	2	2	R	3	1	R	R
Asymmetric valleys related to permafrost	1	2	R	1	2	R	R
Dells	1	2	R	1	2	R	R
Cold-climate varves	2	1	3	3	2	2	R
Lacustrine ice-shove ridges	1	2	3	4	3	3	R
Marine ice-shove ridges	1	2	R	—	—	—	—
Beaded drainage and thaw lakes	1	2	R	R	R	R	R
Loess	2	1	3	4	4	4	4
Dunes	1	2	4	R	3	4	R
Ventifacts	1	2	R	1	2	3	4

12 Environmental reconstructions

I INTRODUCTION

Environmental reconstructions based on periglacial phenomena date back to Łoziński (1909; 1912), the father of periglacial studies. Bertil Högbom's (1914) classic paper on frost action and its effects, and Leffingwell's (1915; 1919) excellent observations on permafrost in Alaska, along with Russian observations, provided much useful information on processes and their climatic associations and thereby aided later reconstructions. Comprehensive reconstructions of this period include those of Soergel (1919) and Kessler (1925). Two decades later the availability of additional field observations and studies, including Troll's (1944; 1958) world-wide survey of periglacial processes and features, permitted improved reconstructions. Among the notable efforts were studies, including maps, by Poser (1947*a*; 1947*b*; 1948*a*; 1950) and Büdel (1951) showing Würm environmental zones in Europe, and a similar effort for Asia by Frenzel and Troll (1952) and Frenzel (1960*a*). The maps, based in part on periglacial phenomena, are very sketchy and subject to considerable modification, but are still among the most noteworthy of their kind.

Far more information for environmental reconstructions is available from Europe (cf. Wright, 1961) and the USSR than from North America. Although fossil periglacial features are widespread in parts of North America, they have been less investigated and there is consequently better stratigraphic control and dating of the European features.

II EUROPE

1 General

Most environmental reconstructions of Europe that are based on periglacial criteria apply to the Continent itself rather than Britain. Nevertheless, periglacial features are also widespread in Britain and are being increasingly studied. Regional summaries include those of Williams (1964; 1965; 1968; 1969), Tufnell (1969), Kelletat (1970*a*; 1970*b*), and Sugden (1971), and there is a wealth of local information becoming available that should favour attempts to reconstruct periglacial environments.

Reconstructions pertaining to the Continent were very prominent several decades ago as reviewed below. Since then much additional local information has accumulated that will permit more refined attempts in the future.

2 Fossil pingos

As noted in the chapter on Frost action, fossil pingos are clear evidence of former permafrost and are becoming increasingly widely known in Europe, including Britain (Frenzel, 1960*a*, 1021, Figure 13; 1967, 29, Figure 17; Maarleveld, 1965; Mitchell, 1971; Mullenders and Gullentops, 1969; Pissart, 1956; 1963; 1965; Svensson, 1969; Watson, 1971; 1972; Watson and Watson, 1972; Wiegand, 1965). Their distribution supports in a general way Poser's reconstruction of the distribution of permafrost during the Würm maximum, which is reviewed below.

3 Depth of frost cracking

In a series of papers, Poser (1947*a*; 1947*b*; 1948*a*) surveyed the distribution of fossil frost cracks and loess or loam wedges in Europe. The wedges he equated with ice-wedge casts – although perhaps some formed without ice wedges as suggested by Péwé (1959) – and he accepted them and the much narrower simple fossil frost cracks as evidence of former permafrost. He assumed that the latter cracks indicate a warmer permafrost climate than the wedges. In plotting the features (Figure 12.1) Poser (1948*a*, 63, Figure 4) noted their depth and width, and found that the deepest and widest were in central Europe, especially central Germany, where maximum depths of wedges were 7–8 m and the greatest widths were 3 m. He considered but, because of regional contrasts, discarded Soergel's (1936, 239, 246) view that the width of wedges is an indicator of age as well as climate. From regional trends Poser (1947*b*, 264; 1948*a*, 64) concluded that during the Würm maximum middle-central Europe was by far the coldest region, followed by southeast Europe and northwest-central Europe, the last with an appreciably warmer winter climate.

Taken alone, the evidence afforded by fossil frost cracks is open to question because so many variables besides temperature affect frost cracking. It may be reasonably assumed that most fossil frost cracks indicate the former presence of permafrost, but to define temperature conditions more closely on the basis of degree of crack development is another matter unless other factors are known to be constant. In addition to terrain and moisture factors, it is not immediately apparent why regional contrasts eliminate age as a variable in the cases considered.

4 Depth of thawing

In the same series of papers in which Poser discussed fossil frost cracks, he reviewed the distribution

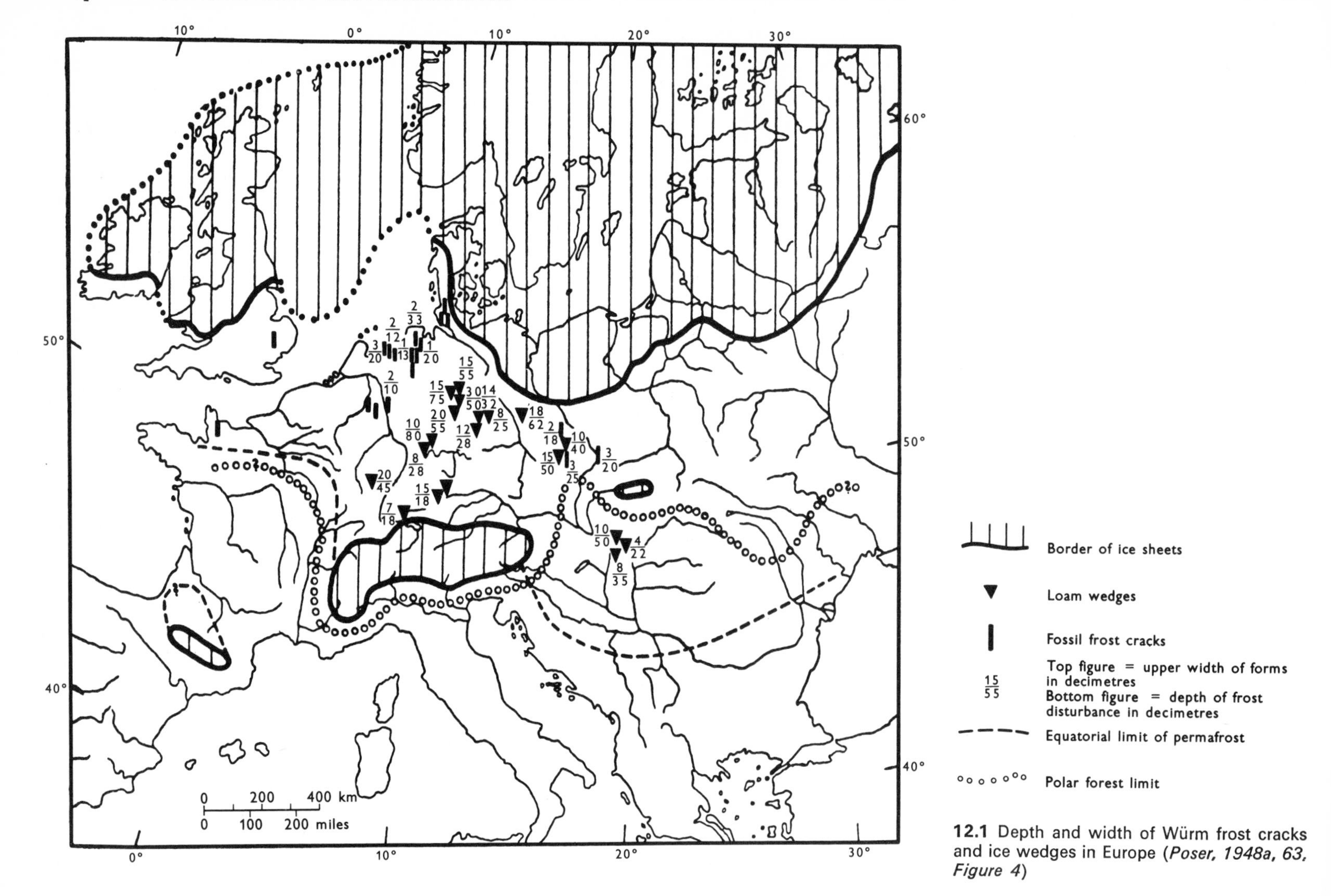

12.1 Depth and width of Würm frost cracks and ice wedges in Europe (*Poser, 1948a, 63, Figure 4*)

of other fossil periglacial features (Figure 12.2). In discussing involutions, he stated that they do not form under present climatic conditions in middle and eastern Europe, where freezing extends to depths of 1·2–1·6 m. Citing Russian workers, especially Sukachëv (1911), Poser concluded that involutions are cryostatic phenomena caused by forces set up in the active layer as downward freezing from the surface approaches the permafrost table. He therefore argued that the depth to which involutions extend approximates the depth of the active layer—i.e. the depth of summer thawing in a permafrost environment. Poser also regarded Stiche as providing information on depth of thawing. He described them as narrow, chevron-like anticlinal deformations of platy bedrock occurring along frost cracks or joints and having an abrupt downward termination. Interpreting Stiche as due to frost action along frost cracks or joints, he accepted their downward termination as controlled by, and therefore indicative of, a former permafrost table. On these premises he reconstructed depth of summer thaw during the Würm Stade (Figure 12.3).

Although accepting most involutions as evidence of permafrost, Kaiser (1960, 133) argued that they could not be used to infer depth of thaw, since there were too many uncertainties involved such as the level of the original surface, contemporaneity of the features, and local climatic influences. Involutions in some places may indeed be associated with permafrost but, as discussed in the chapter on Frost action, they probably also form by seasonal frost action in a non-permafrost environment. Moreover, very similar or identical-appearing features can probably form in environments that have no relation to frost action.

Also it is not demonstrated that Stiche are caused by frost cracking or, if they are, that they terminate at the permafrost table. On the other hand if they are due to frost wedging their depth may merely indicate depth of seasonal freezing and not the presence of permafrost. Pending further investigations, the interpretation of Stiche as permafrost features is problematical. Thus the evidence that Poser adduced for depth of thawing is suspect.

Kaiser (1960, Plate 1 opposite 36) compiled a map of western and southern Europe showing periglacial features he regarded as indicative of permafrost (Figure 12.4). These were ice-wedge casts, involutions, patterned ground, and pingos. He took age differences into account and showed the relationship of the features to various Pleistocene glacial borders. A generalized periglacial map of Europe by Velicko (1972, Figure 3 opposite 68) appeared while this book was in proof.

5 Asymmetric valleys

As noted in the chapter on Fluvial action, the distribution of asymmetric valleys has also been cited as evidence of a periglacial environment. Poser (1948*a*, 54) reported observing two types of asymmetric valleys in the same valley system: (1) Upper parts of valleys where the volume of water was inadequate to carry away all the debris contributed by solifluction from the sunlit

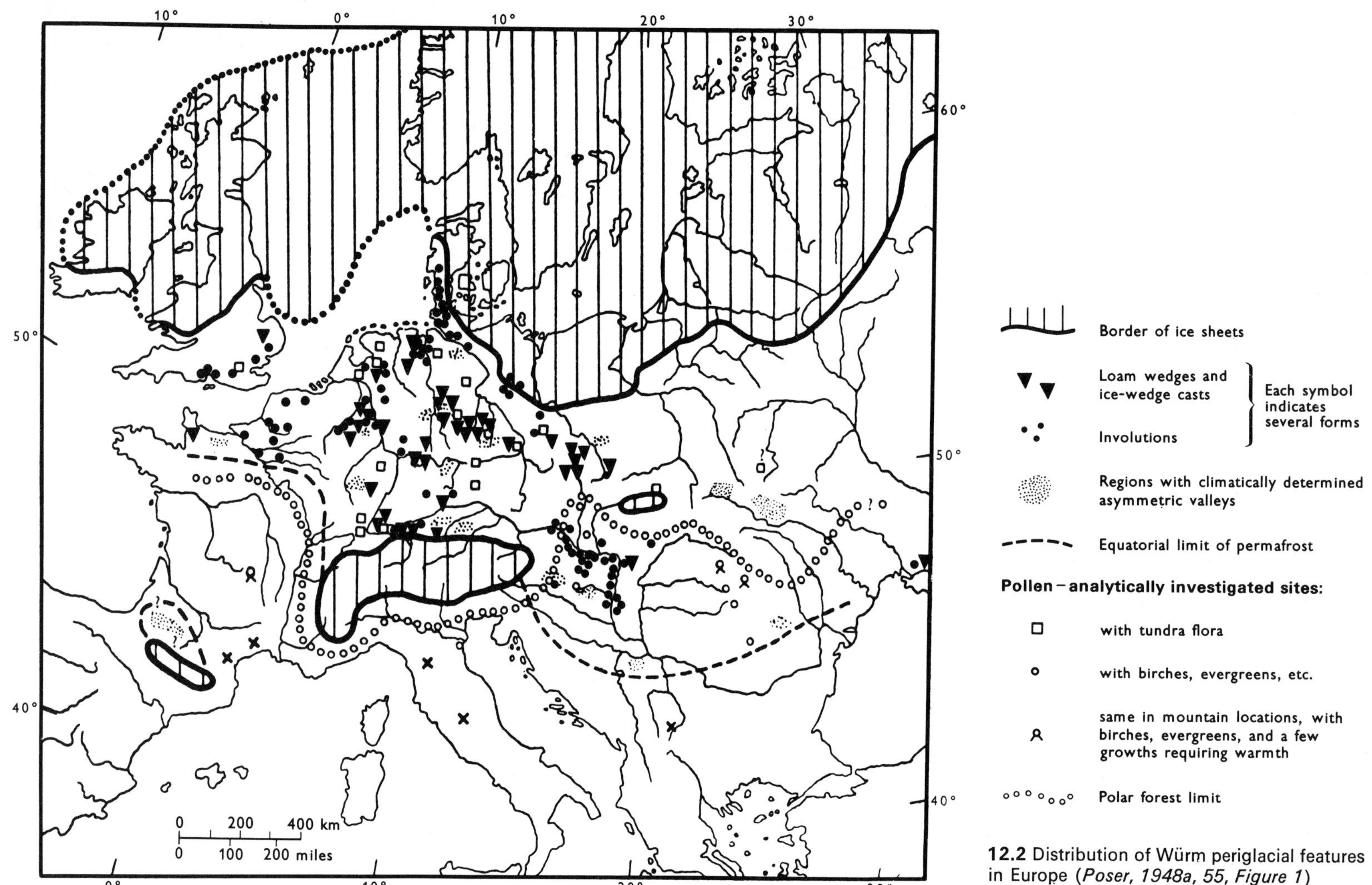

12.2 Distribution of Würm periglacial features in Europe (*Poser, 1948a, 55, Figure 1*)

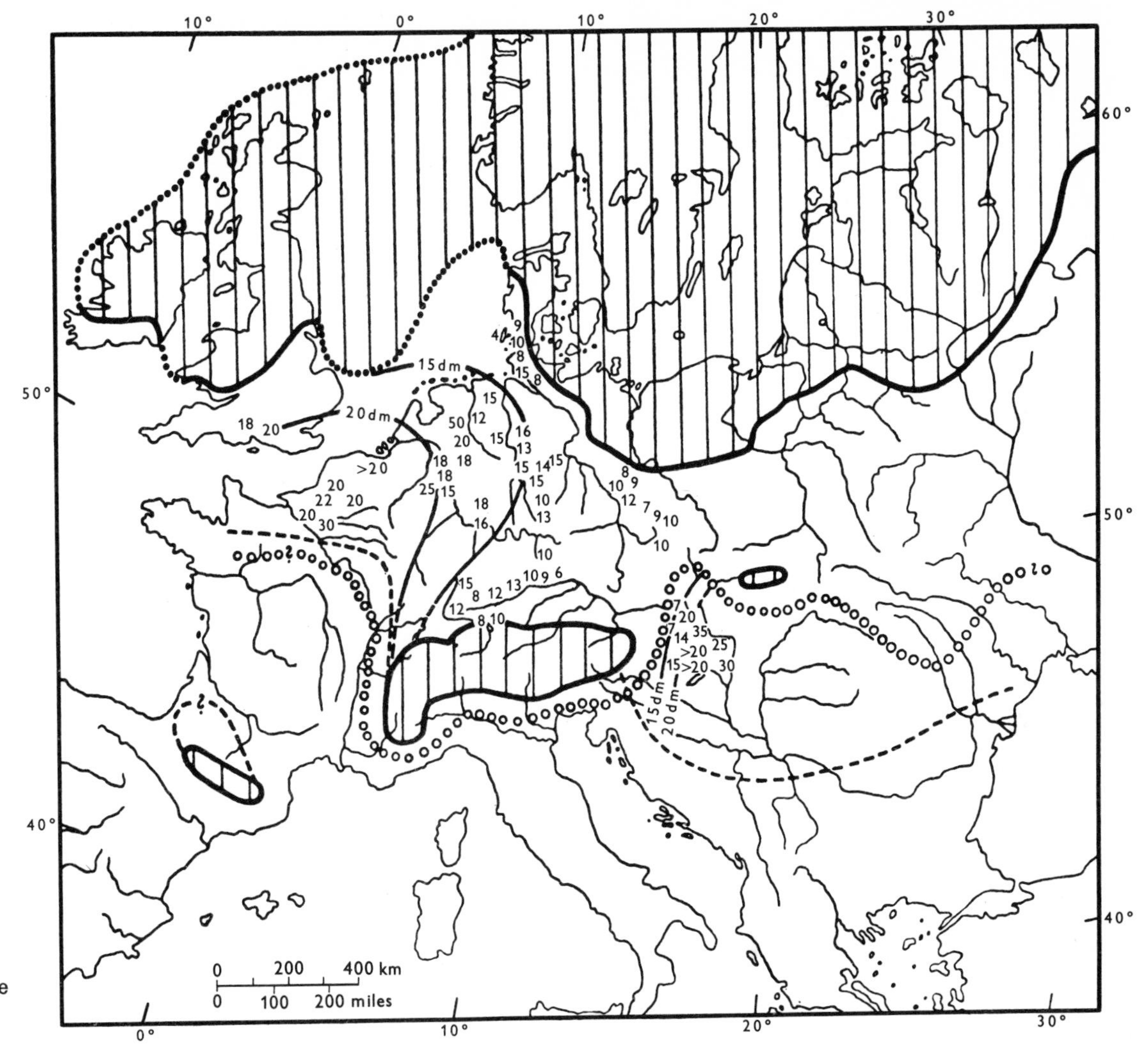

Border of ice sheets

Isolines of thaw depth

Polar forest limit

15 Depth of thaw in decimetres

Equatorial limit of permafrost

12.3 Depth of Würm summer thaw in Europe (*Poser, 1948a, 59, Figure 2*)

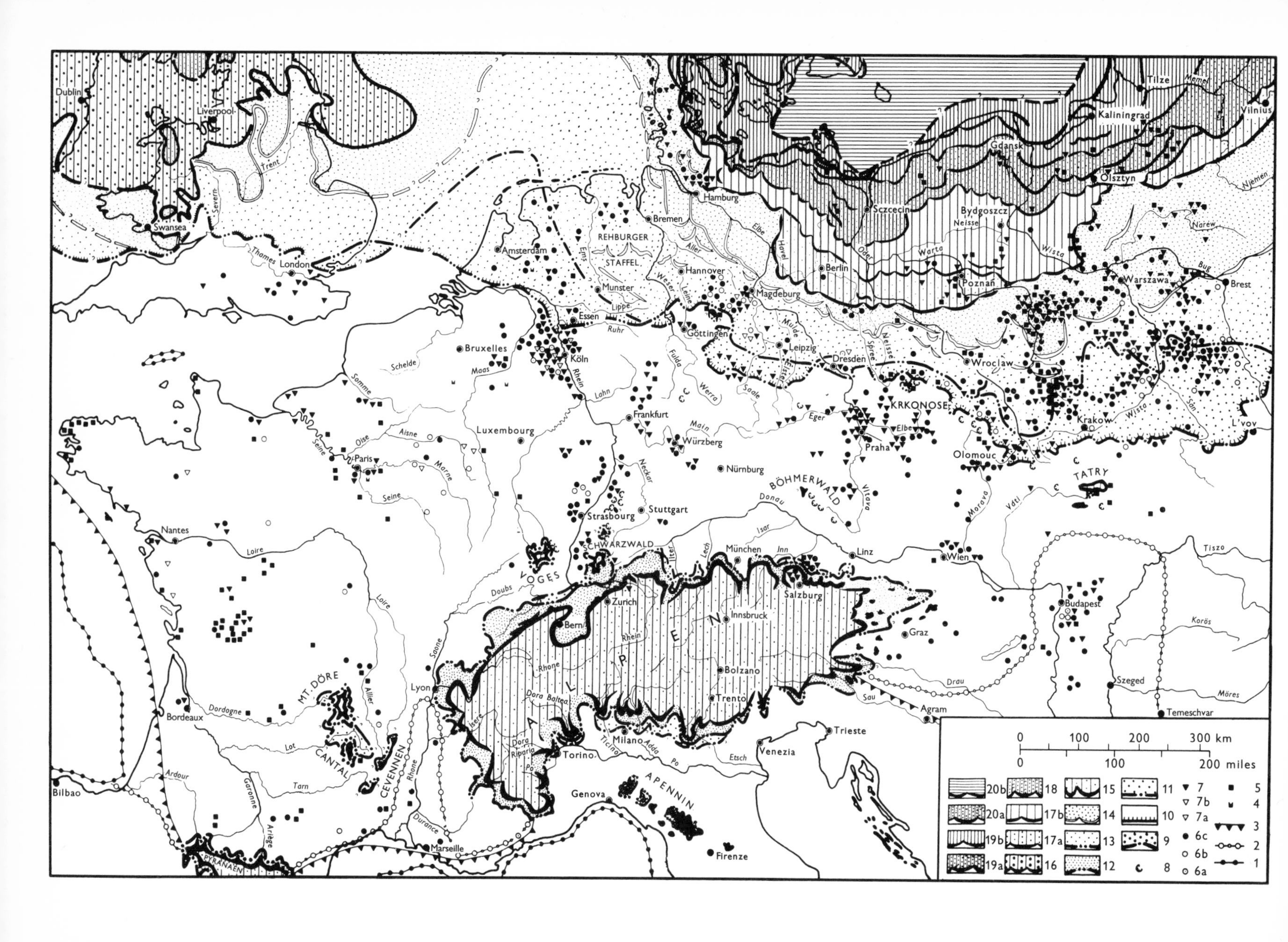

Dublin
Liverpool
Trent
Severn
Swansea
Thames
London
Amsterdam
Ems
REHBURGER
STAFFEL
Munster
Bremen
Hamburg
Aller
Weser
Leine
Elbe
Havel
Hannover
Magdeburg
Berlin
Szczecin
Oder
Warta
Bydgoszcz
Neisse
Wista
Gdansk
Kaliningrad
Tilze
Memel
Vilnius
Olsztyn
Njemen
Narew
Bug
Warszawa
Brest
Poznań
Essen
Lippe
Ruhr
Göttingen
Mulde
Leipzig
Elster
Saale
Dresden
Spree
Neisse
Wroclaw
Bruxelles
Schelde
Maas
Köln
Rhein
Fulda
Werra
Lahn
Frankfurt
Main
Würzberg
Eger
KRKONOSE
Elbe
Praha
Olomouc
Krakow
Wista
San
L'vov
Somme
Oise
Aisne
Seine
Paris
Marne
Seine
Luxembourg
Nürnburg
Neckar
BÖHMERWALD
Vltava
Morava
Váh
TATRY
Stuttgart
Strasbourg
Donau
SCHWARZWALD
Iller
Lech
Isar
München
Inn
Linz
Wien
Tiszo
Nantes
Loire
Loire
VOGES
Doubs
Zurich
Salzburg
Bern
Innsbruck
Budapest
Korös
Rhein
Graz
Rhone
A L P E N
Bolzano
Drau
Szeged
Möres
Temeschvar
Saone
Lyon
MT. DÖRE
Allier
Dordogne
Bordeaux
Dora Baltea
Trento
Sau
Agram
Trieste
Isere
Dora Riparia
Po
Torino
Milano
Ticino
Adda
Po
Etsch
Venezia
CANTAL
CEVENNEN
Lot
Rhone
Tarn
Ardour
Garonne
Bilbao
Genova
APENNIN
Durance
Marseille
Ariège
PYRANAEN
Firenze
0 100 200 300 km
0 100 200 miles
20b
20a
19b
19a
18
17b
17a
16
15
14
13
12
11
10
9
8
7
7b
7a
6c
6b
6a
5
4
3
2
1

slope; consequently the stream was displaced to the opposite side and the gradient of the sunlit slope was preferentially lowered by the mass-wasting. (2) Lower valley sectors where the water volume was sufficient to transport material, and the stream therefore preferentially eroded and steepened the thawed sunlit slope. Poser believed that both types reflect permafrost conditions, and he therefore drew the Pleistocene permafrost boundary in Europe south of them (Figure 12.2). The distribution and significance of asymmetric valleys in southern Germany and eastern Austria have been reviewed by Helbig (1965). As previously discussed, asymmetric valleys are subject to so many interpretations that their value as evidence of permafrost is doubtful; however, this does not deny their climatic significance in places.

6 Distribution of dunes and loess

Poser (1947*b*; 1948*a*) attempted a reconstruction of atmospheric pressure and wind conditions for summer and winter during the Würm maximum (Figures 12.5–12.6). Involved were mainly general theoretical considerations, supplemented by conclusions as to temperature derived from his analysis of depths of frost cracking and thawing, and by information from distribution of dunes (Poser, 1947*b*, 237; 1948*a*, 61).

The reconstruction of atmospheric pressure and wind conditions during Late-Glacial time (Figure 9.1) was directly based on dunes (Poser, 1948*b*; 1950) and supplementary information from loess studies (Poser, 1951). The difficulty of such reconstruction is indicated by varying interpretations of the data (Büdel, 1949*b*, 89–90) and the problems cited by Poser.

7 Summary

Based mainly on conclusions regarding depth of frost cracking and depth of thawing, and on palynological data regarding the nature of the vegetation, Poser (1947*b*; 1948*a*) recognized four climatic provinces in his reconstruction of European environmental conditions for the Würm maximum (Figure 12.7). These were (Poser, 1948*a*, 65–6) (1) Permafrost – tundra climate with subdivisions: (*a*) glacial-maritime province in the west characterized by relatively moist and warm conditions; (*b*) intermediate-glacial province in central Europe characterized by the lowest temperatures and a summer precipitation maximum; (*c*) glacial-continental province farther east characterized by continental conditions the year around but with warmer summers and winters than in central Europe. (2) Continental permafrost – forest province with varied continental conditions. (3) Maritime-tundra province without permafrost, transitional between (1*a*) and (4). (4) Maritime-forest province without permafrost.

Poser was able to draw some conclusions regarding actual temperatures by making the following

A *Pleistocene glaciations of central and western Europe (8–21).* (For further explanation see Kaiser, 1960, 136)
I Würm (Weichsel) borders (15–21):
20 'Gotiglacial': b Rügen line, a Velgast line; 19 'Daniglacial': b Belt line, a Pomeranian line; 18 Scottish Readvance; 17 'Germaniglacial'; 16 Newer Drift; 15 Würm.
II Older glacial borders (9–14):
14 Warthe; 13 Drenthe; 12 Riss; 11 Elster; 10 Maximum extent of Fennoscendian Ice Sheet (Elster and Saale); 9 Mindel and older glaciations.
III Cirque glaciation (8):
8 Cirque glaciers (shown only for Mittelgebirge; predominantly Würm).
B *Periglacial features in central and western Europe (4–7)*
7 Ice-wedge casts: c Würm, b Riss, a Mindel, Günz, or Donau; 6 Involutions: c Würm, b Riss, a Mindel, Günz, or Donau; 5 Patterned ground (predominantly Würm); 4 Pingos (predominantly Würm); 3 Presumed southern limit of continuous permafrost during Pleistocene temperature minimum; 2 Presumed trend of northern forest limit during Pleistocene temperature minimum; 1 Presumed trend of outermost coastline during Pleistocene

12.4 (*Opposite*) **Borders of Pleistocene glaciations and features indicative of permafrost in western and central Europe** (*after Kaiser, 1960, Plate 1 opposite 136*)

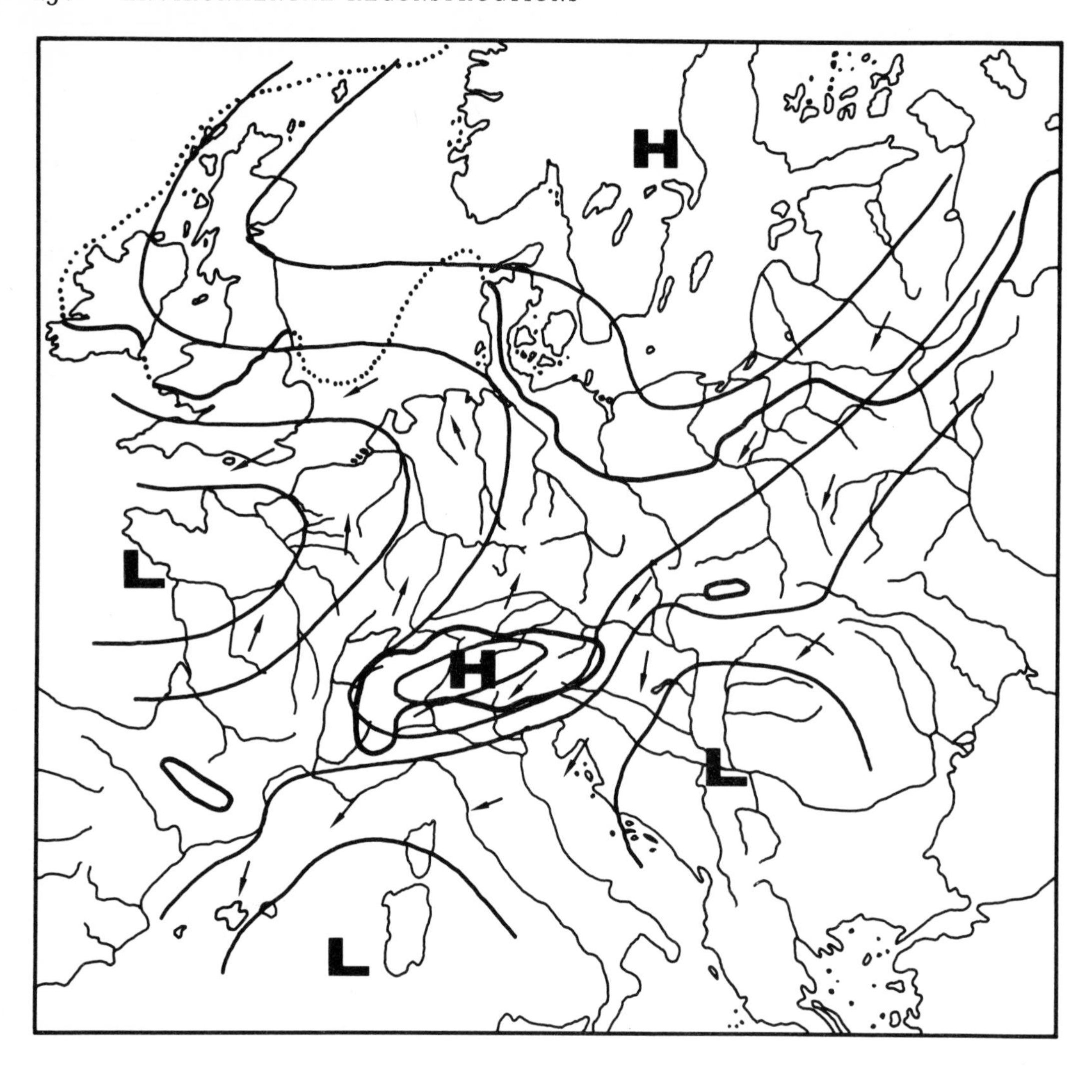

H = highs, L = lows

12.5 Summer atmospheric pressure and wind directions in Europe during Würm Glaciation. (*Poser, 1948a, 61, Figure 3*)

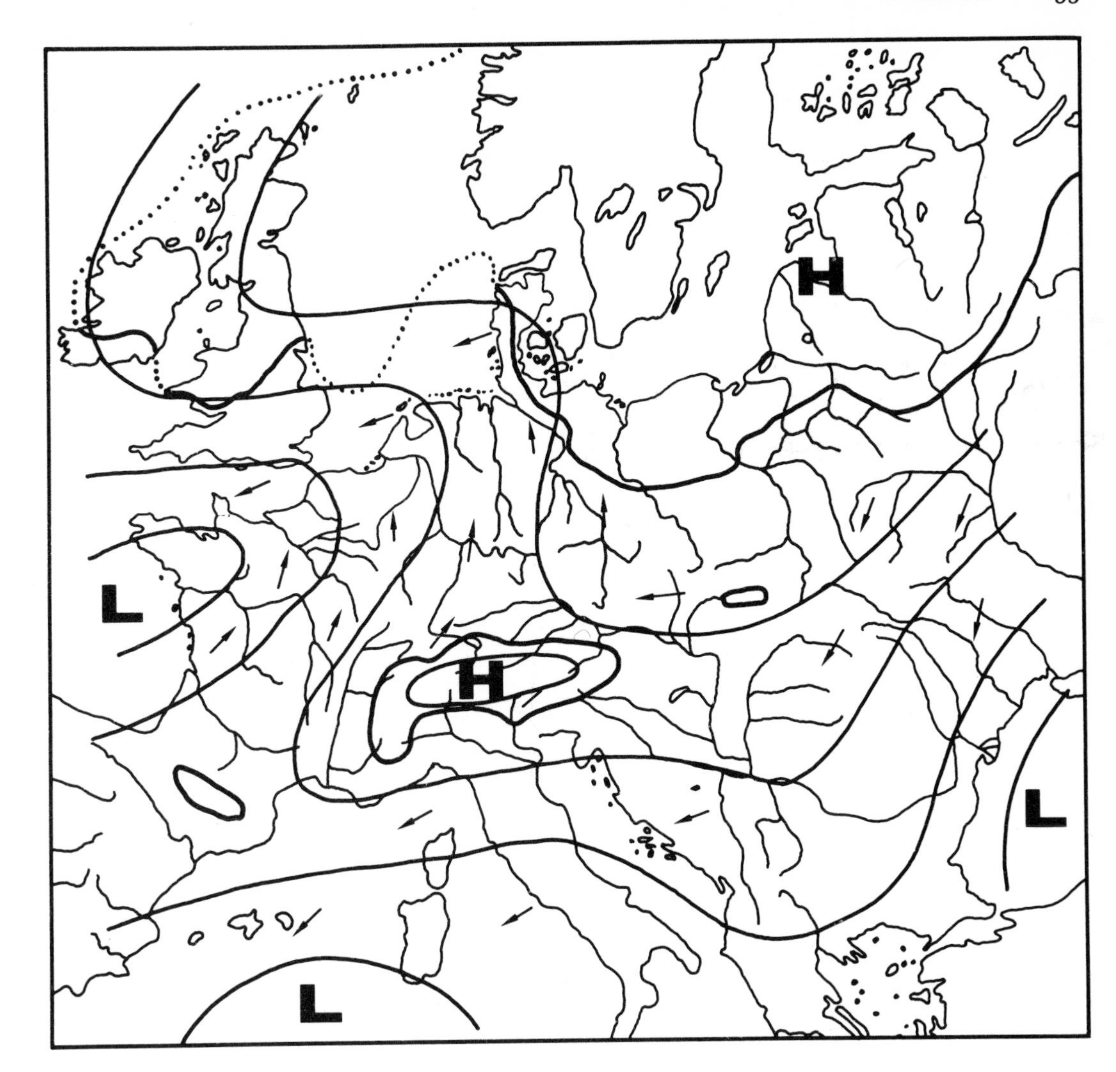

H = highs, L = lows

12.6 Winter atmospheric pressure and wind directions in Europe during Würm Glaciation. (*Poser, 1948a, 65, Figure 5*)

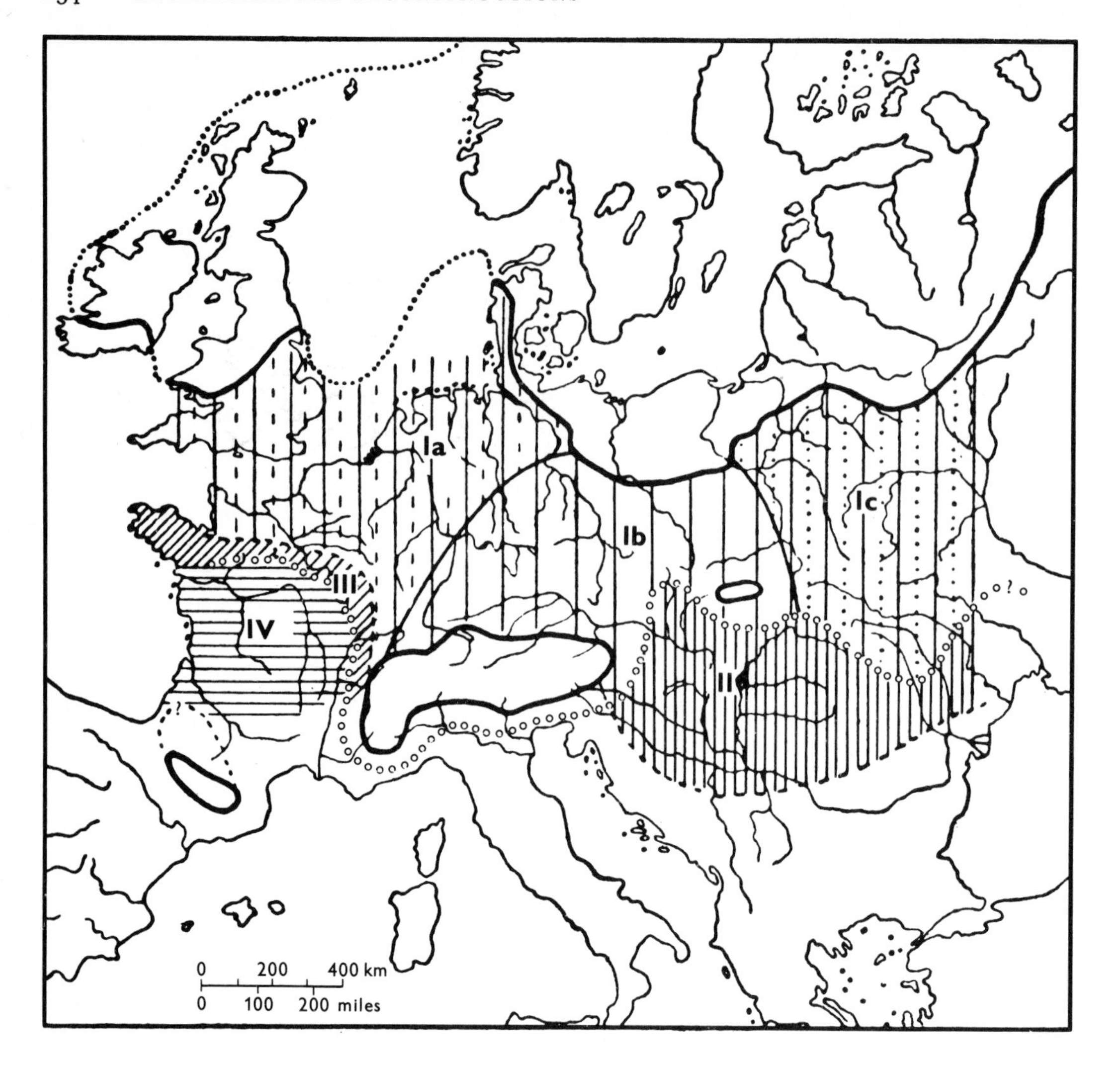

12.7 Climatic regions in Europe during Würm Glaciation (*Poser, 1948a, 65, Figure 6*)

premises: (1) Permafrost today requires a mean annual temperature of $< -2°$. (2) The poleward limit of forest in the Northern Hemisphere approximates the 10° isotherm for July. (3) The distribution of monthly mean temperatures approximates a sine curve. (4) The mean annual temperature approximates the mean temperatures for April and October. (5) The foregoing premises apply to the Würm maximum as well as to the present. (6) The southern limit of permafrost during the Würm maximum is given by the distribution of fossil frost cracks, involutions, Stiche, and asymmetric valleys. (7) Palynological evidence for the position of the forest limit during the Würm maximum is reliable (an unstated premise). All the premises are open to question[1] but (given more reliable data) Poser's pioneering attempt to use periglacial evidence in reconstructing environmental conditions illustrates the possibilities as well as the difficulties involved.

Taking the permafrost limit and the forest limit as providing two isotherms ($-2°$ and $+10°$), Poser found that they crossed in the eastern foreland of the Alps.[2] Here, then, the mean annual temperature (and therefore April and October temperature) was $-2°$ and the July temperature 10°. By virtue of the sine-curve premise, this gave the monthly temperatures in Figure 12.8. Thus the mean summer (June–August) and winter (December–February) temperatures were, respectively, 8·7° and $-12·7°$. Therefore, the mean July and January temperatures during the Würm maximum would have been, respectively, some 8°–9°, and 12° lower than at present (Poser, 1948*a*, 57). Comparison of the position of the Würm permafrost limit and forest limit elsewhere shows that the temperature reduction was different in different places, seasonally as well as regionally (Poser, 1947*a*, 16). Büdel (1953, 250) concluded that the Pleistocene temperature decrease in central Europe must have been as much as 14°, based on mean annual temperatures of $-4·8°$ to $-8·6°$ at the southern boundary of Eurasian permafrost, and thus that the decrease was at least $-5°$ in central Europe where there are ice-wedge casts in areas that now have a mean annual temperature of 8°–9°. Büdel's conclusion for central Europe is very close to Poser's if allowance is made for the fact that Poser assumed $-2°$ for the southern limit of permafrost and that his mean summer and winter temperature reduction would be 10° $\left(\text{i.e.}\ \frac{8+12}{2}\right)$, so that his comparable annual estimate would be a Würm temperature reduction of at least 13°, taking $-5°$ for the southern limit of permafrost. As discussed in the chapter on

[1] Premise 1 suffers from the fact that the mean annual temperature required for permafrost is subject to some regional variation as discussed in the chapter on Frozen ground. Premise 2 does not hold for some regions. The applicability of premises 3–5 to late-Pleistocene conditions was questioned by Butzer (1964, 283, footnote). Premise 6 is based in part on the unreliable evidence of involutions, Stiche, and asymmetric valleys. Premise 7 (with the evidence then available) would not be acceptable to modern palynologists.

[2] Poser's evidence for the southern limit of permafrost east of here is based mainly on the questionable evidence of involutions and asymmetric valleys.

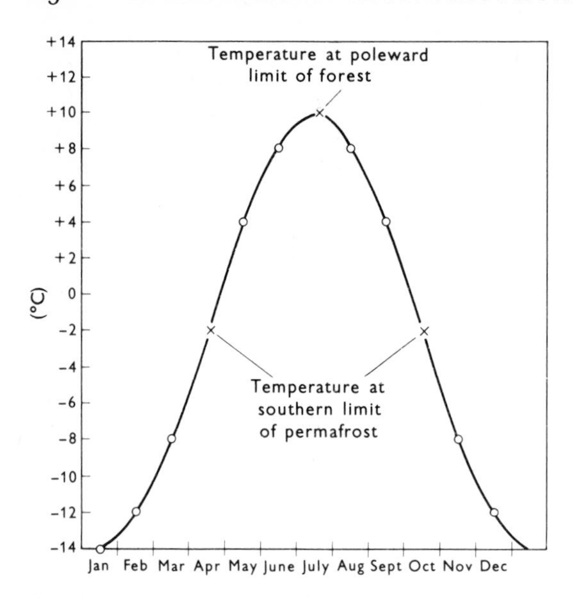

12.8 Sine curve of assumed monthly mean temperatures in Europe where the southern limit of permafrost crossed the northern limit of forest during the Würm maximum (*cf. Poser, 1947a, 16; 1948a, 57)*

Frozen ground, −6° to −8° is perhaps a better temperature requirement for development of ice wedges. If so, the mean annual temperature reduction in central Europe was at least 15°. Mangerud and Skredon (1972, 94) on the evidence of fossil (probably permafrost-crack) wedges suggested that the temperature reduction in western Norway was at least 10°–13°.

Poser's attempt to put quantitative parameters on his reconstructions was a pioneering effort, however uncertain some of the premises and however major the eventual refinements. For instance, Cailleux, Guilcher, and Tricart have mapped fossil frost cracks and/or ice-wedge casts in France as far south as lat. 64°N (Tricart, 1956*b*, 34, Plate III), and therefore permafrost probably existed far south of where it was mapped by Poser and more nearly as it was mapped by Kaiser (Figure 12.4). Continued research on physical periglacial phenomena, combined with biological investigations, will eventually result in more refined reconstructions.

III USSR

Frenzel (1960*a*; 1960*b*; 1967) reviewed the Quaternary environments of northern Eurasia and plotted the distribution of some fossil periglacial features (Frenzel, 1960*a*, 1005, 1010, 1019, 1020, Figures 6–7, 11, 13). He was careful to note that thermokarst phenomena were probably largely caused by human or other non-climatic disturbances rather than by climate change. Although using periglacial evidence, his reconstructions were based mainly on detailed comparisons between the distribution of former and present vegetation and on the assumption that the differences were climatically controlled (Frenzel, 1968).

A map of Pleistocene periglacial features in the USSR, including the former extent of permafrost, was presented by Markov (1961) and Popov (1961) and is reproduced as Figure 1.2. A more recent map by Kost'jaev (1966) shows the southern limit of permafrost much farther north but is based on the dubious assumption that most polygonal systems of soil wedges (Bodenkeile) in Quaternary deposits are due to convection and have no necessary relationship to frost action.

IV NORTH AMERICA

Fossil periglacial features are reported from many places in North America.[1] However, the origin of some features is suggestive or debatable rather than proved (Black, 1969*b*; Flemal, Hinkley, and Hessler, 1969; Galloway, 1970; Malde, 1961; 1964; Prokopovich, 1969; Ruhe, 1969, 177–82), and much remains to be learned. To date, Brunnschweiler (1962; 1964) is the only investigator who has attempted a comprehensive environmental reconstruction based on periglacial features. Field work and a review of the literature originally convinced him that in many places periglacial features extend far south of previous estimates. However, his revised map (Figure 12.9) is more conservative in showing the extent of thermokarst phenomena, although a number of features mapped as probably associated with permafrost remain of debatable origin or significance, including the Mima Mounds of Washington. Consequently the extensive tundra (ET) and other vegetation zones as reconstructed by Brunnschweiler (1964, 230, Figure 8) are suspect to the extent they are based on such evidence. Despite problems of correctly identifying periglacial features, and greater difficulty of dating them than in Europe, periglacial features are beginning to provide scattered evidence regarding former temperature and moisture conditions in North America. For instance, in the Arctic Coastal Plain of Canada, relic permafrost studied by Mackay, Rampton and Fyles (1972) has a mean annual temperature of -7° to -10° at the depth of zero annual amplitude. Since this temperature could not have risen above 0° for a prolonged period without thawing the permafrost, they concluded that it had never been more than 7°–10° higher than now following development of the permafrost, which they believed occurred much more than 40000 years ago.

[1] Antevs, 1932, 51–5; Barsch and Updike, 1971; Berg, 1969; Black, 1965; 1969*a*; 1969*b*; Borns, 1965; Clark, 1968; Clayton and Bailey, 1970; Denny, 1951, 1956, 30–42; Dionne, 1966*b*; 1966*c*; 1969; 1970; Fleisher and Sales, 1971; Flemal, 1972; Galloway, 1970; Goldthwait, 1940, 32–9; 1969; Horberg, 1951, 11–13; Lee, 1957, 2; Mears, 1966; Michalek, 1969; Morgan, 1972; Potter and Moss, 1968; Rapp, 1967, 233–44; Raup, 1951, 113, 114–15, Figures 4–5; Schafer and Hartshorn, 1965, 124; Sevon, 1972; H. T. U. Smith, 1962; Wayne, 1967; and others.

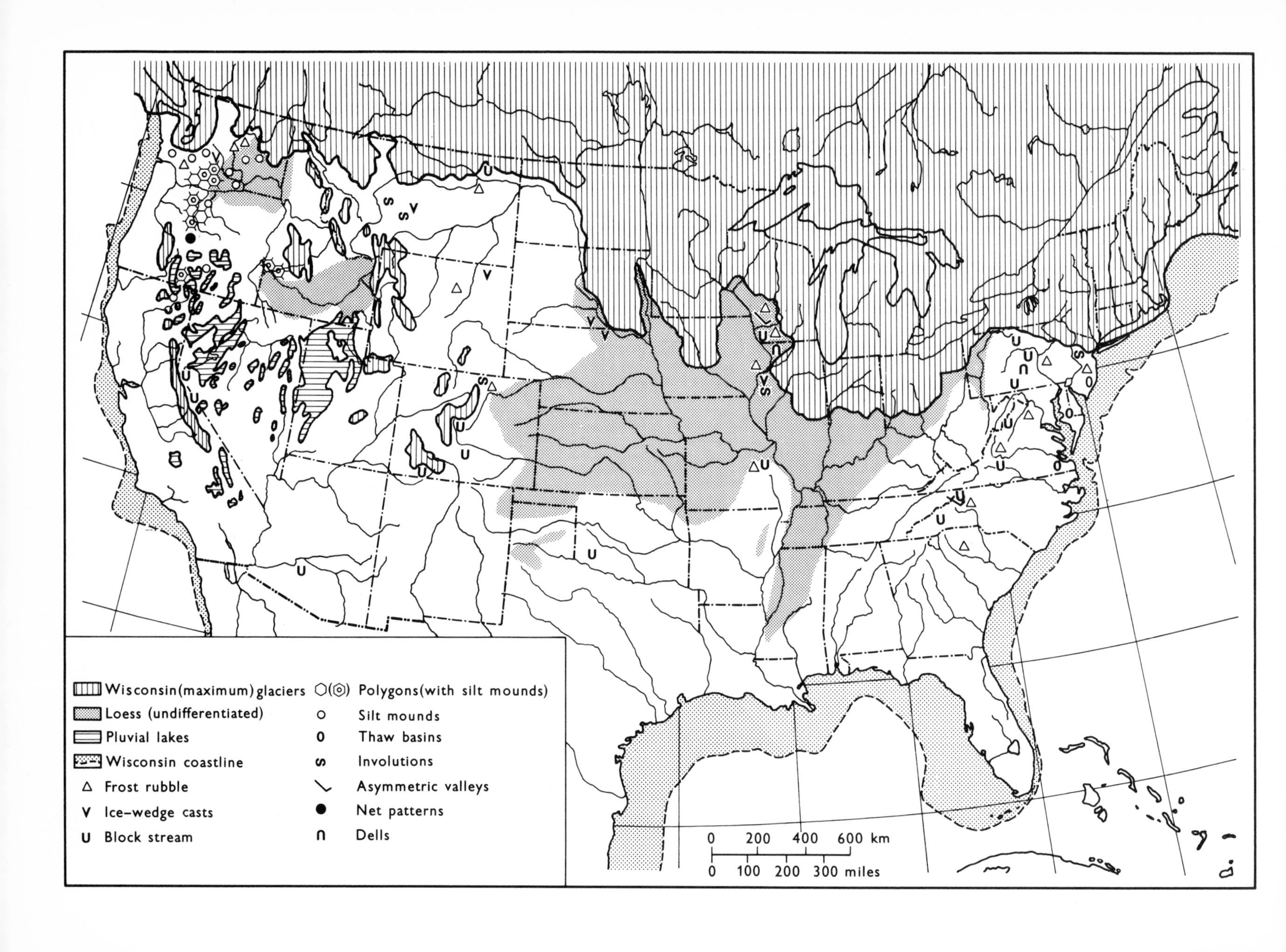

Wisconsin (maximum) glaciers
Loess (undifferentiated)
Pluvial lakes
Wisconsin coastline
Frost rubble
Ice-wedge casts
Block stream
Polygons (with silt mounds)
Silt mounds
Thaw basins
Involutions
Asymmetric valleys
Net patterns
Dells
0 200 400 600 km
0 100 200 300 miles

In the Edmonton, Alberta, area of western Canada, Berg (1969) described fossil sand wedges in Wisconsin sand and gravel beneath till of the last ice advance. The present mean annual precipitation here is 40 cm and the temperature 2° (Berg, 1969, 326). Citing observations that active sand wedges in the Antarctic occur where the annual precipitation is about 16 cm, and Péwé's (1966*b*, 68) conclusion that frost cracking associated with active ice-wedge polygons in Alaska requires at least −5°, Berg (1969, 331–2) concluded that the Edmonton sand wedges record both a drier and colder climate, with precipitation being about half the present and the temperature lower by 7° or more.[1]

In eastern Canada, Morgan (1972) reported ice-wedge casts near Kitchener, Ontario, which he believed originated some 13000–13500 C^{14} years ago. Citing Péwé's (1966*a*, 1966*b*) report that active ice-wedge polygons in Alaska are associated with a mean annual air temperature of −6° to −8°, and comparing this with the present mean annual temperature of 7° in the Kitchener area, Morgan concluded that there has been a temperature rise of at least 13°–14° since the Kitchener polygons formed.

Further east, Dionne (1966*b*; 1969; 1970) reported some 175 ice-wedge casts in Quebec in the vicinity of the Highland Front Moraine south of the St. Lawrence River. They occur mainly in fluvioglacial material. Dionne (1966*b*, 97; 1966*c*, 26) estimated that the former ice wedges required a mean annual temperature of the order of −4° to −5°, based on Péwé's (1966*a*, 78; 1966*b*, 68) observation of active ice wedges at −6° to −8° in Alaska and preliminary observations by Rapp, Gustafsson, and Jobs (1962) who reported probably inactive ice-wedge (?) polygons at −3° to −4° in Swedish Lapland. Subsequent investigations by Rapp and others showed that these polygons are characterized by ice-wedge casts and are indeed inactive but that frost cracking can affect the active layer (Rapp and Annersten, 1969; Rapp and Clark, 1971) and extend into permafrost in exceptionally cold winters (Rapp, personal communication, 1971). Since the present mean annual temperature where Dionne worked is 3° (Dionne, 1969, 308), and assuming that other environmental factors do not negate such comparisons, Dionne's estimate that ice-wedge development requires a mean annual temperature of −4° to −5° would imply that the mean annual temperature of the region was 7°–8° lower than now when the Quebec ice wedges originated some 11500–12500 C^{14} years ago during the St. Antonin glacial episode (Dionne, 1969, 315; 1970, 317–18). Since Dionne cited −3° rather than −3° to −4° for the inactive Lapland occurrences, and considering the −6° to −8° range of Péwé's observations, an 8°–10° lowering may be a better approximation. Apparently ice-wedge casts have not been reported north of the St. Lawrence River, nor

12.9 (*Opposite*) Pleistocene periglacial features in the United States (*Brunnschweiler, 1964, 224, Figure 1*)

[1] Actually Péwé's (1966*b*, 68) −5° temperature referred to the level of zero annual amplitude where the temperature is somewhat warmer than the mean annual air temperature, which in the situation described by Péwé is −6° to −8°, so that the difference if the Alaska comparison is accepted would be at least 8°.

have any significantly younger ones been observed south of it in Quebec (Dionne, 1969, 315).

In the United States, buried ice wedges in silt help to reconstruct the environmental history of the Fairbanks area in central Alaska. Although reported from a number of localities, they have been particularly studied, along with other stratigraphic features, in a 110-m (360-ft) long permafrost tunnel containing a 14-m (45-ft) vertical ventilation shaft, excavated near Fairbanks by the US Army Cold Regions Research Engineering Laboratory (CRREL). Here a lower zone of large ice wedges, 1–2 m (3–6 ft) across, unconformably underlies a zone, some 6 m (20 ft) thick, of smaller wedges that are up to 30 cm (1 ft) across (Figure 12.10) (Sellmann, 1967). Radiocarbon dates of material from the lower zone range from 30700 (+2100 – 1600) to 33700 (+2560 – 1900) C^{14} years BP,[1] and dates from the upper zone range from 8460 (±250) to 14280 (±230) C^{14} years BP (Sellmann, 1967, 17, Table II). Chemical analysis of the cation content of extracted soil water showed a pronounced increase below the unconformity, suggesting that the lower zone had been less subject to thawing and leaching than the upper. Another unconformity separates the upper zone of small wedges from some 3 m (10 ft) of overlying silt. Radiocarbon dates above and below this unconformity bracket its age between 6970 (±135) and 8460 (±250) years BP.

[1] These dates are consistent with dates from below the lower zone, which range from 33200 (± 1900) to > 39000 C^{14} years BP (Sellmann, 1972, 8–9).

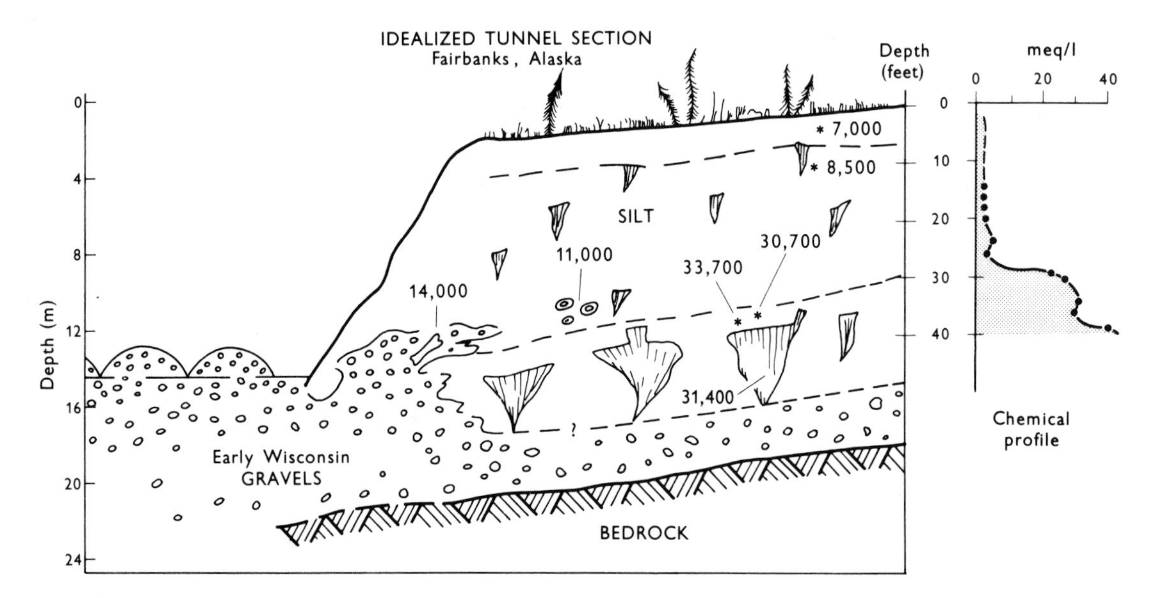

12.10 Idealized section showing distribution of ice wedges, radiocarbon dates, and cation content of soil-water extracts at CRREL permafrost tunnel near Fairbanks, Alaska (*Sellmann, 1967, 15, Figure 13*)

The upper unconformity, which is reported elsewhere in the Fairbanks area as well, is believed to record the lower limit of thawing during the Hypsithermal interval. The lower unconformity, which also appears to occur elsewhere in the area, may represent the lower limit of thawing during a Wisconsin interstadial but confirmation is lacking. The lower-lying large ice wedges suggest epigenetic permafrost (i.e. slow sedimentation rates and a relatively stable land surface) rather than the syngenetic permafrost indicated by the small size and vertical frequency of the overlying wedges.

Ice wedges have also been stratigraphically investigated in other places in the Fairbanks area (Péwé, 1965, 6–36), in the Barrow area (Jerry Brown, 1969, 122) and elsewhere, but much too little has been done to more than suggest the potential significance of such studies for the Quaternary history of Alaska.

Three intersecting generations of ice-wedge casts have been reported from the Laramie Basin of Wyoming by Mears (1966). He considered the oldest and largest ice-wedge casts, some 1·2 m (4 ft) wide at the top and 2·4 m (8 ft) deep, and the youngest ice-wedge casts as recording, respectively, pre-Wisconsin and late-Wisconsin ice-wedges. Because the present mean annual temperature of the area is 5·2° (41·4°F), and Mears believed that the oldest, and maybe all, the ice wedges had formed at a temperature below −6·1° (21°F), he concluded that the mean annual temperature indicated by the oldest was at least 11·1° (20°F) lower than now.

In the mountains of the southwest United States there are slope deposits, believed to be periglacial and to date from the last glaciation, that carry important paleoclimatic implications if correctly interpreted. According to Galloway (1970) this distribution implies a timberline depression of 1300–1400 m, a mean annual temperature 10°–11° lower than today, and a drier rather than wetter (pluvial) climate. Galloway (1970, 256) also concluded that

> unlike snowlines, which are dependent on both warmer temperatures and winter snowfall, the timber line and the lower limit of periglacial action are fairly closely related to summer temperature over a wide range of precipitation and so are potentially more reliable guides to past climates.

According to Barsch and Updike (1971, 112–13), the timberline depression on Kendrick Peak in northern Arizona must have been at least 1000 m during late Pleistocene time.

In Wisconsin the distribution of ice-wedge casts, where the mean annual air temperature is now about 5°, led Black (1969*b*, 234–5) to infer Late Wisconsin mean annual temperatures of −5° to −10°. He argued that the growth of the ice wedges required adequate moisture and temperatures 10°–15° lower than now; their destruction was consequent not on only warmer but also on drier conditions, judging from the fact that the ice-wedge casts consist of eolian sand.

In the eastern United States, repeated measurements spanning a third of a century led

Goldthwait (1969) to conclude that gelifluction features and large-scale sorted patterned ground on Mount Washington (alt. 1917 m), New Hampshire, are inactive. Except for sporadic patches, permafrost is now confined to the summit area above 1825 m. The large-scale periglacial features he studied extend to much lower altitudes where the present mean annual temperature is 0°. Since he believed that the features must have developed under conditions of widespread continuous permafrost, he concluded that the mean annual temperature at that time must have been 6° lower than now. This would be consistent with the −6° to −8° temperature at the southern limit of continuous permafrost in Alaska as reported by Péwé (1966*a*, 78; 1966*b*, 68).

A critical appraisal and reconstruction of North American periglacial environments is clearly becoming increasingly practicable.

References

References have been examined and verified in original except those cited for Figures 6.3–6.5:

AHLMANN, H. W., 1936, Polygonal markings: 7–19 in Scientific results of the Swedish–Norwegian Arctic Expeditions in the summer of 1931 **2**(12): *Geografiska Annaler* **18**, 1–19.

AHNERT, FRANK, 1971, A general and comprehensive theoretical model of slope profile development: *Univ. Maryland Occasional Papers in Geog.* **1**. (95 pp.)

AKADEMIYA NAUK SSSR, 1964, *Fiziko-geograficheskĭ atlas mira:* Moskva. (298 pp.)

ALEKSANDROV, A. I., 1948, Kamennyye morya i reki Urala: *Materialy po geomorfologii Urala, vyp.* **1**, 264–7.

ALESKHOV, A. N., 1936, Über Hochterrassen der Urals: *Zeitschr. Geomorphologie* **9**, 143–9.

AMERICAN METEOROLOGICAL SOCIETY, 1953, Bibliography on frost and frost forecasting: *Meteorol. Abs. and Bibliography* **4**(3), 273–420.

ANDERSEN, B. G., 1963, Preliminary report on glaciology and glacial geology of the Thiel Mountains, Antarctica: *U.S. Geol. Survey Prof. Paper* **475**-B, B140–B143.

ANDERSON, D. L., 1960, The physical constants of sea ice: *Research* **13**, 310–18.

ANDERSON, D. M., 1968, Undercooling, freezing point depression, and ice nucleation of soil water: *Israel J. Chem.* **6**, 349–55.

1970, Phase boundary water in frozen soils: *U.S. Army Corps of Engineers, Cold Regions Research and Engineering Laboratory Research Rept.* **274**. (19 pp.)

1971, Remote analysis of planetary water: *U.S. Army Corps of Engineers, Cold Regions Research Engineering Laboratory Spec. Rept.* **154**. (13 pp.)

ANDERSON, D. M., and TICE, A. R., 1970, Low-temperature phases of interfacial water in clay-water systems: *U.S. Army Corps of Engineers, Cold Regions Research and Engineering Laboratory Research Rept.* **290**. (17 pp.)

ANDERSSON, GUNNAR, and HESSELMAN, HENRIK, 1907, Vegetation ocy flora i Hamra kronopark: Stockholm, *Statens Skogsförsoksanstalt Medd.* **4**(2), 41–110.

ANDERSSON, J. G., 1906, Solifluction, a component of subaerial denudation: *J. Geol.* **14**, 91–112.

ANKETELL, J. M., CEGŁA, JERZY, and DŻUŁYŃSKI, STANISŁAW, 1970, On the deformational structures in systems with reversed density gradients: *Rocznik Polskiego Towarzystwa Geologicznego (Soc. géol. Pologne Annales)* **40**(1), 1–30.

ANTEVS, ERNST, 1925, Retreat of the last ice-sheet in eastern Canada: *Canada Geol. Survey, Mem.* **146**. (142 pp.)

1932, *Alpine zone of Mt. Washington Range:* Auburn, Maine. (118 pp.)

ARCTIC INSTITUTE OF NORTH AMERICA, 1953-, *Arctic bibliography* **1-**.

ARE, FELIX, 1972, The reworking of shores in the permafrost zone: 78–9 in Adams, W. P., and Helleiner, F. M., eds., *International Geography 1972* **1** (Internat. Geog. Cong., 22d, Montreal): Toronto, Univ. Toronto Press. (694 pp.)

AUER, VÄINÖ, 1920, Über die Enstehung der Stränge auf den Torfmooren: *Acta Forestalia Fennica* **12**, 1–145.

AUGHENBAUGH, N. B., 1958, Preliminary report on the geology of the Dufek Massif: *Ohio State Univ. Research Found. Rept.* **825-1**(1) (IGY proj. 4.10, NSF grant Y/4.0/285), 164–208.

BAKLUND, O. O., 1911, Obshchiy obzor deyatelnosti ekspedicii bratiev Kuznetsovykh na Poliarnyy Ural letom 1909 goda: *Zapiski Akademii nauk, Fizikomatematicheskiy otdel, ser. 8,* **28**(1), 70–3.

BAKULIN, F. G., 1958, *L'distost' i osadki pri ottaivanii mnogoletnemerzlykh chetvertichnykh otlozheniĭ Vorkutskogo raĭona:* Moskva, Akad. Nauk SSSR, Inst. Merzlotovedenia im. V. A. Obrucheva. (95 pp.)

BALL, D. F., and GOODIER, R., 1970, Morphology and distribution of features resulting from frost-action in Snowdonia: *Field Studies* **3**, 193–217.

BARANOV, I. Y., 1959, Geograficheskoe rasprostranenie sezonnopromerzayushchikh pochv i mnogoletnemerzlykh gornyk porod: 193–219 (Glava VII) in *Inst. Merzlotovedeniya im. V. A. Obrucheva, Osnovy geokriologii (merzlotovedeniya), Chast' pervaya, Obshchaya geokriologiya:* Moskva, Akad. Nauk SSSR. (459 pp.)

1964, Geographical distribution of seasonally frozen ground and permafrost (Geograficheskoe rasprostranenie sezonnopromerzayushchikh pochv i mnogoletnemerzlykh gornyk porod): *Canada Natl. Research Council Tech. Translation* **1121**. (85 pp.)

BARANOV, I. J., and KUDRYAVTSEV, V. A., 1966, Permafrost in Eurasia: 98–102 in *Permafrost International Conference (Lafayette, Ind., 11–15 Nov. 1963) Proc., Natl. Acad. Sci.–Natl. Research Council Pub.* **1287**. (563 pp.)

BARSCH, DIETRICH, 1969, Studien und Messungen an Blockgletschern in Macun, Unterengadin: *Zeitschr. Geomorphologie, Supp.* **8**, 11–30.

1971, Rock glaciers and ice-cored moraines: *Geografiska Annaler* **53**A, 203–6.

BARSCH, DIETRICH, and UPDIKE, R. G., 1971, Periglaziale Formung am Kendrick Peak in Nord-Arizona während der letzten Kaltzeit: *Geog. Helvetica* **26**, 99–114.

BASHENINA, N. V., 1948, Proiskhozdenie relyefa Yuzhnogo Urala: Moskovskiy ordena Lenina Gosudarstvennyy universitet im. M. V. Lomonosova, *Nauchnoissledovatelskiy institut geografii, OPI* **2**, 1–232.

1967, *Formirovanie sovremennogo relyefa zemnoy poverkhnosti:* Moskva. (194 pp.)

BATTLE, W. R. B., 1960, Temperature observations in bergschrunds and their relationship to frost shattering: 83–95 in Lewis, W. V., ed., Norwegian cirque glaciers, *London, Royal Geog. Soc. Research Ser.* **4**. (104 pp.)

BAULIG, HENRI, 1956, Pénéplaines et pédiplaines: *Soc. belge études géog.* **25**(1), 25–58.

1957, Peneplains and pediplains: *Geol. Soc. America Bull.* **68**, 913–30.

BEHR, F. M., 1918, Über geologisch wichtige Frosterscheinungen in gemässigten Klimaten: *Deutsche geol. Gesell. Zeitschr.* **70**, B. Monatsber., 95–117.

BEHRE, C. H., JR., 1933, Talus behavior above timber in the Rocky Mountains: *J. Geol.* **41**, 622–35.

BELKNAP, R. L., 1941, Physiographic studies in the Holstensborg District of southern Greenland: 199–255 in Hobbs, W. H., ed., Reports of the Greenland Expeditions of the University of Michigan (2), *Michigan Univ. Studies, sci. ser.* **6**. (287 pp.)

BENEDICT, J. B., 1970*a*, Frost cracking in the Colorado Front Range: *Geografiska Annaler* **52A**, 87–93.

1970*b*, Downslope soil movement in a Colorado alpine region: rates, processes, and climatic significance: *Arctic and Alpine Research* **2**, 165–226.

BENNINGHOFF, W. S., 1952, Interaction of vegetation and soil frost phenomena: *Arctic* **5**, 34–44.

1966, Relationships between vegetation and frost in soils: 9–13 in *Permafrost International Conference (Lafayette, Ind., 11–15 Nov. 1963) Proc., Natl. Acad. Sci.–Natl. Research Council Pub.* **1287**. (563 pp.)

BERG, T. E., 1969, Fossil sand wedges at Edmonton, Alberta, Canada: *Biuletyn Peryglacjalny* **19**, 325–33.

BERG, T. E., and BLACK, R. F., 1966, Preliminary measurements of growth of nonsorted polygons, Victoria Land, Antarctica: 61–108 in Tedrow, J. F. C., ed., Antarctic soils and soil-forming processes, *Am. Geophys. Union Antarctic Research. Ser.* **8** (*Natl. Acad. Sci.–Natl. Research Council Pub.* **1418**). (177 pp.)

BERGER, HERFRIED, 1967, Vorgänge und Formen der Nivation in den Alpen, 2 ed.: *Buchreihe des Landemuseums für Kärnten* **17**, Klagenfurt. (89 pp.)

BESCHEL, R. L., 1966, Hummocks and their vegetation in the high arctic: 13–20 in *Permafrost International Conference (Lafayette, Ind., 11–15 Nov. 1963) Proc., Natl. Acad. Sci.–Natl. Research Council Pub.* **1287**. (563 pp.)

BESKOW, G., 1930, Erdfliessen und Strukturböden der Hochgebirge im Licht der Frosthebung: *Geol. Fören. Stockholm, Förh.* **52**, 622–38.

1935, Tjälbildningen och tjällyftningen: *Sveriges Geol. Undersökning Avh. och uppsatser, ser. C* **375** [Årsbok **26**(3)]. (242 pp.)

BESKOW, G., 1947, Soil freezing and frost heaving with special application to roads and railroads (Translated by J. O. Osterberg): *Northwestern Univ., Technol. Inst.* (145 pp.)

BIELORUSOVA, ZH. M., 1963, O nagornykh (soliflukcionnykh) terrasakh ostrova Bolshogo Liakhovskogo: 88–92 in Rutilevskiy, G. L., and Sisko, R. K., *Novosibirskie ostrova*, Trudy Arkticheskogo i Antarkticheskogo nauchnoissledovatelskogo instituta Glavnogo upravleniya severnogo morskogo puti Ministerstva morskogo flota, Leningrad.

BIRD, J. B., 1967, *The physiography of arctic Canada, with special reference to the area south of Parry Channel:* Baltimore, The Johns Hopkins Press. (336 pp.)

BIULETYN PERYGLACJALNY, 1954–, **1**–[An international journal edited by Jan Dylik and published by Łodzkie Towarzystwo Naukowe, Łodz, Poland.]

BLACK, R. F., 1952, Polygonal patterns and ground conditions from aerial photographs: *Photogrammetrical Eng.* **18**, 123–34.

1954, Permafrost – a review: *Geol. Soc. America Bull.* **65**, 839–55.

1963, Les coins de glace et le gel permanent dans le Nord de l'Alaska: *Annales Géog.* **72**(391), 257–71.

1965, Ice-wedge casts of Wisconsin: *Wisconsin Acad. Sci., Arts and Letters, Trans.* **54**, 187–222.

1969*a*, Slopes in southwestern Wisconsin, U.S.A., periglacial or temperate?: *Biuletyn Peryglacjalny* **18**, 69–82.

1969*b*, Climatically significant fossil periglacial phenomena in northcentral United States: *Biuletyn Peryglacjalny* **20**, 225–38.

1969*c*, Thaw depressions and thaw lakes: A review: *Biuletyn Peryglacjalny* **19**, 131–50.

BLACK, R. F., and BARKSDALE, W. L., 1949, Oriented lakes of northern Alaska: *J. Geol.* **57**, 105–18.

BLAGBROUGH, J. W., and BREED, W. J., 1967, Protalus ramparts on Navajo Mountain, southern Utah: *Am. J. Sci.* **265**, 759–72.

BLAGBROUGH, J. W., and FARKAS, S. E., 1968, Rock glaciers in the San Mateo Mountains, south-central New Mexico: *Am. J. Sci.* **266**, 812–23.

BLANCK, E., RIESER, A., and MORTENSEN, H., 1928, Die wissenschaftlichen Ergebnisse einer bodenkundlichen Forschungsreise nach Spitzbergen im Sommer 1926: *Chemie der Erde* **3**, 588–698.

BOBOV, N. G., 1960, Sovremennoe obrazovanie gruntovykh zhil i melkopoligonal'nogo rel'efa na Leno-Viliuiskom mezdurech'e: *Acad. Nauk SSSR, Inst. Merzlotovedeniya im. V. A. Obrucheva, Trudy* **16**, 24–9.

1970, The formation of beds of ground ice: *Soviet Geography, Review and Translation* **11**, 456–563. [From: *Izvestiya Akademii Nauk SSSR, seriya geograficheskaya*, 1969 **6**, 63–8.]

BOCH, S. G., 1946, K voprosu o granice maksimalnogo chetvertichnogo oledeniya v predelakh Uralskogo khrebta v sviazi s nabludeniyami nad nagornymi terrasami: *Chetvertichnoy komisii AN SSSR Byulleten* **8**, 46–72.

BOCH, S. G., and KRASNOV, I. I., 1943, O nagornykh terrasakh i drevnikh poverkhnostiakh vyravnivaniya na Urale i sviazannykh s nimi problemakh: *Izvestiya Vsesoyuznogo geograficheskogo obshchestva* **75**(1), 14–25.

BORNS, H. W., JR., 1965, Late glacial ice-wedge casts in northern Nova Scotia, Canada: *Science* **148**, 1223–5.

BOSTROM, R. C., 1967, Water expulsion and pingo formation in a region affected by subsidence: *J. Glaciology* **6**, 568–72.

BOUT, PIERRE, 1953, Etudes de géomorphologie dynamique en Islande: *Expéditions Polaires Françaises* **3**, Paris, Hermann & Cie, *Actualités Scientifiques et Industrielles* **1197**. (218 pp.)

BOWLEY, W. W., and BURGHARDT, M. D., 1971, Thermodynamics and stones: *EOS, Am. Geophys. Union Trans.* **52**, 4–7.

BOYÉ, MARC, 1950, Glaciaire et périglaciaire de l'Ata Sund nord-oriental, Groenland: *Expéditions Polaires Françaises* **1**, Paris, Hermann & Cie, *Actualités Scientifiques et Industrielles* **1111**. (176 pp.)

BRADLEY, W. C., 1965, Glacial geology, periglacial features and erosion surfaces in Rocky Mountain National Park: *INQUA Cong., 7th, Boulder, Colo. 1965, Guidebook Boulder area, Colorado, USA*, 27–33.

BRETZ, J H., 1935, Physiographic studies in East Greenland: 159–245 in Boyd, L. A., *The fiord region of East Greenland, Am. Geog. Soc. Spec. Pub.* **18**. (369 pp.)

BREWER, MAX, 1958, Some results of geothermal investigations of permafrost in northern Alaska: *Am. Geophys. Union Trans.* **39**, 19–26.

BREWER, R., and HALDANE, A. D., 1957, Preliminary experiments in the development of clay orientation in soils: *Soil Sci.* **84**, 301–9.

BRINK, V. C., *et al.*, 1967, Needle ice and seedling establishment in southwestern British Columbia: *Canadian J. Plant Sci.* **47**, 135–9.

BROCKIE, W. J., 1968, A contribution to the study of frozen ground phenomena – preliminary investigations into a form of miniature stone stripes in East Otago: *New Zealand Geog. Conf., 5th, Proc.*, 191–201.

BROWN, JERRY, 1969, Buried soils associated with permafrost: 115–27 in Pawluk, S., ed., *Symposium on Pedology and Quaternary Research Proc.*, Edmonton, Univ. Alberta. (218 pp.)

BROWN, JERRY, RICKARD, WARREN, and VIETOR, DONALD, 1969, The effect of disturbance on permafrost terrain: *U.S. Army Corps of Engineers, Cold Regions Research and Engineering Laboratory Spec. Rept.* **138**. (13 pp.)

BROWN, R. J. E., 1960, The distribution of permafrost and its relation to air temperature in Canada and the U.S.S.R.: *Arctic* **13**, 163–77.

1966*a*, Influence of vegetation on permafrost: 20–5 in *Permafrost International Conference (Lafayette, Ind., 11–15 Nov. 1963) Proc., Natl. Acad. Sci.–Natl. Research Council Pub.* **1287**. (563 pp.)

1966*b*, Relation between mean annual air and ground temperatures in the permafrost region of Canada: 241–7 in *Permafrost International Conference (Lafayette, Ind., 11–15 Nov .1963) Proc., Natl. Acad. Sci.–Natl. Research Council Pub.* **1287**. (563 pp.)

1967*a*, Permafrost in Canada: *Canada, Geol. Survey Map* **1246** A.

1967*b*, Comparison of permafrost conditions in Canada and the USSR: *Polar Record* **13**, 741–51.

1968, Occurrence of permafrost in Canadian peatlands: *Internat. Peat Cong., 3rd, Quebec 1968, Proc.*, 174–81.

1969, Factors influencing discontinuous permafrost in Canada: 11–53 in Péwé, T. L., ed., *The periglacial environment*, Montreal, McGill–Queen's Univ. Press. (487 pp.)

1970, *Permafrost in Canada*: Toronto, Univ. Toronto Press. (234 pp.)

1972, Permafrost in the Canadian Arctic Archipelago: *Zeitschr. Geomorphologie, Supplementband* **13**, 102–30.

BRÜNING, HERBERT, 1964, Kinematische Phasen und Denudationsvorgänge bei der Fossilation von Eiskeilen: *Zeitschr. Geomorphologie* **8**, 343–50.

BRUNNSCHWEILER, DIETER, 1962, The periglacial realm in North America during the Wisconsin glaciation: *Biuletyn Peryglacjalny* **11**, 15–27.

1964, Der pleistozäne Periglazialbereich in Nordamerika: *Zeitschr. Geomorphologie* **8**, 223–31.

BRYAN, KIRK, 1934, Geomorphic processes at high altitudes: *Geog. Rev.* **24**, 655–6.

1946, Cryopedology – the study of frozen ground and intensive frost-action with suggestions on nomenclature: *Am. J. Sci.* **244**, 622–42.

1951, The erroneous use of *tjaele* as the equivalent of perennially frozen ground: *J. Geol.* **59**, 69–71.

BRYAN, KIRK, and RAY, L. L., 1940, Geologic antiquity of the Lindenmeier site in Colorado: *Smithsonian Misc. Colln.* **99**(2). (76 pp.)

1971, Derived physical characteristics of the Antarctic Ice Sheet: *Univ. Melbourne Meteorol. Dept. Pub.* **18**. (178 pp.)

BUDD, W. F., JENSSEN, D., and RADOK, U., 1970, The extent of basal melting in Antarctica: *Polarforschung* 6, 293–306.

1971, Derived physical characteristics of the Antarctic Ice Sheet: *Univ. Melbourne Meteorol. Dept. Pub.* **18**. (178 pp.)

BÜDEL, JULIUS, 1937, Eiszeitliche und rezente Verwitterung und Abtragung im ehemals nicht vereisten Teil Mitteleuropas: *Petermanns Mitt., Ergänzungsheft* **229**. (71 pp.)

1944, Die morphologischen Wirkungen des Eiszeitklimas im gletscherfreien Gebiet: *Geol. Rundschau* **34**, 482–519.

1948, Die klima-morphologischen Zonen der Polarländer: *Erdkunde* **2**, 22–53.

1949*a*, Die räumliche und zeitliche Gliederung des Eiszeitklimas: *Naturwissenschaften* **36**, 105–12, 133–40.

1949*b*, Neue Wege der Eiszeitforschung: *Erdkunde* **3**, 82–95.

1950, Das System der klimatischen Morphologie: *Deutscher Geographentag München 1948*, **27**(4), 1–36 (65–100 of whole volume).

1951, Die Klimazonen des Eiszeitalters: *Eiszeitalter und Gegenwart* **1**, 16–26.

1953, Die 'periglazial'-morphologischen Wirkungen des Eiszeitklimas auf der ganzen Erde: *Erdkunde* **7**, 249–66.

1959, Periodische und episodische Solifluktion im Rahmen der klimatischen Solifluktionstypen: *Erdkunde* **13**, 297–314.

1960, Die Frostschutt-Zone Südost Spitzbergens: *Colloquium Geographicum* **6**. (105 pp.)

1961, Die Abtragungsvorgänge auf Spitzbergen im Umkreis der Barentsinsel: 337–75 in *Deutscher Geographentag Köln, Tagungsbericht und wissenschaftliche Abhandlungen*, Wiesbaden, Franz Steiner Verlag G.m.b.H. (407 pp.)

1963, Klima-genetische Geomorphologie: *Geog. Rundschau* **15**, 269–85.

1969, Der Eisrinden-Effekt als Motor der Tiefenerosion in der excessiven Talbildungszone: *Würzburger Geog. Arbeiten* **25**. (41 pp.)

BULL, A. J., 1940, Cold conditions and land forms in the South Downs: *Geologist's Assoc., London, Proc.* **51**, 63–71.

BUNTING, B. T., and JACKSON, R. H., 1970, Studies of patterned ground on SW Devon Island, N.W.T.: *Geografiska Annaler* **52**, 194–208.

BUTRYM, J., *et al.*, 1964, New interpretation of 'periglacial structures': *Polska Acad. Nauk Oddzial in Krakowie, Folia Quaternaria* **17**. (34 pp.)

BUTZER, K. W., 1964, *Environment and archeology*: Chicago, Aldine Publishing Co. (524 pp.)

CADY, W. M., *et al.*, 1955, The Central Kuskokwin region, Alaska: *U.S. Geol. Survey Prof. Paper* **268**. (132 pp.)

CAILLEUX, ANDRÉ, 1942, Les actions éoliennes périglaciaires en Europe: *Soc. Géol. France* **21**(1–2), (Mém. 46). (176 pp.)

1947, Caractères distinctifs des coulées de blocailles liées au gel intense: *Geol. Soc. France Comptes rendus*, 323–4.

CAILLEUX, A., and CALKIN, P., 1963, Orientation of hollows in cavernously weathered boulders in Antarctica: *Biuletyn Peryglacjalny* **12**, 147–50.

CAILLEUX, A., and TAYLOR, G., 1954, Cryopédologie, étude des sols gelés: *Actualités Scientifiques et Industrielles* **1203**, *Expéditions Polaires Françaises, Missions Paul-Emile Victor* **IV**, Paris. (218 pp.)

CAINE, N., 1967, The tors of Ben Lomond, Tasmania: *Zeitschr. Geomorphologie* **11**, 418–29.

1968, The blockfields of northeastern Tasmania: *Australian Natl. Univ., Dept. Geog. Pub.* **G/6**. (127 pp.)

1969, A model for alpine talus slope development by slush avalanching: *J. Geol.* **77**(1), 92–100.

CALKIN, P. E., 1971, Glacial geology of the Victoria valley system, Southern Victoria Land, Antarctica: 363–412 in Crary, A. P., ed., Antarctic snow and ice studies—II, *Am. Geophysical Union, Antarctic Research Ser.* **16**. (412 pp.)

CAPPS, S. R., JR., 1910, Rock glaciers in Alaska: *J. Geol.* **18**, 359–75.

CARSON, C. E., and HUSSEY, K. M., 1962, The oriented lakes of Arctic Alaska: *J. Geol.* **70**, 417–39.

CARSON, M. A., and KIRKBY, M. J., 1972, *Hillslope form and process*: Cambridge University Press. (475 pp.)

CASAGRANDE, ARTHUR, 1932, Discussion: 168–72 in Benkelman, A. C., and Olmstead, E. R., A new theory of frost heaving, *Natl. Research Council Highway Research Board, Proc. 11th Ann. Mtg.* (1), 152–77.

CASS, L. A., and MILLER, R. D., 1959, Role of the electric double layer in the mechanism of frost heaving: *U.S. Army Snow Ice and Permafrost Research Establishment Research Rept.* **49**. (15 pp.)

CEGŁA, JERZY, and DŻUŁYŃSKI, STANISŁAW, 1970, Układy niestatecznie warstwowane i ich występowanie w srodowisku peryglacjalnym: *Acta Universitatis Wratislaviensis* **124** (*Studia geograficzne* **13**), 17–42.

CHAIX, ANDRÉ, 1923, Les coulées de blocs du Parc National Suisse d'Engadine: *Le Globe, Mémoires* [*Société de géographie, Genève*] **62**. (38 pp.)

1943, Les coulées de blocs du Parc National Suisse: *Le Globe, Mémoires* [*Société de géographie, Genève*] **82**, 121–8.

CHALMERS, BRUCE, and JACKSON, K. A., 1970, Experimental and theoretical studies of the mechanism of frost heaving: *U.S. Army Corps of Engineers, Cold Regions and Research Engineering Laboratory Research Rept.* **199**. (22 pp.)

CHAMBERS, M. J. G., 1967, Investigations of patterned ground at Signy Island, South Orkney Islands—III, Miniature patterns, frost heaving and general conclusions: *British Antarctic Survey Bull.* **12**, 1–22.

1970, Investigations of patterned ground at Signy Island, South Orkney Islands—IV, Long-term experiments: *British Antarctic Survey Bull.* **23**, 93–100.

CHERNOV, A. A., and CHERNOV, G. A., 1940, *Geologicheskoye stroyenie bassey na reki Kos'-yu v Perhorshkom kraye:* Moskva.

CHURCH, R. E., PÉWÉ, T. L., and ANDRESEN, M. J., 1965, Origin and environmental significance of large-scale patterned ground, Donnelly Dome area, Alaska: *U.S. Army Cold Regions Research and Engineering Laboratory Research Rept.* **159**. (71 pp.)

CLARK, G. M., 1968, Sorted patterned ground: New Appalachian localities south of the glacial border, *Science* **161**, 355–6.

CLAYTON, LEE, and BAILEY, P. K., 1970, Tundra polygons in the northern Great Plains: *Geol. Soc. America, Abs. with Programs* **2**(6), 382.

COOK, F. A., 1956, Additional notes on mud circles at Resolute Bay, Northwest Territories: *Canadian Geographer* **8**, 9–17.

1966, Patterned ground research in Canada: 128–30 in *Permafrost International Conference (Lafayette, Ind., 11–15 Nov. 1963) Proc., Natl. Acad. Sci.–Natl. Research Council Pub.* **1287**. (563 pp.)

COOK, F. A., and RAICHE, V. G., 1962, Freeze–thaw cycles at Resolute, N.W.T.: *Geog. Bull.* **18**, 64–78.

COOKE, R. U., 1970, Stone pavements in deserts: *Assoc. Am. Geographers Annals* **60**, 560–77.

CORTE, A. E., 1961, The frost behavior of soils: laboratory and field data for a new concept—I, Vertical sorting: *U.S. Army Cold Regions Research and Engineering Laboratory Research Rept.* **85**(1). (22 pp.)

1962*a*, Vertical migration of particles in front of a moving freezing plane: *J. Geophys. Research* **67**(3), 1085–90.

1962*b*, Relationship between four ground patterns, structure of the active layer, and type and distribution of ice in the permafrost: *U.S. Army Cold Regions Research and Engineering Laboratory Research Rept.* **88**. (79 pp.)

1962*c*, The frost behavior of soils: laboratory and field data for a new concept—II, Horizontal sorting: *U.S. Army Cold Regions Research and Engineering Laboratory Research Rept.* **85**(2). (20 pp.)

1962*d*, The frost behavior of soils—II, Horizontal sorting: 44–66 in Soil behavior associated with freezing: *Natl. Acad. Sci.–Natl. Research Council Highway Research Board Bull.* **331**.

1962*e*, The frost behavior of soils—I, Vertical sorting: 9–34 in Soil behavior on freezing with and without additives: *Natl. Acad. Sci.–Natl. Research Council Highway Research Board Bull.* **317**. (34 pp.)

1963*a*, Vertical migration of particles in front of a moving freezing plane: *U.S. Army Cold Regions Research and Engineering Laboratory Research Rept.* **105**. (8 pp.)

1963*b*, Particle sorting by repeated freezing and thawing: *Science* **142**, 499–501.

CORTE, A. E., 1963*c*, Relationship between four ground patterns, structure of the active layer, and type and distribution of ice in the permafrost: *Biuletyn Peryglacjalny* **12**, 7–90.

1966*a*, Experiments on sorting processes and the origin of patterned ground: 130–5 in *Permafrost International Conference (Lafayette, Ind., 11–15 Nov. 1963) Proc., Natl. Acad. Sci.–Natl. Research Council Pub.* **1287**. (563 pp.)

1966*b*, Particle sorting by repeated freezing and thawing: *Biuletyn Peryglacjalny* **15**, 175–240.

1967, Soil mound formation by multicyclic freeze–thaw: 1333–8 in Oura, Hirobumi, ed., *Physics of snow and ice, Internat. Conference Low Temperature Science (Sapporo, Japan, 14–19 Aug. 1966) Proc.* **1**(2), Sapporo, Hokkaido Univ., 714–1414.

1969*a*, Geocryology and engineering: 119–85 in Varnes, D. J., and Kiersch, George, eds., *Reviews in engineering geology* **2**, Boulder, Colo., Geol. Soc. America. (350 pp.)

1969*b*, Formacion en el laboratorio de estructuras como pliegues por congelamiento y descongelamiento multiple: *Cuartas Jornadas Geológicas Argentinas* **1**, 215–27.

CORTE, A. E., and HIGASHI, AKIRA, 1964, Experimental research on desiccation cracks in soil: *U.S. Army Cold Regions Research and Engineering Laboratory Research Rept.* **66**. (72 pp.)

COSTIN, A. B., 1955, A note on gilgaies and frost soils: *J. Soil Sci.* **6**, 32–4.

COSTIN, A. B., *et al.*, 1964, Snow action on Mount Twynam, Snowy Mountains, Australia: *J. Glaciology* **5**(38), 219–28.

COUTARD, J.-P., *et al.*, 1970 Gélifraction expérimentale des calcaires de la Campagne de Caen; comparaison avec quelques dépôts périglaciaires de cette région: *Centre de Géomorphologie de Caen Bull.* **6**, 7–44.

CRAWFORD, C. B., and JOHNSTON, G. H., 1971, Construction on permafrost: *Canadian Geotech. J.* **8**, 236–51.

CRAIG, B. G., 1969, photograph, inside front cover: *Papers on Quaternary research in Canada (INQUA Congress, 8th, Paris, 30 Aug.–5 Sept. 1969)*, *Canada Geol. Survey.*

CROSS, C. W., HOWE, ERNEST, and RANSOME, F. L., 1905, Description of the Silverton Quadrangle: *U.S. Geol. Survey, Geol. Atlas*, Folio 120, 25.

CRUICKSHANK, J. G., and COLHOUN, E. A., 1965, Observations on pingos and other landforms in Schuchertdal, Northeast Greenland: *Geografiska Annaler* **47**A, 224–36.

CURREY, D. R., 1964, A preliminary study of valley asymmetry in the Ogotoruk Creek area, northwestern Alaska: *Arctic* **17**, 85–98.

CZEPPE, ZDZISŁAW, 1959, Uwagi o procesie wymarzania głazów: *Czasopismo Geograficzne* **30**, 195–202.

1960, Thermic differentiation of the active layer and its influence upon the frost heave in periglacial regions (Spitsbergen): *Acad. polonaise sciences Bull., sér. sci. géol. et géog.* **8**(2), 149–52.

CZEPPE, ZDZISLAW, 1966, Przebieg głównych procesow morfogenetycznych w południowozachadnim Spitsbergenie: *Uniwersytetu Jagiellońskiego, Zeszyty Naukowe* **127**, *Prace Geograficzne, zeszyt 13, Prace Instytutu Geograficznego, zeszyt* **35**. (129 pp.)

CZUDEK, TADÉĂS, 1964, Periglacial slope development in the area of the Bohemian Massif in Northern Moravia: *Biuletyn Peryglacjalny* **14**, 169–94.

CZUDEK, TADÉĂS, and DEMEK, JAROMÍR, 1961, Výzman pleistocenní kryoplanace na vývoj povrchových tvarů České vysočiny; Symposion o problémech pleistocénu: *Anthropos* **14** (*N.S.* **6**), 57–69.

1970, Thermokarst in Siberia and its influence on the development of lowland relief: *Quaternary Research* **1**, 103–20.

1971, Der Thermokarst im Ostteil des Mitteljakutischen Tieflandes: Scripta Facultatis Scientiarum Naturalium Univ. Purkynianse Brunensis, *Geographia* **1**, 1–19.

CZUDEK, TADÉĂS, DEMEK, JAROMÍR, and STEHLIK, O., 1961, Formy zvětrávání a odnosu pískovců v Hostýnských vršich a Chřibech: *Časopis pro mineralogii a geologii* **6**, 262–9.

DAHL, RAGNAR, 1966*a*, Block fields, weathering pits and tor-like forms in the Narvik mountains, Nordland, Norway: *Geografiska Annaler* **48**A, 55–85.

1966*b*, Block fields and other weathering forms in the Narvik mountains: *Geografiska Annaler* **48**A, 224–7.

DALY, R. A., 1912, Geology of the North American Cordillera at the Forty-ninth Parallel: *Canada Geol. Survey Mem.* **38** (857 pp.) (3 pts.)

DANILOVA, N. S., 1956, Gruntovye zhily i ikh proiskhozhdenie: 109–22 in Meister, L. A., ed., *Materialy k osnovam ucheniya o merzlykh zonakh zemnoi kory, Vypusk—III:* Moskva, Acad. Nauk SSSR, Inst. Merzlotovedeniya im. V. A. Obrucheva. (229 pp.)

1963, Soil wedges and their origin: 90–9 in Meister, L. A., ed., Data on the principles of the study of frozen zones in the earth's crust. *Issue III, Canada Natl. Research Council Tech. Translation* **1088**. (169 pp.)

DAVIES, J. L., 1969, *Landforms of cold climates:* Cambridge, Mass., MIT Press. (200 pp.)

DAVIES, W. E., 1961*a*, Polygonal features on bedrock, North Greenland: *U.S. Geol. Survey Prof. Paper* **424**-D, D218–19.

1961*b*, Surface features of permafrost in arid areas: *Folia Geog. Danica* **9**, 48–56. (Internat. Geog. Cong., 19th, Norden 1960, Symposium SD 2, Physical geography of Greenland.)

DAVISON, CHARLES, 1889, On the creeping of the soilcap through the action of frost: *Geol. Mag.* (Great Britain), decade 3, **6**, 255–61.

DE GEER, GERARD, 1912, A geochronology of the last 12,000 years: *Internat. Geol. Congress, 11th, Compte rendu*, 241–57.

DE GEER, GERARD, 1940, Geochronologia Suecica Principles: *K. Svenska Vetensk. Handl., ser. 3* **18**(6), Stockholm, Almqvist & Wiksells. (367 pp. and atlas.)

DEMEK, JAROMÍR, 1964*a*, Slope development in granite areas of Bohemian Massif: *Zeitschr. Geomorphologie, Supplementband* **5**, 82–106.

1964*b*, Altiplanation terraces in Czechoslovakia and their origin: *J. Czechoslovak Geog. Soc., Cong. Suppl.*, 55–65.

1964*c*, Castle koppies and tors on the Bohemian Highland (Czechoslovakia): *Biuletyn Peryglacjalny* **14**, 192–216.

1964*d*, Formy zvětrávání a odnosu granodioritu v Novohradských horách: *Zprávy Geografického ústavu ČSAV 1964* **9**, 6–15.

1965, Kryoplanační terasy v jihozápadni Anglii: *Sborník Československé společnosti zeměpisné* **70**, 272–7.

1967, Zpráva o studiu kryogenních jevů v Jakutsku: *Sborník Československé společnosti zeměpisné* **72**, 99–114.

1968*a*, Cryoplanation terraces in Yakutia: *Biuletyn Peryglacjalny* **17**, 91–116.

1968*b*, Beschleunigung der geomorphologischen Prozesse durch die Wirkung des Menschen: *Geol. Rundschau* **58**, 111–21.

1969*a*, Cryoplanation terraces, their geographical distribution, genesis and development: *Ceskoslovenské Akademie Věd Rozpravy, Răda Matematických a Přirodních Ved* **79**(4). (80 pp.)

1969*b*, Cryogene processes and the development of cryoplanation terraces: *Biuletyn Peryglacjalny* **18**, 115–25.

DEMEK, JAROMÍR., *et al.*, 1965, *Geomorfologie Českých zemí,* Praha. (335 pp.)

DEMEK, JAROMÍR, and ŠMARDA, J., 1964, Periglaciální jevy v Bulharsku: *Zprávy Geografického ústavu ČSAV 1964* **3**, 1–4.

DENGIN, I. P., 1930, Sledy drevnego oledeneniya v Yablovom khrebte i problema golcovykh terras: *Izvestiya Gosudarstvennogo Geograficheskogo Obshchestva* **62**(2), 153–87.

DENNY, C. S., 1936, Periglacial phenomena in southern Connecticut: *Am. J. Sci. 5th ser.* **32**, 322–42.

1951, Pleistocene frost action near the border of the Wisconsin drift in Pennsylvania: *Ohio J. Sci.* **51**, 116–25.

1956, Surficial geology and geomorphology of Potter County, Pennsylvania: *Geol. Survey Prof. Paper* **288**. (72 pp.)

DERRUAU, M., 1956, Les formes périglaciaires du Labrador-Ungava Central comparées à celles de l'Island Centrale: *Rev. Géomorphologie dynamique* **7**, 12–16.

DIONNE, J.-C., 1966*a*, Un type particulier de buttes gazonnées: *Rev. Géomorphologie dynamique* **16**, 97–100.

DIONNE, J.-C. 1966*b*, Formes de cryoturbation fossiles dans le sud-est du Québec: *Cahiers Géog. Québec.* **10**, 89–100.

1966*c*, Fentes en coin fossiles dans le Québec méridional: *Acad. Sci. [Paris] Comptes rendus* **262**, 24–7.

1968, Observations sur les tourbières reticulées du Lac Saint-Joan: *Mimeo.* (16 pp.)

1969, Nouvelles observations de fentes de gel fossiles sur la côte sud du Saint-Laurent: *Rev. Géog. Montreal* **23**, 307–16.

1970, Fentes en coin fossiles dans la région de Québec: *Rev. Géog. Montréal* **24**, 313–18.

1971*a*, Contorted structures in unconsolidated Quaternary deposits, Lake Saint-Jean and Saguenay regions, Quebec: *Rev. Géog. Montreal* **25**, 5–33.

1971*b*, Vertical packing of flat stones: *Canadian J. Earth Sci.* **8**, 1585–95.

DIONNE, J.-C., and LAVERDIÈRE, CAMILLE, 1967, Sur la mise en place en milieu littoral de cailloux plats posés sur la tranche: *Zeitschr. Geomorphologie* **11**, 262–85.

1972, Ice formed beach features from Lake St. Jean, Quebec: *Canadian J. Earth Sci.*, **9**, 979–90.

DOBROLYUBOVA, T. A., and SOCHKINA, E. D., 1935, Obshaya geologicheskaya karta Evropeyskoy chasti SSSR (Severnyy Ural), list 123: *Trudy Leningradskogo geologo-gidrogeodezicheskogo upravleniya* **8**.

DOLGUSHIN, L. D., 1961, *Geomorfologiya zapadnoy chasti Aldanskogo nagorya:* Moskva, 124–5.

DOMRACHEV, D. V., 1913, Dannye o klimate, pochvakh i rastitelnosti verkhnego techeniya r Tungira Yakutskoĭ obl.: *St. Petersburg, Amurskaya Ekspeditsiya Trudy* **14**.

DOSTOVALOV, B. N., and POPOV, A. I., 1966, Polygonal systems of ice-wedges and conditions of their development: 102–5 in *Permafrost International Conference* (*Lafayette, Ind., 11–15 Nov 1963*) *Proc., Natl. Acad. Sci.–Natl. Research Council Pub.* **1287**. (563 pp.)

DOW, W. G., 1964, The effects of salinity on the formation of mud-cracks: *The Compass of Sigma Gamma Epsilon* **41**, 162–6.

DREW, J. V., *et al.*, 1958, Rate and depth of thaw in arctic soils: *Am. Geophys. Union Trans.* **39**. 697–701.

DRURY, W. H., JR., 1956, Bog flats and physiographic processes in the Upper Kuskokwim River Region, Alaska: *Harvard Univ., Gray Herbarium Contr.* **178**. (130 pp.)

DÜCKER, ALFRED, 1933, 'Steinsohle' oder 'Brodelpflaster'? *Centralbl. Mineralogie*, Jahrg. **1933**(B), 264–7.

1940, Frosteinwirkung auf bindige Böden: *Strassenbau-Jahrb*, 1939/1940, 111–26.

1956, Gibt es eine Grenze zwischen frostsicheren und frostempfindlichen Lockergesteinen?: *Strasse und Autobahn* **7**(3), 78–82.

1958, Is there a dividing line between non-frostsusceptible and frostsusceptible soils?: *Canada Natl. Research Council Tech. Translation* **722**. (12 pp.)

DUMITRASHKO, N. V., 1938, Osnovnye momenty geomorfologii i molodye dvizheniya v rayone reki Verkhney Angary: *Izvestiya AN SSSR, otdeleniye matematicheskikh i estestvenykh nauk*, 399–413.

1948, Osnovnyye problemy geomorfologii i paleogeografif Baykalskoy gornoy oblasti: *Trudy Instituta geografii AN SSSR* **42**; *Materialy pogeomorfologii i paleogeografii SSSR*, 75–141.

DUNN, J. R., and HUDEC, P. P., 1965, The influence of clays on water and ice in rock pores (2), *New York State, Department Public Works Phys. Research Rept.* RR **65**-5. (149 pp.)

1966, Frost deterioration: ice or ordered water? (abs.): *Geol. Soc. America Spec. Paper* **101**, 256.

DUPARC, L., and PEARCE, F., 1905, Sur la présence de hautes terrasses dans l'Oural du Nord: *Soc. Géog. Bull.* **11**, 384–5.

DUPARC, L., PEARCE, F., and TIKANOWICH, M., 1909, Recherches géologiques et pétrographiques sur l'Oural du Nord: *Soc. Physique et histoire naturelle de Genève Mem.* **36**, 149–65.

DURY, G. H., 1965, Theoretical implications of underfit streams: *Geol. Survey Prof. Paper* **452**C. (43 pp.)

1970, General theory of meandering valleys and underfit streams: 264–75 in Dury, G. H., ed., *Rivers and river terraces*: London, Macmillan. (283 pp.)

DYBECK, M. W., 1957, An investigation of some soil polygons in central Iceland: *J. Glaciology* **3**(22), 143–5.

DYLIK, JAN, 1960, Rhythmically stratified slope waste deposits: *Biuletyn Peryglacjalny* **8**, 31–41.

1964*a*, Éléments essentiels de la notion de 'périglaciare': *Biuletyn Peryglacjalny* **14**, 111–32.

1964*b*, The essentials of the meaning of the term of 'periglacial': *Soc. Sci. et Lettres Lødz Bull.* **15**(2), 1–19.

1966, Problems of ice-wedge structures and frost-fissure polygons: *Biuletyn Peryglacjalny* **15**, 241–91.

1967, Solifluxion, congelifluxion and related slope processes: *Geografiska Annaler* **49**A, 167–77.

1969, L'action du vent pendant le dernier âge froid sur le territoire de la Pologne Centrale: *Biuletyn Peryglacjanly* **20**, 29–44.

DYLIK, J., and MAARLEVELD, G. C., 1967, Frost cracks, frost fissures and related polygons: *Mededelingen van de Geol. Stichting, nieuwe ser.* **18**, 7–21.

DŻUŁYŃSKI, STANISŁAW, 1963, Polygonal structures in experiments and their bearing on some 'periglacial phenomena': *Acad. polonaise sci. Bull., sér. sci. géol. et géog.* **11**(3), 145–50.

1965, Experiments on clastic wedges: *Acad. polonaise sci. Bull., sér. sci. géol. et géog.* **13**, 301–4.

1966, Sedimentary structures resulting from convection-like pattern of motion: *Soc. géol. Pologne Annales* **36**(1), 3–21.

EAGER, W. L., and PRYOR, W. T., 1945, Ice formation on the Alaska Highway: *Public Roads* **24**, 55–74, 82.

EAKIN, H. M., 1916, The Yukon–Koyukuk region, Alaska: *U.S. Geol. Survey Bull.* **631**. (88 pp.)
1918, The Cosna–Nowitna region, Alaska: *U.S. Geol. Survey Bull.* **667**, 50–4.
EDELSHTEYN, J. S., 1936, *Instrukciya dlya geomorfologicheskogo izucheniya i kartirovaniya Urala:* Leningrad, Vsesoyuznyi arkticheskiy institut. (91 pp.)
EKBLAW, W. E., 1918, The importance of nivation as an erosive factor, and of soil flow as a transporting agency, in northern Greenland: *Natl. Acad. Sci. Proc.* **4**, 288–93.
ELTON, C. S., 1927, The nature and origin of soil-polygons in Spitsbergen: *Geol. Soc. London Quart. J.* **83**, 163–94.
EMBLETON, CLIFFORD, and KING, C. A. M., 1968, *Glacial and periglacial geomorphology*: London, Edward Arnold; New York, St. Martin's Press. (608 pp.)
EVANS, I. S., 1970, Salt crystallization and rock weathering: A review: *Rev. Géomorphologie dynamique* **19**, 153–77.
EVERETT, D. H., 1961, The thermodynamics of frost damage to porous solids: *Faraday Soc. Trans.* **57**(7), 1541–51.
EVERETT, K. R., 1963, Slope movement in contrasting environment: *Ohio State Univ., Ph.D.* Thesis (251 pp.)
1966, Slope movement and related phenomena: 175–220 in Wilimovsky, N. J., ed., *Environment of the Cape Thompson region, Alaska*, U.S. Atomic Energy Commission, Division of Technical Information. (1250 pp.)
FARRAND, W. R., 1961, Frozen mammoths and modern geology: *Science* **133**, 729–35.
FERRIANS, O. J., 1965, Permafrost map of Alaska: *U.S. Geol. Survey Misc. Geol. Inv. Map* **I–445**.
FERRIANS, O. J., KACHADOORIAN, REUBEN, and GREENE, G. W., 1969, Permafrost and related engineering problems in Alaska: *U.S. Geol. Survey Prof. Paper* **678**. (37 pp.)
FIGURIN, A. E., 1823, Izvlechenie iz zapisok medikokhirurga Firgurina, vedennykh vo vrema opisi beregov Severo-Vostochnoĭ Sibiri: *Gosudarstv. Admiralt. Depart., Zapiski* **5**, 259–328.
FITZE, PETER, 1971, Messungen von Bodenbewegungen auf West-Spitzbergen: *Geog. Helvetica* **26**, 148–52.
FLEISHER, P. J., and SALES, J., 1971, Clastic wedges of periglacial significance, central New York (abs.): *Geol. Soc. America Abs. with Programs* **3**(1), 28–9.
FLEMAL, R. C., 1972, Ice injection origin of the DeKalb mounds, north-central Illinois, U.S.A.: 368–9 in *Internat. Geol. Cong., 24th, Montreal, Abs.* (561 pp.)
FLEMAL, R. C., HINKLEY, KENNETH, and HESSLER, J. L., 1969, Fossil pingo field in north-central Illinois: *Geol. Soc. America Abs. with Programs* **6**, 16.
FLINT, R. F., 1961, Geological evidence of cold climate: 140–55 in Nairn, A. E. M., ed., *Descriptive palaeoclimatology*: New York, Interscience. (300 pp.)
1971, *Glacial and Quaternary geology:* New York, John Wiley. (892 pp.)

FLINT, R. F., and DENNY, C. S., 1958, Quaternary geology of Boulder Mountain, Aquarius Plateau, Utah: *U.S. Geol. Survey Bull.* **1061**-D. (164 pp.)

FORSGREN, BERNT, 1968, Studies of palsas in Finland, Norway and Sweden, 1964–1966: *Biuletyn Peryglacjalny* **17**, 117–23.

FRASER, J. K., 1959, Freeze–thaw frequencies and mechanical weathering in Canada: *Arctic* **12**, 40–53.

FRENCH, H. M., 1971*a*, Ice cored mounds and patterned ground, southern Banks Island, western Canadian Arctic: *Geografiska Annaler* **53**A, 32–8.

1971*b*, Slope asymmetry of the Beaufort Plain, Northwest Banks Island, N.W.T., Canada: *Canadian J. Earth Sci.* **8**, 717–31.

1972, The role of wind in periglacial environments, with special reference to northwest Banks Island, western Canadian Arctic: 82–4 in Adams, W. P., and Helleiner, F. M., eds., *International Geography 1972*, **1** (Internat. Geog. Cong., 22d, Montreal), Toronto, Univ. Toronto Press. (694 pp.)

FRENZEL, BURKHARD, 1960*a*, Die Vegetations- und Landschaftszonen Nord-Eurasiens während der letzten Eiszeit und während der postglazialen Wärmezeit—I, Teil: *Allgemeine Grundlagen, Akad. Wiss. u. Lit. Mainz Abh., Kl. Math.-Naturwiss.* **1959**(13), 937–1099.

1960*b*, Die Vegetations- und Landschaftszonen Nord-Eurasiens während der letzten Eiszeit und während der postglazialen Wärmezeit—II, Teil: *Rekonstruktionsversuch der letzteiszeitlichen und wärmezeitlichen Vegetation Nord-Eurasiens, Akad. Wiss. u. Lit. Mainz Abh., Kl. Math.-Naturwiss.* **1960**(6), 287–453.

1967, *Die Klimaschwankungen des Eiszeitalters*, Braunschweig, Friedr. Vieweg. (296 pp.)

1968, Grundzüge der pleistozänen Vegetationsgeschichte Nord-Eurasiens: *Erdwissenschaftliche Forschung* **1**: Wiesbaden, Franz Steiner. (326 pp.)

FRENZEL, BURKHARD, and TROLL, CARL, 1952, Die Vegetationszonen des nördlichen Eurasiens während der letzten Eiszeit: *Eiszeitalter und Gegenwart* **2**, 154–67.

FRIEDMAN, J. D., *et al.*, 1971, Observations on Icelandic polygon surfaces and palsa areas. Photo interpretation and field studies: *Geografiska Annaler* **53**A, 115–45.

FRISTRUP, BØRGE, 1953, Wind erosion within the arctic deserts: *Geog. Tidsskrift* **52**, 51–65.

FRÖDIN, JOHN, 1918, Über das Verhältnis zwischen Vegetation und Erdfliessen in den alpinen Regionen des schwedischen Lappland: *Lunds Univ. Årssk., N. F. Avd. 2* **14**(24). (30 pp.)

FURRER, GERHARD, 1954, Solifluktionsformen im schweizerischen Nationalpark: *Schweizer. naturf. Gesell. Ergebnisse der wissenschaftlichen Untersuchungen des schweizerischen Nationalparks* **4** (Neue Folge) (29), 201–75.

1955, Bodenformen aus dem subnivalen Bereich: *Die Alpen* (*Les Alpes*) **31**, 146–51.

FURRER, GERHARD, 1965, Die Höhenlage von subnivalen Bodenformen: *Univ. Zürich, Habilitationsschrift, Philosophischen Fakultät—II.* (89 pp.)

1968, Untersuchungen an Strukturböden in Ostspitzbergen, ihre Bedeutung für die Erforschung rezenter und fossiler Frostmusterformen in den Alpen bzw. im Alpenvorland: *Polarforschung* **6**, 202–6.

FURRER, GERHARD, and BACHMANN, FRITZ, 1968, Die Situmetrie (Einregelungsmessung) als morphologische Untersuchungsmethode: *Geog. Helvetica* **23**, 1–14.

FYLES, J. G., 1963, Surficial geology of Victoria and Stefansson Islands, District of Franklin: *Canada Geol. Survey Bull.* **101**. (38 pp.)

GAKKELYA, G. J., and KOROTKEVICH, E. S., 1962, Severnaya Yakutiya (Fizikogeograficheskaya kharakteristika): *Trudy Arkticheskogo i antarkticheskogo nauchno-issledovatelskogo Instituta* **236**, 79.

GALLOWAY, R. W., 1961, Periglacial phenomena in Scotland: *Geografiska Annaler* **43**, 348–53.

1970, The full-glacial climate in the southwestern United States: *Assoc. Am. Geographers Annals* **60**, 245–56.

GANYESHIN, G. S., 1949, O nagornykh terrasakh v nizhnem Priamurye: *Izvestiya Veseoyuznogo geograficheskogo obshchestva* **81**(2), 254–6.

GARDNER, JIM, 1969, Snowpatches: their influence on mountain wall temperatures and the geomorphic implications: *Geografiska Annaler* **51**A, 114–20.

GEIGER, RUDOLF, 1965, *The climate near the ground*: Cambridge, Mass., Harvard Univ. Press. (611 pp.)

GEIKIE, JAMES, 1894, *The great ice age and its relation to the antiquity of man*, 3 ed.: London, Edward Stanford. (850 pp.)

GERASIMOV, I. P., and MARKOV, K. K., 1968, Permafrost and ancient glaciation: *Canada Defence Research Board Translation T499R*, 11–19.

GERDEL, R. W., 1969, Characteristics of the cold regions: *U.S. Army Corps of Engineers, Cold Regions Research and Engineering Laboratory, Cold Regions Science and Engineering Mon.* **1**A. (53 pp.)

GLEN, J. W., 1958, The flow law of ice: 171–83 in Symposium of Chamonix, *Internat. Assoc. Sci. Hydrology pub.* **47**. (394 pp.)

GOLDTHWAIT, R. P., 1940, Geology of the Presidential Range: *New Hampshire Acad. Sci. Bull.* **1**. (43 pp.)

1969, Patterned soils and permafrost on the Presidential Range (abs.): *INQUA Cong., 8th, Paris 1969, Résumés des Communications*, 150.

GORCHAKOVSKIY, P. L., 1954, Vysokogornaya vegetaciya Yaman-Tau, samoy vysokoy gory yuzhnogo Urala: *Botanicheskiy zhurnal* **39**, 827–41.

GOW, A. J., UEDA, H. T., and GARFIELD, D. E., 1968, Antarctic ice sheet: Preliminary results of first core hole to bedrock: *Science* **161**, 1011–13.

GRADWELL, M. W., 1954, Soil frost studies at a high country station—1: *New Zealand J. Sci. and Technol., sec. B* **36**, 240–57.

GRADWELL, M. W., 1957, Patterned ground at a high-country station: *New Zealand J. Sci. and Technol.* **38**(B), 793–806.

GRAVE, N. A., 1956, An archaeological determination of the age of some hydrolaccoliths (pingos) in the Chuckchee Peninsula: *Canada Defence Research Board Translation* T**218**R. (3 pp.)

1967, Temperature regime of permafrost under different geographical and geological conditions: 1339–43 in Oura, Hirobumi, ed., *Physics of snow and ice, Internat. Conference Low Temperature Science (Sapporo, Japan, 14–19 Aug. 1966) Proc.* **1**(2), Sapporo, Hokkaido Univ., 714–1414.

1968*a*, Merzlye tolshchi zemli: *Priroda,* 1968 **1**, 46–53.

1968*b*, The earth's permafrost beds: *Canada Defense Research Board Translation* T**499**R, 1–10.

GRAWE, O. R., 1936, Ice as an agent of rock weathering: A discussion: *J. Geol.* **44**, 173–82.

GRAVIS, G. F., 1964, *Stadiynost v razvitii nagornykh terras (na primere khrebta Udokan):* Voprosy geografii Zabaykalskogo Severa, 133–42.

GREGORY, K. J., 1966, Aspect and landforms in north east Yorkshire: *Biuletyn Peryglacjalny* **15**, 115–20.

GRIM, R. E., 1952, Relation of frost action to the clay-mineral composition of soil materials: 167–72 in Frost action in soils: a symposium, *Natl. Acad. Sci.–Natl. Research Council Highway Research Board Spec. Rept.* **2**. (385 pp.)

GUILCHER, A., 1950, Nivation, cryoplanation et solifluction quaternaires dans les colins de Bretagne Occidentale et du Nord de Devonshire: *Rev. Géomorphologie dynamique,* 53–78.

GUILLIEN, YVES, 1951, Les grèzes litées de Charente: *Rev. Géog. Pyrenées Sud-Ouest* **22**, 154–62.

GUILLIEN, YVES, and LAUTRIDOU, J.-P., 1970, Recherches de gélifraction expérimentale du Centre de Géomorphologie—1. Calcaires des Charentes: *Centre de Géomorphologie de Caen Bull.* **5**. (45 pp.)

GULLENTOPS, F., *et al.*, 1966, Observations géologiques et palynologiques dans la Vallée de la Lienne: *Acta Geog. Lovanensia* **4**, 192–204.

GUSSOW, W. C., 1954, Piercement domes in Canadian Arctic: *Am. Assoc. Petroleum Geologists Bull.* **38**, 2225–6.

HABRICH, WULF VON, 1968, Vegetationshöcker auf steilgeneigten Terrassenhängen in der Frostschuttzone Nordostkanadas: *Polarforschung* **6**, 212–15.

HACK, JOHN T., 1960, Origin of talus and scree in northern Virginia (abs.): *Geol. Soc. America Bull.* **71**(12), part 2, 1877–8.

HAEFELI, R., 1953, Creep problems in soils, snow and ice: *Intern. Conf. Soil Mechanics and Found. Eng., 3d, Switzerland 1953, Proc.* **3**, 238–51.

HAEFELI, R., 1954, Kriechprobleme im Boden, Schnee und Eis: *Wasser- und Energiewirtschaft* **46**(3), 51–67.

HALLSWORTH, E. G., ROBERTSON, G. K., and GIBBONS, R. F., 1955, Studies in pedogenesis in New South Wales—VII, The 'gilgai' soils: *J. Soil Sci.* **6,** 1–31.

HAMBERG, AXEL, 1915, Zur Kenntnis der Vorgänge im Erdboden beim Gefrieren und Auftauen sowie Bemerkungen über die erste Kristallisation des Eises in Wasser: *Geol. Fören. Stockholm, Förh.* **37**, 583–619.

HAMELIN, L. E., 1957, Les tourbières réticulées du Quebec–Labrador subarctique: interprétation morphoclimatique: *Cahiers Géog. de Quebec* **12**(3), 87–106.

1969, Le glaciel de Iakoutie, en Sibérie nordique: *Cahiers Géog. de Quebec* **13**(29), 205–16.

HAMELIN, L.-E., and COOK, F. A., 1967, Le périglaciaire par l'image; Illustrated glossary of periglacial phenomena: *Quebec, Les Presses de l'Université Laval.* (237 pp.)

HAMILTON, A. B., 1966, Freezing shrinkage in compacted clays: *Canadian Geotechnical J.* **3**, 1–17.

HANSEN, SIGURD, 1940, Varvighed i danske og skaanske senglaciale Aflejringer. Med. saerlig Hensyntagen til Egernsund Issøsystemet: *Danmarks Geol. Undersøgelse, 2d Raekke* **63**. (478 pp.)

HANSON, H. C., 1950, Vegetation and soil profiles in some solifluction and mound areas in Alaska: *Ecology* **31**, 606–30.

HARRIS, C., 1972, Processes of soil movement in turf-banked solifluction lobes, Okstindan, northern Norway: 155–174 in Price, R. J. and Sugden, D. E., compilers, *Polar geomorphology, Inst. British Geographers Spec. Pub.* **4**. (215 pp.)

HARRISON, S. S., 1970, Note on the importance of frost weathering in the disintegration and erosion of till in east-central Wisconsin: *Geol. Soc. America Bull.* **81**, 3407–9.

HAY, THOMAS, 1936, Stone stripes: *Geog. J.* **87**, 47–50.

HEINTZ, A. E., and GARUTT, V. E., 1965, Determination of the absolute age of the fossil remains of mammoth and woolly rhinoceros from the permafrost in Siberia by the help of radio carbon (C_{14}): *Norsk geol. tidsskr.* **45**, 73–9.

HELAAKOSKI, A. R., 1912, Havaintoja jäätymisilmiöiden geomorfologisista vaikutuksista: 1–108 in *Finland Geografiska Fören. Medd., Julkaisuja* **9**.

HELBIG, KLAUS, 1965, Asymmetrische Eiszeittäler in Süddeutschland und Ostösterreich: *Würzburger Geog. Arbeiten* **14**. (108 pp.)

HENDERSON, E. P., 1959*a*, Surficial geology of Sturgeon Lake map-area, Alberta: *Canada Geol. Survey Mem.* **303**. (108 pp.)

1959*b*, A glacial study of central Quebec–Labrador: *Canada Geol. Survey Bull.* **50**. (94 pp.)

1968, Patterned ground in southeastern Newfoundland: *Canadian J. Earth Sci.* **5**, 1443–53.

HENOCH, W.-E.-S., 1960, String-bogs in the Arctic 400 miles north of the tree-line: *Geog. J.* **126**, 335–9.

HEYWOOD, W. W., 1957, Isachsen area, Ellef Ringnes Island, District of Franklin, Northwest Territories: *Canada Geol. Survey Paper* **56-8**. (36 pp.)

HIGASHI, AKIRA, and CORTE, A. E., 1971, Solifluction: A model experiment: *Science* **171**, 480–2.

1972, Growth and development of perturbations on the soil surface due to the repetition of freezing and thawing: *Hokkaido Univ., Faculty of Eng. Mem.* **13** *suppl.*, 49–63.

HIGHWAY RESEARCH BOARD, 1948, Bibliography on frost action in soils, annotated: *Natl. Acad. Sci.–Natl. Research Council Highway Research Board Bibliography* **3**. (57 pp.)

1952*a*, Frost action in roads and airfields (*HRB Spec. Rept.* **1**): *Natl. Acad. Sci.–Natl. Research Council Pub.* **211**. (287 pp.)

1952*b*, Frost action in soils, a symposium (*HRB Spec. Rept.* **2**): *Natl. Acad. Sci.–Natl. Research Council Pub.* **213**. (385 pp.)

1957, Fundamental and practical concepts of soil freezing (*HRB Bull.* **168**): *Natl. Acad. Sci.–Natl. Research Council Pub.* **528**. (205 pp.)

1959, Highway pavement design in frost areas, a symposium: Part 1, Basic considerations (*HRB Bull.* **225**): *Natl. Acad. Sci.–Natl. Research Council Pub.* **685**. (131 pp.)

1962, Soil behavior associated with freezing (*HRB Bull.* **331**): *Natl. Acad. Sci.–Natl. Research Council Pub.* **1013**. (115 pp.)

1970, Frost action: Bearing, thrust, stabilization, and compaction (*HRB Record* **304**): *Natl. Acad. Sci.–Natl. Acad. Eng.* (51 pp.)

HOBBS, W. H., 1931, Loess, pebble bands, and boulders from glacial outwash of the Greenland continental glacier: *J. Geol.* **39**, 381–5.

HOEKSTRA, PIETER, 1969, Water movement and freezing pressures: *Soil Sci. Amer. Proc.* **33**, 512–18.

HOEKSTRA, PIETER, and MILLER, R. D., 1965, Movement of water in a film between glass and ice: *U.S. Army Corps of Engineers, Cold Regions Research Engineering Laboratory Research Rept.* **153**. (8 pp.)

HÖGBOM, BERTIL, 1910, Einige Illustrationen zu den geologischen Wirkungen des Frostes auf Spitzbergen: *Uppsala Univ., Geol. Inst. Bull.* **9**, 41–59.

1914, Über die geologische Bedeutung des Frostes: *Uppsala Univ., Geol. Inst. Bull.* **12**, 257–389.

HÖGBOM, IVAR, 1923, Ancient inland dunes of northern and middle Europe: *Geografiska Annaler* **5**, 113–243.

HÖLLERMANN, P. W., 1964, Rezente Verwitterung, Abtragung und Formenschatz in den Zentralalpen am Beispiel des oberen Suldentales (Ortlergruppe): *Zeitschr. Geomorphologie, Supplementband* **4**. (257 pp.)

1967, Zur Verbreitung rezenter periglazialer Kleinformen in den Pyrenäen und Ostalpen: *Göttinger Geog. Abh.* **40**. (198 pp.)

HOLMES, G. E., HOPKINS, D. M., and FOSTER, H. L., 1968, Pingos in central Alaska: *U.S. Geol. Survey Bull.* **1241**-H. (40 pp.)

HOLMQUIST, P. J., 1898, Ueber mechanische Störungen und chemische Umsetzungen in dem Bänderthon Schwedens: *Uppsala Univ., Geol. Inst. Bull.* **3**, 412–32.

HOPKINS, D. M., 1949, Thaw lakes and thaw sinks in the Imuruk Lake area, Seward Peninsula, Alaska: *J. Geol.* **57**, 119–31.

HOPKINS, D. M., and SIGAFOOS, R. S., 1951, Frost action and vegetation patterns on Seward Peninsula, Alaska: *U.S. Geol. Survey Bull.* **974**-C, 51–100.

1954, Role of frost thrusting in the formation of tussocks: *Am. J. Sci.* **252**, 55–9.

HOPKINS, D. M., *et al.*, 1955, Permafrost and ground water in Alaska: *U.S. Geol. Survey Prof. Paper* **264**-F, 113–46.

HORBERG, LELAND, 1951, Intersecting minor ridges and periglacial features in the Lake Agassiz Basin, North Dakota: *J. Geol.* **59**, 1–18.

HÖVERMANN, J., 1953, Die Periglazial-Erscheinungen im Harz: *Göttingen Geog. Abh.* **14**, 1–39.

1960, Über Strukturböden im Elburs (Iran) und zur Frage des Verlaufs der Strukturbodengrenze: *Zeitschr. Geomorphologie* **4**, 173–4.

HOWE, ERNEST, 1909, Landslides in the San Juan Mountains, Colorado: including a consideration of their causes and their classification: *U.S. Geol. Survey Prof. Paper* **67**. (58 pp.)

HOWE, JOHN, 1971, Temperature readings in test bore holes: *Mt. Washington Observatory News Bull.* **12**(2), 37–40.

HOWELL, J. V., Chm., 1960, Glossary of geology and related sciences, 2 ed.: Washington, D.C., *Am. Geol. Inst.* (325 pp. with supplement.)

HRÁDEK, M., 1967, Drobné tvary v pegmatitu Čertových kamenů v Hrubém Jeseníku: *Zprávy Geografichkého ústavu ČSAV* **3**, 1–8.

HUGHES, O. L., 1969, Distribution of open-system pingos in central Yukon Territory with respect to glacial limits: *Canada Geol. Survey Paper* **69-34**. (8 pp.)

HUGHES, T. M., 1884, On some tracks of terrestrial and freshwater animals: *Geol. Soc. London Quart. J.* **40**, 178–86.

HUME, J. D., and SCHALK, MARSHALL, 1964, The effects of ice-push on arctic beaches: *Am. J. Sci.* **262**, 267–73.

HUNT, C. B., and WASHBURN, A. L., 1966, Patterned ground: B104–33 in Hunt, C. B. *et al.*, Hydrologic basin, Death Valley, California: *U.S. Geol. Survey Prof. Paper* **494**-B. (138 pp.)

HUSSEY, K. M., 1962, Ground patterns as keys to photointerpretation of arctic terrain: *Iowa Acad. Sci.* **69**, 332–41.

HUSSEY, K. M., and MICHELSON, R. W., 1966, Tundra relief features near Point Barrow, Alaska: *Arctic* **19**, 162–84.

HUXLEY, J. S., and ODELL, N. E., 1924, Notes on surface markings in Spitsbergen: *Geog. J.* **63**, 207–29.

ILYIN, R. S., 1934, Nagornyye terasy i kurumy: *Izvestiya Gosudarstvenogo Geograficheskogo obshchestva* **66**(4), 621–5.

INGLIS, D. R., 1965, Particle sorting and stone migration by freezing and thawing: *Science* **148**, 1616–17.

INSTITUT MERZLOTOVEDENIYA im. V. A. OBRUCHEVA, 1956, *Osnovnyye ponyatiya i terminy geokriologii (merzlotovedeniya)*: Moskva, Akad. Nauk, SSSR. (16 pp.)

1960, Fundamental concepts and terms in geocryology (permafrost studies) (Osnovnyye ponyatiya i terminy geokriologii (merzlotovedeniya)): *U.S. Army Corps of Engineers, Arctic Construction and Frost Effects Laboratory Translation* **28**. (11 pp.)

IVAN, A., 1965, Zpráva o výzkumu kryoplanačních teras v severozápadni části Rychlebských hor: *Zprávy Geografického ústavu ČSAV* **7**(**146**-B), 1–3.

IVES, J. D., 1966, Block fields, associated weathering forms on mountain tops and the Nunatak hypothesis: *Geografiska Annaler* **48**A, 220–3.

IVES, J. D., and FAHEY, B. D., 1971, Permafrost occurrence in the Front Range, Colorado Rocky Mountains, U.S.A.: *J. Glaciology* **10**, 105–11.

IVES, R. L., 1940, Rock glaciers in the Colorado Front Range: *Geol. Soc. America Bull.* **51**, 1271–94.

JÄCKLI, HEINRICH, 1957, Gegenwartsgeologie des bündnerischen Rheingebietes: *Beiträge zur Geologie der Schweiz, Geotechnische Ser.* **36**. (136 pp.)

JACKSON, K. A., and CHALMERS, BRUCE, 1956, Study of ice formation in soils: *U.S. Army Corps of Engineers New England Division, Arctic Construction and Frost Effects Laboratory Tech. Rept.* **65**. (29 pp.)

JACKSON, K. A., and UHLMANN, D. R., 1966, Particle sorting and stone migration due to frost heave: *Science* **152**, 545–6.

JAHN, ALFRED, 1948*a*, Badania nad strukturą i temperaturą gleb w Zachodniej Grenlandii: *Polska Akad. Umiejętności, Rozprawy Wydž., matem.-przyr.* **72**A, (6). (121 pp., 63–183 of whole volume.)

1948*b*, Badania nad struktura i temperaturą gleb w Grenlandii zachodniej – Research on the structure and temperature of the soils in western Greenland: *Acad. polonaise sci. et lettres Bull., cl. sci. math. et nat.–Ser. A, sci. math. N° sommaire A*, 1940–1946, 50–9.

1960, Some remarks on evolution of slopes on Spitsbergen: *Zeitschr. Geomorphologie, Supplementband* **1** (*Morphologie des versants*), 49–58.

1961*a*, Quantitative analysis of some periglacial processes in Spitsbergen: *Universytet Wrocławski im. Bolesława Bieruta, zeszyty naukowe, nauki przyrodnicze, ser. B* **5** (Nauka o Ziemi II), 1–34.

1961*b*, Problemy geograficzne Alaski wo świetle podrózi naukowej odbytej v 1960 roku: *Czasopisno Geograficzne* **32**(2), 115–81.

JAHN, ALFRED, 1965, Formy i procesy stokowe w Karkonoszach: *Opera Corcontica* **2**, 7–16.

1966, *Alaska:* Warszawa, Panstwowe Wydawanictwo Naukowe. (498 pp.)

1970, Zagadnienia strefy peryglacjalnej: Warszawa, Panstwowe Wydawnictwo Naukowe. (202 pp.)

JAHN, ALFRED, and CZERWIŃSKI, JANUSZ, 1965, The rôle of impulses in the process of periglacial soil structure formation: *Acta Universitatis Wratislaviensis* **44** (*Studia Geograficzne* **7**), 1–13.

JESSUP, R. W., 1960, The stony tableland soils of the southeastern portion of the Australian arid zone and their evolutionary history: *J. Soil Sci.* **11**(2), 188–96.

JOHANSSON, SIMON, 1914, Die Festigkeit der Bodenarten bei verschiedenem Wassergehalt nebst Vorschlag zu einer Klassifikation: *Sveriges Geol. Undersökning Årsbok* **7** (3) (avh. och uppsatser, ser. C, 256). (110 pp.)

JOHNSSON, G., 1959, True and false ice-wedges in southern Sweden: *Geografiska Annaler* **41**, 15–33.

1962, Periglacial phenomena in southern Sweden: *Geografiska Annaler* **44**, 378–404.

JÜNGST, H., 1934, Zur geologischen Bedeutung der Synärese: *Geol. Rundschau* **25**, 312–25.

KACHURIN, S. P., 1959, Kriogennye fiziko-geologicheskie yavleniya v raionakh s mnogoletnemerzlymi porodami: 365–98 (Glava XI) in *Inst. Merzlotovedeniya im. V. A. Obrucheva, Osnovy geokriologii (merzlotovedeniya), Chast' pervaya, Obshchaya geokriologiya:* Moskva, Akad. Nauk SSSR. (459 pp.)

1961, *Termakarst na territorii SSSR* (Thermokarst on the territory of the USSR): Moscow, Acad. Nauk. (291 pp.)

1962, Thermokarst within the territory of the U.S.S.R.: *Biuletyn Peryglacjalny* **11**, 49–55.

1964, Cryogenic physico-geological phenomena in permafrost regions (Kriogennye fiziko-geologicheskie yavleniya v raionakh s mnogoletnemerzlymi porodami): *Canada Natl. Research Council Tech. Translation* **1157**. (91 pp.)

KAISER, KARLHEINZ, 1960, Klimazeugen des periglazialen Dauerfrostbodens in Mittel- und Westeuropa: *Eiszeitalter und Gegenwart* **11**, 121–41.

KALETSKAYA, M. W., and MIKLUKHO-MAKLAY, A. D., 1958, Nekotorye cherty chetvertichnoy istorii vostochnoy chasti pechorskogo basseyna i zapadnogo sklona Poliarnogo Urala: *Trudy Instituta geografii AN SSSR* **76**; *Materialy po geomorfologii i paleografii SSSR* **20**, 1–67.

KAPLAR, C. W., 1965, Stone migration by freezing of soil: *Science* **149**, 1520–21.

1969, Phenomena and mechanism of frost heaving: *Highway Research Board Annual Mtg., 49th, Washington, D.C. 1970, preprint.* (44 pp.)

1970, Phenomenon and mechanism of frost heaving: 1–13 in *Highway Research Board, Frost action: Bearing, thrust, stabilization, and compaction* (*HRB Record* **304**): Natl. Acad. Sci.–Natl. Acad. Eng. (51 pp.)

KAPLAR, C. W., 1971, Experiments to simplify frost susceptibility testing of soils: *U.S. Army Corps of Engineers, Cold Regions Research Engineering Laboratory Tech. Rept.* **223**. (23 pp.)

KAPLINA, T. N., 1965, Kriogennye sklonovye protsessy: Moskva, Izdatel'stvo 'Nauka'. (296 pp.)

KATASONOV, E. M., 1972, Regularities in cryogenic phenomena development: 34–5 in Adams, W. P., and Helleiner, F. M., eds., *International Geography 1972* **1** (Internat. Geog. Cong., 22, Montreal), Toronto, Univ. Toronto Press. (694 pp.)

KATASONOV, E. M., and SOLOV'EV, P. A., 1969, Guide to trip round Central Yakutia: *Internat. Symposium, Paleogeography and Periglacial Phenomena of Pleistocene*, Yakutsk. (88 pp.)

KELLER, B., 1910, Po dolinam i goram Altaya: *Trudy pochvennobotanicheskoy ekspedicii po issledovaniyam Aziatskoy Rosii, Chast II., Botanicheskie issledovaniya*, 233–4.

KELLETAT, DIETER, 1969, Verbreitung und Vergesellschaftung rezenter Periglazialerscheinungen im Appenin: *Göttinger Geog. Abh.* **48**. (114 pp.)

1970*a*, Rezente Periglazialerscheinungen im Schottischen Hochland. Untersuchungen zu ihrer Verbreitung und Vergesellschaftung: *Göttinger Geog. Abh.* **51**, 67–140.

1970*b*, Zum Problem der Verbreitung, des Alters und der Bildungsdauer alter (inaktiver) Periglazialerscheinungen im Schottischen Hochland: *Zeitschr. Geomorphologie* **14**, 510–19.

KENDREW, W. G., 1941, *The climates of the continents*, 3 ed.: London, Oxford Univ. Press. (473 pp.)

KENNEDY, BARBARA, and MELTON, M. A., 1967, Stream-valley asymmetry in an arctic–subarctic environment: Conditions governing the geomorphic processes: *Arctic Inst. North Am. Research Paper* **42**. (41 pp.)

1972, Valley assymetry and slope forms of a permafrost area in the Northwest Territories, Canada: 107–121 in Price, R. J. and Sugden, D. E., compilers, *Polar geomorphology, Inst. British Geographers Spec. Pub.* **4**. (215 pp.)

KERFOOT, D. E., 1972, Thermal contraction cracks in an arctic tundra environment: *Arctic*, **25**, 142–50.

KESSELI, J. E., 1941, Rock streams in the Sierra Nevada, California: *Geog. Rev.* **31**, 203–27.

KESSLER, PAUL, 1925, *Das eiszeitliche Klima und seine geologischen Wirkungen im nicht vereisten Gebiet*: Stuttgart, E. Schweizerbart'sche. (204 pp.)

KHESTHOVA, T. N., *et al.*, 1961, Issledovanie sloya seznnogo Promerzaniya i protaivaniya pochv (Gornykh porod): 44–69 in *Polevye geokriologicheskie (merzlotnye) issledovaniya.* Chast' I. *Geokriologicheskaya s'emka:* Moskva, Akad. Nauk SSSR, Inst. Merzlotovedeniya im. V. A. Obrucheva. (423 pp.)

1969, The layer of seasonal freezing and thawing of soil (rock): 3–24 in *Permafrost investigations in the field: Canada Natl. Research Council Tech. Translation* **1358**. (227 pp.)

KINDLE, E. M., 1917, Some factors affecting the development of mud-cracks: *J. Geol.* **25**, 135–44.

KING, C. A. M., and HIRST, R. A., 1964, The boulder-fields of the Åland Islands: *Fennia* **89**(2). (41 pp.)

KING, R. B., 1971, Boulder polygons and stripes in the Cairngorm Mountains, Scotland: *J. Glaciology* **10**, 375–86.

KLATKA, TADEUSZ, 1961*a*, Problèmes des sols striés de la partie septentrionale de la presqu'île de Sorkapp (Spitsbergen): *Biuletyn Peryglacjalny* **10**, 291–320.

1961*b*, Indices de structure et de texture des champs de pierres des Lysógory: *Soc. Sci. et Lettres Łodz Bull.* **12**(10), 1–21.

1962, Geneza i wiek głoborzy Łysogorskich: *Geographica Lodziendzia Acta* **12**. (129 pp.)

1968, Microrelief of slopes in the coastal area south of Hornsund, Vestspitsbergen: 265–81 in Birkenmajer, K., ed., *Polish Spitsbergen Expeditions 1957–1960*, Warsaw, Polish Academy of Sciences. (466 pp.)

KNECHTEL, M. M., 1951, Giant playa-crack polygons in New Mexico compared with arctic tundra-crack polygons (abs.): *Geol. Soc. America Bull.* **62**, 1455.

1952, Pimpled plains of eastern Oklahoma: *Geol. Soc. America Bull.* **63**, 689–700.

KOERNER, R. M., 1961, Glaciological observations in Trinity Peninsula, Graham Land, Antarctica: *J. Glaciology* **3**, 1063–74.

KÖPPEN, W., 1936, Das geographische System der Klimate: C1–44 in Köppen, W., and Geiger, R., eds., *Handbuch der Klimatologie* **1**, Berlin, Gebrüder Borntraeger.

KÖPPEN-GEIGER, 1954, *Klima der Erde – Climate of the earth* (map – 1:16,000,000): Darmstadt, Germany, Justus Perthes.

KORNILOV, B. A., 1962, *Relyef yugovostochnoy okrainy Aldanskogo nagorya*, Moskva. (96 pp.)

KOROFIEYEV, N. V., 1939, K voprosu genezisa nagornykh terras: *Problemy Arktiky* **6**, 89–91.

KORZHUYEV, S. S., 1959, Geomorfologiya severozapadnoy chasti Stanovogo khrebta i yeye yuzhnogo obramleniya: *Trudy Instituta geografii AN SSSR* **78**, 74–123.

KOSTYAEV [KOST'JAEV], A. G., 1966, Über die Grenze der unterirdischen Vereisung und die Periglazialzone im Quartär: *Petermanns Mitt.* **110**(4), 253–61.

1969, Wedge- and fold-like diagenetic disturbances in Quaternary sediments and their paleogeographic significance: *Biuletyn Peryglacjalny* **19**, 231–70.

KOZLOV, M. T., 1966, K voprosu ob obrazovanii nagornykh terras: *Formirovanie relyera i chetvertichnykh otlozheniy Kolskogo Poluostrova*, 126–32.

KOZMIN, N. M., 1890, O lednikovykh yavleniyakh v Olekminsko-Vitimskoy gornoy strane i o svyazi ikh s obrazovaniyem zolotnoshykh rossypey: *Izvestiya Vostochno – Sibirskogo otdela imperatorskogo Russkogo Geograficheskogo obshchestva* **21**(1).

KRÁL, V., 1968, Geomorfologie vrcholové oblasti Krušných hor a problém paroviny: *Rozpravy ČSAV, řada MPV* **78**(9), 1–65.

KRASNOV, I. I., and KOZLOVSKAYA, S. F., 1966, *Geologiya Sibirskoy platformy*, Moskva, 399–419.

KROPACHEV, A. M., and KROPACHEVA, T. S., 1956, Nagornyye terrasy odnogo iz rayonov vostochnogo Zapoliarya: *Nauchnye trudy, Ministerstvo vysshego obrazovaniya SSSR, Molotovksoy Gornyy institut, Sbornik* **1**, 126–35.

KRUMME, OSKAR, 1935, Frost und Schnee in ihrer Wirkung auf den Boden im Hochtaunus: *Rhein-Mainische Forschungen* **13**. (73 pp.)

KUNSKÝ, J., and LOUČEK, D., 1956, Stone stripes and thufurs in the Krkonoše: *Biuletyn Peryglacjalny* **4**, 345–9.

KUSHEV, S. L., 1957, Relyef i prirodnyye usloviya Tuvinskoy avtonomnoy oblasti: *Trudy Tuvinskoy kompleksnoy ekspedicii* **3**, 11–45.

LABA, J. T., 1970, Lateral thrust in frozen granular soils caused by temperature change: 27–37 in *Highway Research Board, Frost action: Bearing, thrust, stabilization, and compaction* (*HRB Record* **304**): Natl. Acad. Sci.–Natl. Acad. Eng. (51 pp.)

LACHENBRUCH, A. H., 1959, Periodic heat flow in a stratified medium with application to permafrost problems: *U.S. Geol. Survey Bull.* **1083**-A, 1–36.

1961, Depth and spacing of tension cracks: *J. Geophys. Res.* **66**(12), 4273–92.

1962, Mechanics of thermal contraction cracks and ice-wedge polygons in permafrost: *Geol. Soc. America Spec. Paper* **70**. (69 pp.)

1966, Contraction theory of ice wedge polygons: A qualitative discussion: 63–71 in *Permafrost International Conference* (*Lafayette, Ind., 11–15 Nov. 1963*) *Proc., Natl. Acad. Sci.–Natl. Research Council Pub.* **1287**. (563 pp.)

1970*a*, Some estimates of the thermal effects of a heated pipeline in permafrost: *U.S. Geol. Survey Circ.* **632**. (23 pp.)

1970*b*, Thermal considerations in permafrost: J1–2 and discussion; J2–5 in Adkison, W. L., and Borsge, M. M., eds., *Geological Seminar on the North Slope of Alaska Proc.*: Los Angeles, Am. Assoc. Petroleum Geologists, Pacific Sec., A1–R10.

LACHENBRUCH, A. H., and MARSHALL, B. V., 1969, Heat flow in the Arctic: *Arctic* **22**, 300–11.

LAMAKINY, V. V., and LAMAKINY, N. V., 1930, Sayano-Dzhidinskoye nagorye: *Zemlevedenie* **32**(1–2), 21–54.

LAMBE, T. W., and KAPLAR, C. W., 1971, Additives for modifying the frost susceptibility of soils (1): *U.S. Army Corps of Engineers, Cold Regions Research and Engineering Laboratory Tech. Rept.* **123**(1). (41 pp.)

LANG, W. B., 1943, Gigantic drying cracks in Animas Valley, New Mexico: *Science*, New Ser. **98**, 583–4.

LASCA, N. P., 1969, The surficial geology of Skeldal, Mesters Vig, Northeast Greenland: *Medd. om Grønland* **176**(3). (56 pp.)

LAUTRIDOU, J. P., 1971, Conclusions generales des expériences de gélifraction expérimentale: *Centre de Géomorphologie, Bull.* **10**, 65–84.

LAZAREV, P. A., 1961, Kratkiy geomorfologicheskiy ocherk khrebta Tuora-Sis: *Voprosy geografii Yakutii* **1**, 5–11.

LEE, H. A., 1957, Surficial geology of Fredericton, York and Sunbury counties, New Brunswick: *Canada Geol. Survey Paper* **56-2**. (11 pp.)

LEFFINGWELL, E. DE K., 1915, Ground-ice wedges; the dominant form of ground-ice on the north coast of Alaska: *J. Geol.* **23**, 635–54.

1919, The Canning River Region, northern Alaska: *U.S. Geol. Survey Prof. Paper* **109**. (251 pp.)

LEGGET, R. F., BROWN, R. J. E., and JOHNSTON, G. H., 1966, Alluvial fan formation near Aklavik, Northwest Territories, Canada: *Geol. Soc. America Bull.* **77**, 15–29.

LEWIS, C. A., 1966, The nivational landforms and the reconstructed snowline of Slaettaratindur, Faeroe Islands: *Biuletyn Peryglacjalny* **15**, 293–302.

LEWIS, W. V., 1939, Snowpatch erosion in Iceland: *Geog. J.* **94**, 153–61.

LINELL, K. A., and KAPLAR, C. W., 1959, The factor of soil and material type in frost action: *Natl. Acad. Sci.–Natl. Research Council Highway Research Board Bull.* **225**, 81–128.

LINTON, D. L., 1969, The abandonment of the term 'periglacial': 65–70 in van Zinderen Bakker, E. M., ed., *Palaeoecology of Africa and of the surrounding islands and Antarctica* **5**: Cape Town, Balkema. (240 pp.)

LONGWELL, C. R., 1928, Three common types of desert mud-cracks: *Am. J. Sci.*, 5th ser. **15**, 136–45.

LONGWELL, C. R., FLINT, R. F., and SANDERS, J. E., 1969, *Physical geology*: New York, John Wiley. (685 pp.)

LONGWELL, C. R., KNOPF, A., and FLINT, R. F., 1939, *A textbook of geology, part I–Physical geology*, 2 (revised) ed.: New York, John Wiley. (543 pp.)

LOUIS, HERBERT, 1930, Morphologische Studien in Südwest-Bulgarian: *Geog. Abh.* [*Pencks*], 3rd ser. **2**. (119 pp.)

LÖVE, DORIS, 1970, Subarctic and subalpine: Where and what?: *Arctic and Alpine Research* **2**, 63–73.

LOVELL, C. W., JR., 1957, Temperature effects on phase composition and strength of partially-frozen soil: *Natl. Acad. Sci.–Natl. Research Council, Highway Research Board Bull.* **168**, 74–95.

LOW, P. F., HOEKSTRA, PIETER, and ANDERSON, D. M., 1967, Some thermodynamic relationships for soils at or below the freezing point **2**: Effects of temperature and pressure on unfrozen soil water: *U.S. Army Cold Regions Research and Engineering Laboratory Research Rept.* **222**. (5 pp.)

ŁOZIŃSKI, W., 1909, Über die mechanische Verwitterung der Sandsteine im gemässigten Klima: *Acad. sci. cracovie Bull. internat., cl. sci. math. et naturelles* **1**, 1–25.

LOZIŃSKI, W., 1912, Die periglaziale Fazies der mechanischen Verwitterung: *Internat. Geol. Congress, 11th, Stockholm 1910, Compte rendu*, 1039–53.

LUNDQVIST, G., 1949, The orientation of the block material in certain species of earth flow: 335–47 in *Glaciers and Climate*: *Geographiska Annaler, 1949* **1–2**.

LUNDQVIST, JAN, 1962, Patterned ground and related frost phenomena in Sweden: *Sveriges Geol. Undersökning Årsbok* **55**(7) (*Avh. och uppsatser, ser C* **583**). (101 pp.)

1969, Earth and ice mounds: a terminological discussion: 203–15 in Péwé, T. L., *The periglacial environment*: Montreal, McGill–Queen's Univ. Press. (487 pp.)

LYFORD, W. H., GOODLETT, J. C., and COATES, W. H., 1963, Landforms, soils with fragipans, and forest on a slope in the Harvard Forest: *Harvard Forest Bull.* **30**. (68 pp.)

LYUBIMOVA, E. L., 1955, Botaniko-geograficheskie issledovaniya yuzknoy chasti Pripoliarnogo Urala: *Trudy Instituta geografii AN SSSR* **64**, 201–41.

MAARLEVELD, G. C., 1965, Frost mounds, a summary of the literature of the past decade: *Mededelingen van de Geologische Stichting, Nieuwe Serie* **17**. (16 pp.)

MACCARTHY, G. R., 1952, Geothermal investigations on the arctic slope of Alaska: *Am. Geophys. Union Trans.* **33**(4), 589–93.

MACKAY, J. R., 1953, Fissures and mud circles on Cornwallis Island, N.W.T.: *The Canadian Geographer* **3**, 31–7.

1958*a*, Arctic 'vegetation arcs': *Geog. J.* **124**, 294–5.

1958*b*, The Anderson River Map-Area, N.W.T.: *Canada, Dept. Mines and Technical Surveys, Geog. Branch Mem.* **5**. (137 pp.)

1962, Pingos of the Pleistocene Mackenzie River delta area: *Geog. Bull.* **18**, 21–63.

1963*a*, The Mackenzie Delta area, N.W.T.: *Canada, Mines and Tech. Surveys, Geog. Branch Mem.* **8**. (202 pp.)

1963*b*, Origin of the pingos of the Pleistocene Mackenzie Delta area: *Canada Natl. Research Council, Associate Comm. Soil and Snow Mechanics, Tech. Memo.* **76**, 79–83.

1963*c*, Progress of break-up and freeze-up along the Mackenzie River: *Geog. Bull.* **19**, 103–16.

1965, Gas-domed mounds in permafrost, Kendall Island, N.W.T.: *Geog. Bull.* **7**, 105–15.

1966*a*, Pingos in Canada: 71–6 in *Permafrost International Conference (Lafayette, Ind., 11–15 Nov. 1963) Proc., Natl. Acad. Sci.–Natl. Research Council Pub.* **1287**. (563 pp.)

1966*b*, Segregated epigenetic ice and slumps in permafrost, Mackenzie Delta area, N.W.T.: *Geog. Bull.* **8**, 59–80.

1970, Disturbances to the tundra and forest tundra environment of the western Arctic: *Canadian Geotech. J.* **7**, 420–32.

MACKAY J. R., 1971*a*, Ground ice in the active layer and the top portion of permafrost: 26–30 in Brown, R. J. E., ed. *Proceedings of a seminar on the permafrost active layer, 4 and 5 May 1971, Canada Natl. Research Council Tech. Memo.* **103**. (63 pp.)

1971*b*, The origin of massive icy beds in permafrost, western arctic coast, Canada: *Canadian J. Earth Sci.* **8**, 397–422.

1972*a*, The world of underground ice: *Assoc. Am. Geog. Annals* **62**, 1–22.

1972*b*, Some observations on growth of pingos: 141–7 in Kerfoot, D. E., ed., *Mackenzie Delta area monograph:* Internat. Geog. Cong., 22d, Montreal, 1972: Brock Univ. (174 pp.)

1972*c*, Some observations on ice-wedges, Garry Island, N.W.T.: 131–9 in Kerfoot, D. E., ed., *Mackenzie Delta area monograph:* Internat. Geog. Cong., 22d, Montreal, 1972: Brock Univ. (174 pp.)

MACKAY, J. R., RAMPTON, V. N., and FYLES, J. G., 1972, Relic Pleistocene permafrost, Western Arctic, Canada: *Science* **176**, 1321–3.

MACKAY, J. R., and STAGER, J. K., 1966, The structure of some pingos in the Mackenzie Delta area, N.W.T.: *Geog. Bull.* **8**(4), 360–8.

MACKIN, J. H., 1947, Altitude and local relief of Bighorn area during the Cenozoic: *Wyoming Geol. Assoc. Field Conf., Bighorn Basin, Guidebook*, 103–20.

MAKEROV, J., 1913, Nagornyye terrasy v Sibirii i proiskhozhdenie yikh: *Izvestiya geologicheskogo komiteta* **32**(8), 761–801.

MALAURIE, JEAN, 1952, Sur l'asymétrie des versants dans l'île de Disko, Groenland: *Acad. Sci. [Paris] Comptes rendus* **234**, 1461–2.

1968, Thèmes de recherche géomorphologique dans le Nord-Ouest du Groenland: *Centre de Recherches et Documentation Cartographiques et Géographiques* [*Paris*], *Mémoires et Documents, Numéro hors série*. (495 pp.)

MALAURIE, JEAN, and GUILLIEN, YVES, 1953, Le modelé cryo-nival des versants meubles de Skansen (Disko, Groenland). Interprétation générale des grèzes litées: *Soc. Géol. France Bull., ser. 6* **3**, 703–21.

MALDE, H. E., 1961, Patterned ground of possible solifluction origin at low altitude in the western Snake River Plain, Idaho: *U.S.G.S. Prof. Paper* **424**-B, B-170–3.

1964, Patterned ground in the western Snake River Plain, Idaho, and its possible cold-climate origin: *Geol. Soc. America Bull.* **75**, 191–208.

MANGERUD, JAN, and SKREDEN, S. A., 1972, Fossil ice wedges and ground wedges in sediments below the till at Voss, western Norway: *Norsk Geol. Tidsskr.* **52**, 73–96.

MARBUT, C. F., and WOODWORTH, J. B., 1896, The clays about Boston: 989–98 in Shaler, N. S., Woodworth, J. B., and Marbut, C. F., Glacial brick clays of Rhode Island and southeastern Massachusetts: *U.S. Geol. Survey 17th Ann. Rept.* (1). (1076 pp.)

MARKOV, K. K., 1961, Sur les phénomènes périglaciaires du Pléistocène dans le territoire de l'U.R.S.S.: *Biuletyn Peryglacjalny* **10**, 75–85.

MARKOV, K. K., and BODINA, E. L., 1961, Karta periglayatsial'yx obrazovaniy Antarktidy: 53–60 in *Antarktika, Doklady Komissii 1960, Acad. Nauk SSSR, Mezhdyvedomstvennaya Komissiya po Izychenyu Antarktiki.* (88 pp.)

1966, Map of periglacial formations in Antarctica: 49–59 in *Antarctica, Commission Reports 1960, Acad. Sci. USSR. Interdepartmental Commission on Antarctic Research.* (103 pp.) (Israel Program for Scientific Translations, Jerusalem.)

MARUSZCZAK, H., 1961, Phénomènes périglaciaires dans le Pirin et sur la Vitocha (Bulgaria): *Biuletyn Peryglacjalny* **10**, 225–34.

MATHER, K. F., GOLDTHWAIT, R. P., and THIESMEYER, L. R., 1942, Pleistocene geology of western Cape Cod, Massachusetts: *Geol. Soc. America Bull.* **53**, 1127–74.

MATHEWS, W. H., and MACKAY, J. R., 1963, Snowcreep studies, Mount Seymour, B.C.: Preliminary field investigations, *Geog. Bull.* **20**, 58–75.

MATTHES, F. E., 1900, Glacial sculpture of the Bighorn Mountains, Wyoming: *U.S. Geol. Survey 21st Ann. Rept.* (2), 173–90.

MATTHEWS, B., 1962, Frost-heave cycles at Schefferville, October 1960–June 1961 with a critical examination of methods used to determine them: 112–24 in Field Research in Labrador–Ungava: *McGill Sub-Arctic Research Lab. Ann. Rept. 1960–61, McGill Sub-Arctic Research Papers* **12**. (137 pp.)

MAYO, L. R., 1970, Classification and distribution of aufeis deposits, Brooks Range and Arctic Slope, Alaska (abs.): *Alaska Sci. Conf., 20th, College, Alaska 1969, Proc.*, 310.

MCARTHUR, D. S., and ONESTI, L. J., 1970, Contorted structures in Pleistocene sediments near Lansing, Michigan: *Geografiska Annaler* **52**A, 186–93.

MCCABE, L. H., 1939, Nivation and corrie erosion in west Spitsbergen: *Geog. J.* **94**, 447–65.

MCCALLIEN, W. J., RUXTON, B. P., and WALTON, B. J., 1964, Mantle rock tectonics. A study in tropical weathering at Accra, Ghana: *[Great Britain] Overseas Geol. Surveys, Overseas Geol. and Mineral Resources* **9**(3), 257–94.

MCCRAW, J. D., 1967, Some surface features of McMurdo Sound region, Victoria Land, Antarctica: *New Zealand J. Geol. and Geophys.* **10**, 394–417.

MCDOWALL, I. C., 1960, Particle size reduction of clay minerals by freezing and thawing: *New Zealand J. Geol. and Geophys.* **3**(3), 337–43.

MEARS, BRAINERD, JR., 1966, Ice-wedge pseudomorphs in the Laramie Basin, Wyoming (abs.): *Geol. Soc. America Spec. Papers* **87**, 295.

MEINARDUS, WILH., 1912*a*, Beobachtungen über Detritussortierung und Strukturboden auf Spitzbergen: *Gesell. Erdkunde Berlin Zeitschr.*, **1912**, 250–59.

MEINARDUS, WILH., 1912*b*, Über einige charakteristische Bodenformen auf Spitzbergen: *Naturh. Ver. Preuss. Rheinlande u. Westfalens, Medizinisch-naturwiss. Gesell. Münster Sitzungsber., Sitzung* **26**, 1–42.

1923, Meteorologische Ergebnisse der Kerguelenstation 1902–1903: 341–435 in E. v. Dryglaski, *Deutsche Südpolar-Expedition 1901–1903* **3** (Meteorologie, Teilbd. 1, 1 Hälfte): Berlin and Leipzig. (578 pp.)

1930, Arktische Böden: 27–96 in Blanck, E., *Handbuch der Bodenlehre* **3**, Berlin, Julius Springer. (550 pp.)

MELLOR, MALCOLM, 1970, Phase composition of pore water in cold rocks: *U.S. Army Corps of Engineers, Cold Regions Research and Engineering Laboratory Research Rept.* **292**. (61 pp.)

MERTIE, J. B., Jr., 1937, The Yukon–Tanana Region, Alaska: *U.S. Geol. Survey Bull.* **872**. (276 pp.)

MICHALEK, D. D., 1969, *Fanlike features and related periglacial phenomena of the southern Blue Ridge*, Univ. North Carolina (Chapel Hill), Ph.D. Thesis. (198 pp.)

MILLER, R. D., 1966, Phase equilibria and soil freezing: 193–7 in *Permafrost International Conference (Lafayette, Ind., 11–15 Nov. 1963) Proc., Natl. Acad. Sci.–Natl. Res. Council Pub.* **1287**. (563 pp.)

MILORADOVICH, B. V., 1936, Geologicheski ocherk severo-vostochnogo poberezhya severnogo ostrova Novoy Zemli: *Trudy Arkticheskogo instituta* **38**, 51–121.

MITCHELL, G. F., 1971, Fossil pingos in the south of Ireland: *Nature* **230**, 43–4.

MOHAUPT, WILLI, 1932, *Beobachtungen über Bodenversetzungen und Kammeisbildungen aus dem Stubai und dem Grödener Tal:* Univ. Hamburg Thesis., Hamburg, Hans Christians Druckerei und Verlag. (54 pp.)

MORAWETZ, S. O., 1932, Beobachtungen an Schutthalden, Schuttkegeln und Schuttflecken: *Zeitschr. Geomorphologie* **7** (1932–3), 25–43.

MORGAN, A. V., 1972, Late Wisconsinan ice-wedge polygons near Kitchener, Ontario, Canada: *Canadian Jour. Earth Sci.*, **9**, 607–17.

MORTENSEN, HANS, 1932, Über die physikalische Möglichkeit der 'Brodel'-Hypothese: *Centralbl. Mineralog.* **1932**(B), 417–22.

1933, Die 'Salzsprengung' und ihre Bedeutung für die regionalklimatische Gliederung der Wüsten: *Petermanns Mitt.* **79**, 130–5.

MÜCKENHAUSEN, E., 1960, Eine besondere Art von Pingos am Hohen Venn/Eifel: *Eiszeitalter und Gegenwart* **11**, 5–11.

MULLENDERS, WILLIAM, and GULLENTOPS, FRANS, 1969, The age of the pingos of Belgium: 321–35 in Péwé, T. L., ed., *The periglacial environment*: Montreal, McGill–Queen's Univ. Press. (487 pp.)

MÜLLER, FRITZ, 1954, *Frostbodenerscheinungen in NE- und N-Groenland:* Zürich Univ., Philos. Fakultät II, Diplomarbeit für das Höhere Lehramt in Geographie. (221 pp.) (Thesis.)

MÜLLER, FRITZ, 1959, Beobachtungen über Pingos. Detailuntersuchungen in Ostgrönland und in der kanadischen Arktis: *Medd. om Grønland* **153**(3). (127 pp.)

1963, Observations on pingos (Beobachtungen über Pingos): *Canada Natl. Research Council Tech. Translation* **1073**. (117 pp.)

1968, Pingos, modern: 845–7 in Fairbridge, R. W., ed., *The encyclopedia of geomorphology*, Reinhold Book Corp. (1295 pp.)

MULLER, S. W., 1947, *Permafrost or permanently frozen ground and related engineering problems:* Ann Arbor, Mich., J. W. Edwards. (231 pp.)

NAKANO, YOSHISUKE, and BROWN, JERRY, 1972, Mathematical modeling and validation of the thermal regimes in tundra soils, Barrow, Alaska: *Arctic and Alpine Research* **4**, 19–38.

NANSEN, FRIDTJOF, 1922, *Spitzbergen*, 3 edn.: Leipzig, F. A. Brockhaus. (327 pp.)

NATIONAL ACADEMY OF SCIENCES–NATIONAL RESEARCH COUNCIL, 1966, *Permafrost International Conference Proc., Natl. Acad. Sci.–Natl. Research Council Pub.* **1287**. (563 pp.)

NEAL, JAMES, 1965, Giant desiccation polygons of Great Basin playas: *Air Force Cambridge Research Laboratories, Environmental Research Paper* **123**. (30 pp.)

NEAL, J. T., LANGER, A. M., and KERR, P. F., 1968, Giant desiccation polygons of Great Basin playas: *Geol. Soc. America Bull.* **79**, 69–90.

NICHOLS, R. L., 1953, Marine and lacustrine ice-pushed ridges: *J. Glaciology* **2**, 172–5.

1966, Geomorphology of Antarctica: 1–59 in Tedrow, J. C. F., ed., *Antarctic soils and soil forming processes, Am. Geophysical Union, Antarctic Research Ser.* **8**. (117 pp.)

1971, Glacial geology of the Wright Valley, McMurdo Sound: 293–340 in Quam L. O., ed., *Research in the Antarctic, Am. Assoc. Advancement Science Pub.* **93**. (768 pp.)

NIKOLSKAYA, V. V., and CHICHAGOV, V. P., 1962, Drevniye periglacialnyye yavleniya v basseyne Amura: *Voprosy kriologii pri izuchenii chetvertichnykh otlozheniy*, 45–52.

NIKOLSKAYA, V. V., and SHCHERBAKOV, I. N., 1956, Znaki byvshego oledeniya gor Tukuringra-Dzhagdy: *Izvestiya AN SSSR, seria geograficheskaya* **2**, 58–65.

NIKOLSKAYA, V. V., TIMOFIEYEV, D. A., and CHICHAGOV, V. P., 1964, Zonalnye tipy pedimentov basseyna Amura: *Zapisky Zabaykalskogo otdela geograficheskogo obshchestva SSSR* **24**, 67–86.

NOBLES, L. H., 1961, Surface features of the ice-cap margin, northwestern Greenland: 752–67 in Raasch, G. O., ed., *Geology of the Arctic*, 2 vols., Toronto, Univ. Toronto Press. (1196 pp.)

1966, Slush avalanches in northern Greenland and the classification of rapid mass movements: *Internat. Assoc. Sci. Hydrol. pub.* **69**, 267–72.

NØRVANG, AKSEL, 1946, Nogle Forekomster af Arktisk Strukturmark (Brodelboden) bevarede i danske Istidsaflejringer: *Danmarks Geol. Undersøgelse—II, Raekke* **74**. (65 pp.)

O'BRIEN, ROBERT, 1971, Observations on pingos and permafrost hydrology in Schuchert Dal., N. E. Greenland: *Medd. om Gronland* **195**(1). (20 pp.)

O'BRIEN, R., ALLEN, C. R., and DODSON, B., 1968, Geomorphology: 3–16 in *Scoresby Land Expedition, 1968:* Univ. Dundee. (61 pp.)

OBRUCHEV, S. V., 1937, Soliflukcionnye (nagornyye) terrasy i yikh genezis na osnovanii rabot v Chukotskom kraye: *Problemy Arktiki* **3**, 27–48; **4**, 57–83.

OHLSON, BIRGER, 1964, Frostakitivität, Verwitterung und Bodenbildung in den Fjeldgegenden von Enontekiö, Finnisch–Lappland: *Fennia* **89**(3), 1–180.

ØSTREM, GUNNAR, 1971, Rock glaciers and ice-cored moraines, a reply to D. Barsch: *Geografiska Annaler* **53**A, 207–13.

OUTCALT, S. I., 1970, *A study of needle ice events at Vancouver, Canada, 1961–68:* Univ. British Columbia, Ph.D. Thesis. (135 pp.)

1971, An algorithm for needle ice growth: *Water Resources Research* **7**, 394–400.

OWENS, E. H., and MCCANN, S. B., 1970, The role of ice in the arctic beach environment with special reference to Cape Ricketts, southwest Devon Island, Northwest Territories, Canada: *Am. J. Sci.* **268**, 397–414.

PADALKA, G., 1928, O vysokykh terrasakh na Severnom Urale: *Vestnik geologicheskogo komiteta III* **4**, 9–15.

PANOŠ, V., 1960, Příspěvek k poznání geomorfologie krasové oblasti 'Na Pomezi' v Rychlebských horách: *Sborník Vlastivědného ústavu v Olomouci, oddíl A, Přírodni vědy* **4**, 33–88.

PANOV, D. G., 1937, Geomorfologicheskiy ocherk Polyarnych Uralid i zapadnoy chasti polyarnogo chelfa: *Trudy instituta geografii AN SSSR* **26**.

PATALEEV, A. V., 1955, Morozoboĭnye treshchiny v gruntakh: *Priroda* **44**(12), 84–5.

PECK, R. B., HANSON, W. E., and THORNBURN, T. H., 1953, *Foundation engineering:* New York, John Wiley. (410 pp.)

PÉCSI, M., 1963, Die periglazialen Erscheinungen in Ungarn: *Petermanns Mitt.* **107**, 161–82.

1964, *Ten years of physicogeographic research in Hungary*, Budapest, 61–3.

1965, Les principaux problèmes des recherches géomorphologiques dans les montagnes hongroises moyennes: *Geographia Polonica* **9**, 87–99.

PELTIER, L. C., 1950, The geographic cycle in periglacial regions as it is related to climatic geomorphology: *Assoc. Am. Geog. Annals* **40**, 214–36.

PENNER, E., 1968, Particle size as a basis for predicting frost action in soils: *Soils and Foundations* **8**(4), 21–9.

PEROV, V. F., 1959, O nablyudeniyach nad processami nivacii i soliflyukcii v Khibinskikh gorach: *Voprosy fizicheskoy geografii polyarnych stran* **2**, 83–6.

PESSL, FRED, JR., 1969, Formation of a modern ice-push ridge by thermal expansion of lake ice in southeastern Connecticut: *U.S. Army Corps of Engineers, Cold Regions Research and Engineering Laboratory Research Rept.* **259**. (15 pp.)

PÉWÉ, T. L., 1959, Sand-wedge polygons (Tesselations) in the McMurdo Sound Region, Antarctica – A progress report: *Am. J. Sci.* **257**, 545–52.

1964, New type large-scale sorted polygons near Barrow, Alaska (abs.): *Geol. Soc. America Spec. Paper* **76**, 301.

1965, Fairbanks area: 6–36 in Péwé, T. L. (Organizer), *Guidebook for field conference F:* INQUA Congress, 7th, Boulder, Col. 1965. (141 pp.)

1966*a*, Ice-wedges in Alaska – classification, distribution, and climatic significance: 76–81 in *Permafrost International Conference (Lafayette, Ind., 11–15 Nov. 1963) Proc., Natl. Acad. Sci.–Natl. Research Council Pub.* **1287**. (563 pp.)

1966*b*, Paleoclimatic significance of fossil ice wedges: *Biuletyn Peryglacjalny* **15**, 65–73.

1966*c*, *Permafrost and its effect on life in the North:* Corvallis, Oregon State Univ. Press. (40 pp.)

1969, The periglacial environment: 1–9 in Péwé, T. L., ed., *The periglacial environment*: Montreal, McGill–Queen's Univ. Press. (487 pp.)

1970, Altiplanation terraces of early Quaternary age near Fairbanks, Alaska: *Acta Geographica Lodziensia* **24**, 357–63.

1971, Permafrost and environmental-engineering problems in Arctic (abs.): *Internat. Symposium on Arctic Geology, 2d, San Francisco 1971, Program abstracts*, 44.

PÉWÉ, T. L., CHURCH, R. E., and ANDRESEN, M. J., 1969, Origin and paleoclimatic significance of large-scale patterned ground in the Donnelly Dome Area, Alaska: *Geol. Soc. America Special Paper* **103**. (87 pp.)

PHILBERTH, KARL, 1960, Sur une explication de la régularité dans des sols polygonaux: *Acad. Sci. [Paris] Comptes rendus* **251**, 3004–6.

1964, Recherches sur les sols polygonaux et striés: *Biuletyn Peryglacjalny* **13**, 99–198.

PIHLAINEN, J. A., BROWN, R. J. E., and LEGGET, R. F., 1956, Pingo in the Mackenzie Delta, Northwest Territories, Canada: *Geol. Soc. America Bull.* **67**, 1119–22.

PISSART, A., 1956, L'origine périglaciare des viviers des Hautes Fagnes: *Soc. géol. Belgique Annales* **79**, 1955–6, B119–31.

1958, Les dépressions fermées dans la région parisienne: Le probleme de leur origine: *Rev. Géomorphologie dynamique* **9**, 73–83.

1963, Les traces de 'pingos' du Pays de Galles (Grande-Bretagne) et du plateau des Hautes Fagnes (Belgique): *Zeitschr. Geomorphologie, N.F.* **2**, 147–65.

1964, Contribution expérimentale à la genèse des sols polygonaux: *Soc. géol. Belgique Annales* **87** (1963–4), *Bull.* **7**, B214–23.

1965, Les pingos des Hautes Fagnes: Les problèmes de leur genèse: *Soc. géol. Belgique Annales* **88**, 277–89.

PISSART, A., 1966*a*, Expériences et observations à propos de la genèse des sols polygonaux triés: *Rev. belge Géographie* **90**(1), 55–73.

1966*b*, Le role géomorphologique du vent dans la région de Mould Bay (Ile Prince Patrick–N.W.T.–Canada): *Zeitschr. Geomorphologie* **10**, 226–36.

1966*c*, Étude de quelques pentes de l'île Prince Patrick: *Soc. géol. Belgique* **89**, 377–402.

1967, Les pingos de l'Ile Prince-Patrick (76°N–120°W): *Geog. Bull.* **9**, 189–217.

1968, Les polygones de fente de gel de l'Ile Prince Patrick (Arctique Canadien – 76° lat. N.): *Biuletyn Peryglacjalny* **17**, 171–80.

1969, Le méchanism périglaciaire dressant les peirres dan le sol. Resultats d'expériences: *Acad. Sci. [Paris] Comptes rendus* **268**, 3015–17.

1970*a*, Les phénomènes physiques essentielles liés au gel, les structures périglaciaires qui en résultent et leur signification climatique: *Soc. géol. Belgique Annales* **93**, 7–49.

1970*b*, The pingos of Prince Patrick Island (76°N–120°W): *Canada Natl. Research Council Tech. Translation* **1401**. (46 pp.)

1972, Mouvements de sols gelés subissant des variations de température sous 0°: résultats de mesures dilatométriques: 124–6 in Adams, W. P., and Helleiner, F. M., eds., *International Geography 1972* **1** (Internat. Geog. Cong., 22d, Montreal), Toronto, Univ. Toronto Press. (694 pp.)

PLASCHEV, A. V., 1956, Vzryv ledyanogo bugra: *Priroda* **45**(9), 113.

POPOV, A. I., 1961, Cartes des formations périglaciaires actuelles et Pléistocènes en territoire de l'U.R.S.S.: *Biuletyn Peryglacjalny* **10**, 87–96.

1967, *Merzlotnyye yavleniya v zemnoy kore (kriolitologiya)*: Moscow, Izdatel'stvo Moskovskogo Universiteta. (304 pp.)

PORSILD, A. E., 1938, Earth mounds in unglaciated arctic northwestern America: *Geog. Rev.* **28**, 46–58.

PORTER, S. C., 1966, Pleistocene geology of Anaktuvuk Pass, Central Brooks Range, Alaska: *Arctic Inst. No. America Tech. Paper* **18**. (100 pp.)

POSER, HANS, 1931, Beiträge zur Kenntnis der arktischen Bodenformen: *Geol. Rundschau* **22**, 200–231.

1932, Einige Untersuchungen zur Morphologie Ostgrönlands: *Medd. om Grønland* **94**(5). (55 pp.)

1947*a*, Dauerfrostboden und Temperaturverhältnisse während der Würm-Eiszeit im nicht vereisten Mittel- und Westeuropa: *Naturwissenschaften* **34**, 10–18.

1947*b*, Auftautiefe und Frostzerrung im Boden Mitteleuropas während der Würm-Eiszeit: *Naturwissenschaften* **34**, 323–8, 262–7.

POSER, HANS, 1948*a*, Boden- und Klimaverhältnisse in Mittel- und Westeuropa während der Würmeiszeit: *Erdkunde* **2**, 53–68.

1948*b*, Äolische Ablagerungen und Klima des Spätglazials in Mittel- und Westeuropa: *Naturwissenschaften* **9**, 269–75, 307–12.

1950, Zur Rekonstruktion der spätglazialen Luftdruckverhältnisse in Mittel- und Westeuropa auf Grund der vorzeitlichen Binnendünen: *Erdkunde* **4**, 81–8.

1951, Die nördliche Lössgrenze in Mitteleuropa und das spätglaziale Klima: *Eiszeitalter und Gegenwart* **1**, 27–55.

1954, Die Periglazial–Erscheinungen in der Umgebund der Gletscher des Zemmgrundes (Zillertaler Alpen): *Göttinger Geog. Abh.* **15**, 125–80.

POSER, HANS, and MÜLLER, THEODOR, 1951, Studien an den asymmetrischen Tälern des Niederbayerischen Hügellandes: *Akad. Wiss. Göttingen Nachrichten, Math.–Phys. Klasse IIb* **1951**(1), 1–32.

POTTER, NOEL, JR., 1969, *Rock glaciers and mass-wastage in the Galena Creek area, northern Absaroka Mountains:* Univ. Minnesota, Ph.D. Thesis. (150 pp.)

1972, Ice-cored rock glacier, Galena Creek, northern Absaroka Mountains, Wyoming: *Geol. Soc. America Bull,* **83**, 3025–57.

POTTER, NOEL, JR., and MOSS, J. H., 1968, Origin of the Blue Rocks block field and adjacent deposits, Berks County, Pennsylvania: *Geol. Soc. America Bull.* **79**, 255–62.

POTTS, A. S., 1970, Frost action in rocks: Some experimental data: *Inst. British Geographers Trans.* **49**, 109–24.

POWERS, W. E., 1936, The evidences of wind abrasion: *J. Geol.* **44**, 214–19.

PREOBRAZHENSKIY, V. S., 1959, Alpiyskie i golcovyye yavleniya v prirode khrebtov Stanovogo nagorya (Kodar i Udokan): *Izvestiya AN SSSR, ser. geograficheskaya* **4**, 67–72.

1962, *Relyef i istoriya yego razvitiya:* Moskva, Prirodnyye usloviya osvoyeniya Severa Chitinskoy oblasti, 1–126.

PRICE, L. W., 1970, Up-heaved blocks: A curious feature of instability in the tundra: *Assoc. Am. Geographers Proc.* **2**, 106–10.

1971, Vegetation, microtopography, and depth of active layer on different exposures in subarctic alpine tundra: *Ecology* **52**, 638–47.

1972, The periglacial environment, permafrost, and man: *Assoc. Am. Geographers, Comm. on College Geog. Resource Paper* **14**. (88 pp.)

PRICE, W. A., 1968, Oriented lakes: 784–796 in Fairbridge, R. W., ed., *The encyclopedia of geomorphology*: New York, Reinhold Book Corp. (1295 pp.)

PRINDLE, L. M., 1905, The gold placers of the Fortymile, Birch Creek and Fairbanks regions: *U.S. Geol. Survey Bull.* **251**, 89.

PROKOPOVICH, N. P., 1969, Pleistocene permafrost in California's Central Valley?: *Geol. Soc. America, Abs. with Programs* **5**, 66.

PRYALUKHINA, A. F., 1958, O rastitelnosti golcov i podgolcovoy polosy Bikino-Imanskogo vodorazdela: *Botanicheskiy zhurnal* **43**, 92–6.

PULINA, MARIAN, 1968, Gleby poligonalne w jaskini Czarnej, Tatry Zachodnie (Les sols polygonaux dans la grotte Czarna, les Tatras Occidentales): *Speleologia* **3**(2), 99–104.

RABOTNOV, T. A., 1937, Vegetaciya vysokogornogo poyassa basseyna verkhovyev rek Aldana i Timptona: *Izvestiya Gosudarstvennogo Geograficheskogo Obshchestva* **69**, 585–605.

RAPP, ANDERS, 1960*a*, Recent development of mountain slopes in Kärkevagge and surroundings, northern Scandinavia: *Geografiska Annaler* **42**(2–3), 65–200.

1960*b*, Talus slopes and mountain walls at Tempelfjorden, Spitsbergen: *Norsk Polarinstitutt Skr.* **119**. (96 pp.)

1967, Pleistocene activity and Holocene stability of hillslopes, with examples from Scandinavia and Pennsylvania: 229–44 in L'évolution des versants, *Congrès et Colloques de l'Université de Liège* **40**. (384 pp.)

RAPP, ANDERS, and ANNERSTEN, LENNART, 1969, Permafrost and tundra polygons in northern Sweden: 65–91 in Péwé, T. L., ed., *The periglacial environment*: Montreal, McGill–Queen's Univ. Press. (487 pp.)

RAPP, ANDERS, and CLARK, G. M., 1971, Large nonsorted polygons in Padjelanta National Park, Swedish Lappland: *Geografiska Annaler* **53**A, 71–85.

RAPP, ANDERS, GUSTAFSSON, KJELL, and JOBS, PER, 1962, Iskilar i Padjelanta?: *Ymer* **82**, 188–202.

RAUP, H. M., 1951, Vegetation and cryoplanation: *Ohio J. Sci.* **51**(3), 105–16.

1965, The structure and development of turf hummocks in the Mesters Vig district, Northeast Greenland: *Medd. om Grønland* **166**. (113 pp.)

1969, Observations on the relation of vegetation to mass-wasting processes in the Mesters Vig District, Northeast Greenland: *Medd. om Grønland* **176**(6). (216 pp.)

1971*a*, The vegetational relations of weathering, frost action, and patterned ground processes, in the Mesters Vig District, Northeast Greenland: *Medd. om Grønland* **194**(1). (92 pp.)

1971*b*, Miscellaneous contributions on the vegetation of the Mesters Vig District, Northeast Greenland: *Medd. om Grønland* **194**(2). (105 pp.)

REID, CLEMENT, 1887, On the origin of dry chalk valleys and of Coombe rock: *Geol. Soc. London Quart. J.* **43**, 364–73.

REID, J. R., 1970*a*, Report on formation of a 'frost boil', Umiat, Alaska, 1953: Written communication.

1970*b*, Ground patterns and frost contraction polygons in North Dakota, Written communication.

RICHMOND, G. M., 1952, Comparison of rock glaciers and block streams in the La Sal Mountains, Utah (abs.): *Geol. Soc. America Bull.* **63**, 1292–3.

1962, Quaternary stratigraphy of the La Sal Mountains, Utah: *U.S. Geol. Survey Prof. Paper* **324**. (135 pp.)

1964, Glaciation of Little Cottonwood and Bells Canyons, Wasatch Mountains, Utah: *U.S. Geol. Survey Prof. Paper* **454D**. (41 pp.)

RICHTER, H., 1965, Die periglazialen Zonen ausserhalb des Jungmöranengebietes: 230–42 in Gellert, J. F., ed., *Die Weichsel-Eiszeit im Gebiet der DDR*, Berlin.

RICHTER, H., HAASE, G., and BARTHEL, H., 1963, Die Golezterrassen: *Petermanns Mitt.* **3**, 183–92.

RICHTER, KONRAD, 1951, Die stratigraphische Bewertung periglazialer Umlagerungen im nördlichen Niedersachsen: *Eiszeitalter und Gegenwart* **1**, 130–42.

ROBITAILLE, B., 1960, Géomorphologie du Sud-Est de l'Île Cornwallis, Territoires du Nord-Ouest: *Cahiers Géog. de Québec* **4**, 359–65.

ROHDENBURG, H., and MEYER, B., 1969, Zur Deutung Pleistozäner Periglazialformen in Mitteleuropa: *Göttinger Bodenkindliche Berichte* **7**, 49–70.

RÖMKENS, M. J. M., 1969, *Migration of mineral particles in ice with a temperature gradient*, Cornell Univ., Ph.D. Thesis. (109 pp.)

ROSS, JOHN, 1835, *Narrative of a second voyage in search of a North-West Passage and of a residence in the arctic regions during the years 1829, 1830, 1831, 1832, 1833*, London, A. W. Webster. (740 pp.)

RUDAVIN, V. V., 1967, Kriogennyye obrazovaniya i processy v Yuzhno-muyskom khrebte: *Geokriologicheskye usloviya Zabaykalya i Pribaykalya*, 175–82.

RUDBERG, STEN, 1958, Some observations concerning mass movement on slopes in Sweden: *Geol. Fören. Stockholm, Förh.* **80**, 114–25.

1962, A report on some field observations concerning periglacial geomorphology and mass movement on slopes in Sweden: *Biuletyn Peryglacjalny* **11**, 311–23.

1964, Slow mass movement processes and slope development in the Norra Storfjäll area, southern Swedish Lappland: *Zeitschr. Geomorphologie, Supplementband* **5** (*Fortschritte der internationalen Hangforschung*), 192–203.

1970, Naturgeografiska uppsatser vid Göteborgs universitet höstterminen 1959—vårterminen 1969: *Geografiska Fören. i Göteborg Medd.* **10**.

1972, Periglacial zonation—a discussion: *Göttinger Geog. Abh.* **60**, 221–33.

RUHE, R. V., 1969, *Quaternary landscapes in Iowa:* Ames, Iowa, Iowa State Univ. Press. (255 pp.)

RUSANOV, B. S., *et al.*, 1967, *Geomorfologiya Vostochnoy Yakutii:* Yakutsk. (375 pp.)

RUSSELL, R. J., 1933, Alpine land forms in western United States: *Geol. Soc. America Bull.* **44**, 927–50.

1943, Freeze–thaw frequencies in the United States: *Am. Geophys. Union Trans.* **24**, 125–33.

RYZHOV, B. V., 1961, K voprosu o geomorfologii i stroyenii chetvertichnogo pokrova verkhovyev seti Shilkinskogo-Argunskogo mezhdurechya v svyazi s usloviami zaleganiya kassiteronosnykh rossypey: *Materialy Vsesoyuznogo soveshchaniya po izuchenii chetvertichnogo perioda* **3**, 277–82.

SAMUELSSON, CARL, 1927, Studien über die Wirkungen des Windes in den kalten und gemässigten Erdteilen: *Uppsala Univ., Geol. Inst. Bull.* **20**, 57–230.

SANGER, F. J., 1966, Degree-days and heat conduction in soils: 253–62 in *Permafrost International Conference (Lafayette, Ind., 11–15 Nov. 1963) Proc., Natl. Acad. Sci.–Natl. Research Council Pub.* **1287**. (563 pp.)

SAVEL'EV, B. A., 1960, Peculiarities of the ice-thawing process of the ice cover and in frozen ground: 160–67 in *Problems of the North* (Translation of *Problemy Severa* **1**, 1958), Canada Natl. Research Council. (376 pp.)

SCHAFER, J. P., and HARTSHORN, J. H., 1965, The Quaternary of New England: 113–28 in Wright, H. E., Jr., and Frey, D. G., eds., *The Quaternary of the United States*: Princeton, N.J., Princeton Univ. Press. (922 pp.)

SCHEIDEGGER, A. E., 1970, *Theoretical geomorphology* 2d (revised) ed., Berlin–Heidelberg–New York, Springer-Verlag. (435 pp.)

SCHENK, ERWIN, 1955*a*, Die Mechanik der periglazialen Strukturböden: *Hessischen Landesamtes für Bodenforschung Abh.* **13**. (92 pp.)

1955*b*, Die periglazialen Strukturbodenbildungen als Folgen der Hydratationsvorgänge im Boden: *Eiszeitalter und Gegenwart* **6**, 170–84.

1966, Origin of string bogs: 155–9 in *Permafrost International Conference (Lafayette, Ind., 11–15 Nov. 1963) Proc., Natl. Acad. Sci.–Natl. Research Council Pub.* **1287**. (563 pp.)

SCHMERTMANN, J. H., and TAYLOR, R. S., 1965, Quantitative data from a patterned ground site over permafrost: *U.S. Army Cold Regions Research and Engineering Laboratory Research Rept.* **96**. (76 pp.)

SCHMID, JOSEF, 1955, *Der Bodenfrost als morphologischer Faktor*, Heidelberg, Dr. Alfred Hüthig Verlag. (144 pp.)

1958, Rezente und fossile Frosterscheinungen im Bereich der Gletscherlandschaft der Gurgler Ache (Ötztaler Alpen): *Schlern-Schriften* **190**, 255–64.

SCHOSTAKOWITSCH, W. B., 1927, *Der ewig gefrorene Boden Siberiens:* Gesell, Erdkunde Berlin Zeitschr., 394–427.

SCHOTT, CARL, 1931, Die Blockmeere in den deutschen Mittelgebirgen: *Forschungen zur Deutschen Landes- und Volkskunde* **29**(1), 1–78.

SCHRAMM, J. R., 1958, The mechanism of frost heaving of tree seedlings: *Am. Philosophical Soc. Proc.* **102**(4), 333–50.

SCHULTZ, C. B., and FRYE, J. C. (eds.), 1968, Loess and related eolian deposits of the world: *INQUA Congress, 7th, Boulder, Col. 1965, Proc.* **12**. (367 pp.)

SCOTT, B. W., 1965, The ecology of the alpine tundra on Trail Ridge: *INQUA Congress, 7th, Boulder, Colo. 1965, Guidebook Boulder area*, 13–16.

SCOTT, R. F., 1969, The freezing process and mechanics of frozen ground: *U.S. Army Cold Regions Research and Engineering Laboratory, Cold Regions Science and Engineering Mon.* **II**-D1. (67 pp.)

SEGERSTROM, KENNETH, 1950, Erosion studies at Parícutin, State of Michoacán, Mexico: *U.S. Geol. Survey Bull.* **965**-A, 1–151.

SEKYRA, JOSEF, 1960, Působení mrazu na půdu; Kryopedologie se zvláštním zřetelem k ČSSR (Frost action on the ground; Cryopedology with special reference to Czechoslovakia): *Geotechnica* **27**. (164 pp.)

1964, Kvarterně-geologické a geomorfologické problémy krkonošského krystalinika: *Opera Corcontica* **1**, 7–24.

SEIDENFADEN, GUNNAR, 1931, Moving soil and vegetation in East Greenland: *Medd. om Grønland* **87**(2). (21 pp.)

SELLMANN, P. V., 1967, Geology of the USA CRREL permafrost tunnel Fairbanks, Alaska: *U.S. Army Cold Regions Research and Engineering Laboratory Tech. Rept.* **199**. (22 pp.)

SELZER, GEORG, 1959, 'Erdkegel' als heutige Frostboden-Bildungen an Rutschhangen im Saarland: *Eiszeitalter und Gegenwart* **10**, 217–23.

SEVON, W. D., 1972, Late Wisconsinan periglacial boulder deposits in northeastern Pennsylvania (abs.): *Geol. Soc. America Abs. with Programs* **4**(1), 43–4.

SHARP, R. P., 1942*a*, Soil structures in the St. Elias Range, Yukon Territory: *J. Geomorphology* **5**, 274–301.

1942*b*, Periglacial involutions in northeastern Illinois: *J. Geology* **50**, 113–33.

SHARPE, C. F. S., 1938, *Landslides and related phenomena:* New York, Columbia Univ. Press. (137 pp.) (Also: New Jersey, Pageant Books, 1960.)

SHEARER, J. M., *et al.*, 1971, Submarine pingos in the Beaufort Sea: *Science* **174**, 816–18.

SHREVE, R. L., 1968*a*, The Blackhawk landslide: *Geol. Soc. America Spec. Paper* **108**. (47 pp.)

1968*b*, Sherman landslide: 395–401 in The great Alaska earthquake of 1964. *Hydrology* A: *Natl. Acad. Sci. pub.* **1603**. (441 pp.)

SHROCK, R. R., 1948, *Sequence in layered rocks:* New York, McGraw-Hill. (507 pp.)

SHUMSKII, P. A., 1959, Podzemnye l'dy: 274–327 (Glava IX) in *Inst. Merzlotovedeniya im. V. A. Obrucheva, Osnovy geokriologii (merzlotovedeniya), Chast' pervaya, Obshchaya geokriologiya*, Moskva, Akad. Nauk SSSR. (459 pp.)

SHUMSKII, P. A., 1964*a*, Ground (subsurface) ice (Podzemnye l'dy): *Canada Natl. Research Council Tech. Translation* **1130**. (118 pp.)

1964*b*, *Principles of structural glaciology:* New York, Dover Publications. (497 pp.) (Translated from the Russian by David Kraus.)

SHVETSOV, P. F., 1959, Obshchie zakonomernosti vozniknoveniya i razvitiya mnogoletneĭ kriolitozony: 77–107 (Glava IV) in *Inst. Merzlotovedenya im. V. A. Obrucheva, Osnovy geokriologii (merzlotovedniya), Chast'pervaya, Obshchaya geokriologiya,* Moskva, Akad. Nauk SSSR. (459 pp.)

1964, General mechanisms of the formation and development of permafrost (Obshchie zakonomernosti vozniknoveniya i razvitiya mnogolerneĭ kriolitozony) *: Canada Natl. Research Council Tech. Translation* **1117**. (91 pp.)

SIGAFOOS, R. S., and HOPKINS, D. M., 1952, Soil instability on slopes in regions of perennially-frozen ground: 176–92 in *Frost action in soils: a symposium, Natl. Acad. Sci.–Natl. Research Council Highway Research Board Spec. Rept.* **2**. (385 pp.)

SMITH, D. I., 1961, Operation Hazen – The geomorphology of the Lake Hazen region, N.W.T.: *McGill Univ., Geog. Dept. Misc. Paper* **2**. (100 pp.)

SMITH, H. T. U., 1949*a*, Physical effects of Pleistocene climatic changes in nonglaciated areas: Eolian phenomena, frost action, and stream terracing: *Geol. Soc. America Bull.* **60**, 1485–516.

1949*b*, Periglacial features in the driftless area of southern Wisconsin: *J. Geol.* **57**, 196–215.

1962, Periglacial frost features and related phenomena in the United States: *Biuletyn Peryglacjalny* **11**, 325–42.

1964, Periglacial eolian phenomena in the United States: 177–86 in *INQUA Congress, 6th, Warsaw 1961, Rept.* **4**, Lodz, Poland. (596 pp.)

1965, Dune morphology and chronology in central and western Nebraska: *J. Geol.* **73**, 557–78.

1968, 'Piping' in relation to periglacial boulder concentrations: *Biuletyn Peryglacjalny* **17**, 195–204.

SOCHAVA, V. B., 1930, Gora Standukhina na kraynem severovostocke Azii: *Priroda* **11–12**.

SOERGEL, WOLFGANG, 1919, *Lösse, Eiszeiten, and palaolithische Kulturen:* Jena, G. Fischer. (177 pp.)

1936, Diluviale Eiskeile: *Zeitschr. Deutsche Geol. Gesell.* **88**, 223–47.

SOFRONOV, G. P., 1945, K geomorfologii Voykarskogo rayona-Poliarnyy Ural: *Izvestiya AN SSSR, ser. geologicheskaya* **4**.

SOLOVIEV, P. A., 1962, Alasnvy relyef Centralnoj Jakutii i ego proiskhozhdeniye: 38–53 in *Mnogoletnemerzlyye porody i soputstvuyushchiye yim Yavleniya na territorii JASSR,* Moskva, Izdatelstvo AN SSSR.

SOLOVYEV, V. V., 1961, Sledy drevnego oledeniya i periglyacialnykh usloviy v Yuzhnom Primorye: *Trudy Vsesoyuznogo geologicheskogo Instituta, novaya seriya* **64**, 141–8.

SOONS, J. M., and GREENLAND, D. E., 1970, Observations on the growth of needle ice: *Water Resources Research* **6**, 579–93.

SØRENSEN, THORVALD, 1935, Bodenformen und Pflanzendecke in Nordostgrönland: *Medd. om Grønland* **93**(4). (69 pp.)

SOUCHEZ, R., 1967, Gélivation et évolution des versants en bordure de l'Inlandsis d'Antarctide orientale: 291–8 in L'évolution des versants: *Congrès et Colloques de l'Université de Liège* **40**. (384 pp.)

SPÖNEMANN, J., 1966, Geomorphologische Untersuchungen an Schicht-kämmen des Niedersächsischen Berglandes: *Göttinger Geog. Abh.* **36**. (167 pp.)

SPRINGER, M. E., 1958, Desert pavement and vesicular layer of some soils of the desert of the Lahontan Basin, Nevada: *Soil Sci. Soc. America Proc.* **22**, 63–6.

STEHLÍK, O., 1960, Skalni tvary ve východni části Moravskoslezských Beskyd: *Dějepis a zemèpis ve škole* **3**, 46–7.

ST-ONGE, DENIS, 1965, La géomorphologie de l'Île Ellef Ringnes, Territoires du Nord-Ouest, Canada: *Canada, Ministère des Mines et des Relevés techniques, Direction de la Géographie Étude géog.* **38**. (46 pp.)

1969, Nivation landforms: *Geol. Survey Canada Paper* **69-30**. (12 pp.)

STAGER, J. K., 1956, Progress report on the analysis of the characteristics and distribution of pingos east of the Mackenzie Delta: *Canadian Geographer* **7**, 13–20.

STEARNS, S. R., 1966, Permafrost (perennially frozen ground): *U.S. Army Cold Regions Research and Engineering Laboratory, Cold Regions Science and Engineering* **1**(A2). (77 pp.)

STECHE, HANS, 1933, Beiträge zur Frage der Strukturböden: *Sächsischen Akad. Wiss. Leipzig Berichte, Math.–phys. Kl.* **85**, 193–272.

STEINEMANN, SAMUEL, 1953, Kammeis, eine anomale Wachstumsform der Eiskristalle: *Zeitschr. angew. Mathematik und Physik (ZAMP)* **4**, 500–6.

1955, Mushfrost, an anomalous growth form of the ice crystal (Kammeis, eine anomale Wachstrumsform der Eiskristalle): *Canada Natl. Research Council Tech. Translation* **528**. (11 pp.)

STEPHENSON, P. J., 1961, Patterned ground in Antarctica (correspondence): *J. Glaciology* **3**, 1163–4.

STINGL, HELMUT, 1969, Ein periglazialmorphologisches Nord-Süd-Profil durch die Ostalpen: *Göttinger Geog. Abh.* **40**. (115 pp.)

1971, Zur Verteilung von Gross- und Miniaturformen von Strukturböden in den Ostalpen: *Akad. Wiss Göttingen Nachr., Math.-Phys. Kl.* **2**, 25–40.

STRAHLER, A. N., 1969, *Physical geography*, 3 ed.: New York, John Wiley. (733 pp.)

STREIFF-BECKER, RUDOLPH, 1946, Strukturböden in den Alpen: *Geographica Helvetica* **1**, 150–7.

STRIGIN, V. M., 1960, Vysotnaya fiziko-geograficheskaya poyasnost Denezhkina Kamnya na Severnom Urale: *Vorprosy fizicheskoy geografii Urala (Moskovskoe obshchestvo isptateley prirody, Geograficheskaya sekciya)*, 113–18.

STROCK, CLIFFORD, and KORAL, R. L., eds., 1959, *Handbook of air conditioning, heating and ventilating*, 2 ed.: New York, The Industrial Press. (1472 pp.)

STRUGOV, A. S., 1955, Vzryv gidrolakkolita: *Priroda* **44**(6), 117.

SUGDEN, D. E., 1971, The significance of periglacial activity on some Scottish mountains: *Geog. J.* **137**, 388–92.

SUKACHËV, V., 1910, Rastitelnost verkhney chasti basseyna reki Tungira, Olekminskogo okruga Yakutskòy oblasti: *Trudy Amurskoy ekspedicii* **16** (Botanicheskie issledovaniya), 265.

1911, K voprosu o vliianii merzloty na pochvu: *Akad. Nauk Izv., ser. 6* **5**, 51–60.

[SUKATSCHEW, W.], 1912, Die Vegetation des oberen Einzugsgebietes des Flusses Tungir im Kreise Olekminsk im Bezirk Jakutsk: *Arbeiten der auf Allerhochsten Befehl ausgefuhrten Amur-Expedition* **1** (16), St. Petersburg.

SUZDALSKIY, O. V., 1952, Po povodu nagornyhk terras Visherskogo Urala: *Izvestiya Veseoyuznogo geograficheskogo obshchestva* **84**, 102–3.

SVENSSON, HARALD, 1969, A type of circular lake in northernmost Norway: *Geografiska Annaler* **51**A(1–2), 1–12.

SWINZOW, G. K., 1969, Certain aspects of engineering geology in permafrost: *Eng. Geol.* **3**, 177–215.

TABER, STEPHEN, 1918, Ice forming in clay soils will lift surface weights: *Eng. News-Rec.* **80**(6), 262–3.

1929, Frost heaving: *J. Geol.* **37**, 428–61.

1930*a*, The mechanics of frost heaving: *J. Geol.* **38**, 303–17.

1930*b*, Freezing and thawing of soils as factors in the destruction of road pavements: *Public Roads* **11**, 113–32.

1943, Perennially frozen ground in Alaska: its origin and history: *Geol. Soc. America Bull.* **54**, 1433–548.

1950, Intensive frost action along lake shores: *Am. J. Sci.* **248**, 784–93.

1952, Geology, soil mechanics, and botany: *Science* **115**, 713–14.

1953, Origin of Alaska silts: *Am. J. Sci.* **251**, 321–36.

TANTTU, ANTTI, 1915, Ueber die Enstehung der Bülten und Stränge der Moore: *Acta Forestalia Fennica* **4**, 1–24.

TARNOGRADSKIY, G. S., 1963, Reliktovye nagornye terasy na zapadnom sklone Severnogo Urala: *Izvestiya Vsesoyuznogo obshchestva* **95**(4), 358–60.

TARR, R. S., 1897, Rapidity of weathering and stream erosion in the arctic latitudes: *Am. Geologist* **19**, 131–6.

TAYLOR, GRIFFITH, 1922, *The physiography of the McMurdo Sound and Granite Harbour region: British Antarctic (Terra Nova) Expedition 1910–1913:* London, Harrison. (246 pp.)

TEDROW, J. C. F., 1969, Thaw lakes, thaw sinks and soils in northern Alaska: *Biuletyn Peryglacjalny* **20**, 337–44.

1970, Soil investigations in Inglefield Land, Greenland: *Medd. om Grønland* **188**(3). (93 pp.)

TEICHERT, C., 1935, Bedeutung des Windes in arktischen Gegenden: *Natur und Volk* **65**, 619–28.

1939, Corrasion by wind-blown snow in polar regions: *Am. J. Sci.* **237**, 146–8.

TE PUNGA, M. T., 1956, Altiplanation terraces in Southern England: *Biuletyn Peryglacjalny* **4**, 331–8.

TERZAGHI, KARL, 1952, Permafrost: *Boston Soc. Civil Engineers J.* **39**, 1–50.

THOMAS, W. N., 1938, Experiments on the freezing of certain building materials: *Great Britain, Department Scientific and Industrial Research, Building Research Tech. Paper* **17**. (146 pp.)

THOMPSON, W. F., 1962, Preliminary notes on the nature and distribution of rock glaciers relative to true glaciers and other effects of the climate on the ground in North America: *Internat. Assoc. Sci. Hydrology, Symposium of Obergurgl, Pub.* **58**, 212–19.

1968, New observations on alpine accordances in the western United States: *Assoc. Am. Geographers Annals.* **58**, 650–69.

THORARINSSON, SIGURDUR, 1951, Notes on patterned ground in Iceland, with particular reference to the Icelandic 'flás': *Geografiska Annaler* **33**, 144–56.

THORODDSEN, TH., 1913, Polygonboden und 'thufur' auf Island: *Petermanns Mitt.* **59**(2), 253–5.

1914, An account of the physical geography of Iceland with special reference to the plant life: 187–343 in Kolderup-Rosenvinge, L., and Warming, Eugene, eds., 1912–18, *The botany of Iceland* **1**: Copenhagen, J. Frimodt; London, John Weldon. (675 pp.)

THORSTEINSSON, R., 1961, The history and geology of Meighen Island, Arctic Archipelago: *Canada Geol. Survey Bull.* **75**, 19.

TIMOFIEYEV, D. A., 1965, *Srednaya i nizhnaya Olekma:* Moskva. (137 pp.)

TOLL, EDUARD V., 1895, Die fossilen Eislager und ihre Beziehungen zu den Mammuthleichen: *Wissenschaftliche Resultate der von der Kaiserlichen Akademie der Wissenschaften zur Erforschung des Janalandes und der Neusibirischen Inseln in den Jahren 1885 und 1886 ausgesandten Expedition—Abt. III, L'Acad. Impériale des Sciences de St.-Pétersbourg, VII ser.* **42**(13). (86 pp.)

TOLMACHEV, I. P., 1903, Geologicheskaya poyezdka v Kuznecki Alatau letom 1902 goda: *Izvestiya Imperatorskogo Russkogo Geograficheskogo obshchestva* **39**, 390–436.

TRICART, JEAN, 1956*a*, Étude expérimentale du problème de la gélivation: *Biuletyn Peryglacjalny* **4**, 285–318.

1956*b*, Cartes des phénomènes périglaciaires quaternaires en France: *Carte géologique détaillée de la France, Mémoires.* (40 pp.)

TRICART, JEAN, 1963, *Géomorphologie des régions froides:* Paris, Presses Universitaires de France. (289 pp.)

1966, Un chott dans le désert Chilien: La Pampa del Tamarugal: *Rev. Géomorphologie dynamique* **16**, 12–22.

1967, Le modelé des régions périglaciaires: Tricart, J., and Cailleux, A., *Traité de Géomorphologie* **2**, Paris, SEDES. (512 pp.)

1969, *Geomorphology of cold environments* (Translated by Edward Watson): London, Macmillan; New York, St. Martin's Press. (320 pp.)

1970, Convergence de phénomènes entre l'action du gel et celle du sel: *Acta Geographica Lodziensia* **24**, 425–36.

TROELSEN, J. C., 1952, An experiment on the nature of wind erosion, conducted in Peary Land, North Greenland: *Dansk. geol. Foren. Medd.* **12**, 221–2.

TROLL, CARL, 1944, Strukturböden, Solifluktion und Frostklimate der Erde: *Geol. Rundschau* **34**, 545–694.

1947, Die Formen der Solifluktion und die periglazialer Bodenabtragung: *Erdkunde* **1**, 162–75.

1958, Structure soils, solifluction, and frost climates of the earth (Strukturböden, Solifluktion, und Frostklimate der Erde): *U.S. Army Snow Ice and Permafrost Research Establishment Translation* **43**. (121 pp.)

1969, Inhalt, Probleme und Methoden geomorphologischer Forschung (mit besonderer Berücksichtigung der klimatischen Fragestellung): *Geol. Jahrbuch Beihandlung* **80**, 225–57.

TSKHURBAYEV, F. J., 1966, Geomorfologiya, chetvertichnyye otlozheniya i zolotonosnyye rossypy Nerskogo ploskogorya: *Geologiya rossypey zolota i zakonimernosti yikh razmeshcheniya v centralnoy chasti Yanokolymskogo skladchatogo poiasa*, 129–60.

TSVETAYEV, A. A., 1960, Klimaticheskie osobennosti gornogo rayona Iremel: *Voprosy fizicheskoy geografii Urala*, 101–12.

TSYTOVICH, N. A., 1957, The fundamentals of frozen ground mechanics: 116–19 in *Internat. Conf. Soil Mech. and Found. Eng., 4th, London 1957, Proc.* **1**(28). (466 pp.)

1958, Comments: 92–3 in *Internat. Conf. Soil Mech. and Found. Eng., 4th, London 1957, Proc.* **3**. (291 pp.)

TSYTOVICH, N. A., *et al.*, 1959, O fizicheskikh yavleniyakh i protsessakh v promerzayushchikh, merzlykh i protaivayushchikh gruntakh: 108–52 (Glava V) in *Inst. Merzlotovedeniya im. V. A. Obrucheva, Osnovy geokriologii (merzlotovedeniya), Chast'pervaya, Obshchaya geokriologiya:* Moskva, Akad. Nauk SSSR. (459 pp.)

1964, Physical phenomena and processes in freezing, frozen and thawing soils (O fizicheskikh yavleniyakh i protsessakh v promerzayushchikh, merzlykh i protaivayushchikh gruntakh): *Canada Natl. Research Council Tech. Translation* **1164**. (109 pp.)

TUFNELL, LANCE, 1969, The range of periglacial phenomena in northern England: *Biuletyn Peryglacjalny* **19**, 291–323.

1972, Ploughing blocks with special reference to north-west England: *Biuletyn Peryglacjalny* **21**, 237–70.

TYULINA, L. O., 1931, O yavleniyakh sviazannykh s pochvennoy merzlotoy i moroznym vyvetrivaniyem na gore Iremel (Yuzhnyy Ural): *Izvestiya Gosudarstvennogo geograficheskogo obshchestva* **63**(2–3), 124–44.

TYULINA, L. N., 1936, O lesnoy restitelnosti Anadyrskogo kraya i yeye vzaimootnoshenii s tundroy: *Trudy Arkticheskogo instituta* **40**, 7–212.

1948, O sledakh oledeneniya na severovostochnom poberezhye Baykala: *Problemy fizicheskoy geografii* **13**, 77–90.

TYUTYUNOV, I. A., 1964, *An introduction to theory of the formation of frozen rocks* (Translated from the Russian by J. O. H. Muhlhaus; translation edited by N. Rast): Oxford, Pergamon; New York, Macmillan. (94 pp.)

U.S. ARMY ARCTIC CONSTRUCTION AND FROST EFFECTS LABORATORY, 1958, Cold room studies, third interim report of investigations **1**: *Corps of Engineers, New England Div. Tech. Rept.* **43**. (46 pp.)

U.S. ARMY COLD REGIONS RESEARCH and ENGINEERING LABORATORY, 1951–Bibliography on snow, ice, and permafrost: *CRREL Rept.* **12**.

U.S. ARMY WATERWAYS EXPERIMENT STATION, 1948, Trafficability of soils. Laboratory tests to determine effects of moisture content and density variations: *U.S. Army Corps of Engineers Tech. Memo.* **3-240**. (First supp. 28 pp.)

1953, The unified soil classification system: *U.S. Army Corps of Engineers Tech. Memo.* **3-357**. (30 pp.)

U.S. FOREST SERVICE, 1968, Snow avalanches: *U.S. Dept. Agriculture Handbook* **194** (revised). (84 pp.)

VARNES, D. B., 1958, Landslide types and processes: 20–47 in Eckel, E. G., ed., Landslides and engineering practice: *Natl. Acad. Sci.–Natl. Research Council Pub.* **544** (*Highway Research Board Spec. Rept.* **29**). (232 pp.)

VARSANOFYEVA, V. A., 1929, Geomorfologicheskiy ocherk basseyna reki Ylykha: *Trudy Instituta po izuchenii severa* **42**. (120 pp.)

1932, Geomorfologicheskiye nablyudeniya v severnom Urale: *Izvestiya Gosudarstvennogo Geograficheskogo Obshchestva* **64**(2–3), 105–71.

VELIČKO, A. A., 1972, La morphologie cryogène relicte: caractères fondamentaux et cartographie: *Zeitschr. Geomorphologie, Supplementband* **13**, 59–72.

VILBORG, L., 1955, The uplift of stones by frost: *Geografiska Annaler* **37**, 164–9.

WAHRHAFTIG, CLYDE, 1965, Physiographic divisions of Alaska: *U.S. Geol. Survey Prof. Paper* **482**. (52 pp.)

WAHRHAFTIG, CLYDE, and COX, ALLAN, 1959, Rock glaciers in the Alaska Range: *Geol. Soc. America Bull.* **70**, 383–436.

WALKER, H. J., and ARNBORG, L., 1966, Permafrost and ice-wedge effect on riverbank erosion: 164–71 in *Permafrost International Conference (Lafayette, Ind., 11–15 Nov. 1963) Proc., Natl. Acad. Sci.–Natl. Research Council Pub.* **1287**. (563 pp.)

WARD, W. H., and ORVIG, S., 1953, The glaciological studies of the Baffin Island Expedition, 1950—Part IV, The heat exchange at the surface of the Barnes Ice Cap during the ablation period: *J. Glaciology* **2**, 158–72.

WASHBURN, A. L., 1947, Reconnaisance geology of portions of Victoria Island and adjacent regions, Arctic Canada: *Geol. Soc. America Mem.* **22**. (142 pp.)

1950, Patterned ground: *Rev. Canadienne Géographie* **4**(3–4), 5–59.

1951, Geography and arctic lands: 267–87 in Taylor, Griffith, ed., *Geography in the twentieth century:* New York, Philosophical Library; London, Methuen. (630 pp.)

1956*a*, Unusual patterned ground in Greenland: *Geol. Soc. America Bull.* **67**, 807–10.

1956*b*, Classification of patterned ground and review of suggested origins: *Geol. Soc. America Bull.* **67**, 823–65.

1967, Instrumental observations of mass-wasting in the Mesters Vig district, Northeast Greenland: *Medd. om Grønland* **166**(4). (318 pp.)

1969*a*, Weathering, frost action, and patterned ground in the Mesters Vig district, Northeast Greenland: *Medd. om Gronland* **176**(4). (303 pp.)

1969*b*, Patterned ground in the Mesters Vig district, Northeast Greenland: *Biuletyn Peryglacjalny* **18**, 259–330.

1970, An approach to a genetic classification of patterned ground: *Acta Geographica Lodziensia* **24**, 437–46.

WASHBURN, A. L., and GOLDTHWAIT, R. P., 1958, Slushflows (abs.): *Geol. Soc. America Bull.* **69**, 1657–8.

WASHBURN, A. L., SMITH, D. D., and GODDARD, R. H., 1963, Frost cracking in a middle-latitude climate: *Biuletyn Peryglacjalny* **12**, 175–89.

WATERS, R. S., 1962, Altiplanation terraces and slope development in West-Spitsbergen and South-West England: *Biuletyn Peryglacjalny* **11**, 89–101.

WATSON, EDWARD, 1969, The slope deposits in the Nant Iago valley near Cader Idris, Wales: *Biuletyn Peryglacjalny* **18**, 95–113.

1971, Remains of pingos in Wales and the Isle of Man: *Geol. J.* **7**(2), 381–92.

1972, Pingos of Cardiganshire and the latest ice limit: *Nature* **236**, 343–4.

WATSON, EDWARD, and WATSON, SYBIL, 1971, Vertical stones and analogous structures: *Geografiska Annaler* **53**A, 107–14.

1972, Investigations of some pingo basins near Aberystwyth, Wales: 212–23 in *Quaternary geology: Internat. Geol. Cong., 24th, Montreal, Proc. sec.* **12**. (226 pp.)

WAYNE, W. J., 1967, Periglacial features and climatic gradient in Illinois, Indiana, and western Ohio, east-central United States: 393–414 in Cushing, E. J., and Wright, H. E. Jr., eds., *Quaternary paleoecology*: New Haven, Yale Univ. Press. (433 pp.)

WEBB, P. N., and MCKELVEY, B. D., 1959, Geological investigations in South Victoria Land, Antarctica—Part 1, Geology of Victoria Dry Valley: *New Zealand J. Geol. and Geophys.* **2**, 120–36.

WELLMAN, A. W., and WILSON, A. T., 1965, Salt weathering, a neglected geological erosive agent in coastal and arid environments: *Nature* **205**, 1097–8.

WENDLER, GERD, 1970, Some measurements of the extinction coefficients of river ice: *Polarforschung* **6**(1), 1969, 253–6.

WHITE, E. M., 1972, Soil-desiccation features in South Dakota depressions: *Geol. J.* **80**, 106–11.

WHITE, E. M., and AGNEW, A. F., 1968, Contemporary formation of patterned ground by soils in South Dakota: *Geol. Soc. America Bull.* **79**, 941–4.

WHITE, E. M., and BONESTEEL, R. G., 1960, Some gilgaied soils in South Dakota: *Soil Sci. Soc. Am. Proc.* **24**, 305–9.

WHITE, S. E., 1971, Rock glacier studies in the Colorado Front Range, 1967 to 1968: *Arctic and Alpine Research* **3**, 43–64.

1972, Alpine subnival boulder pavements in Colorado Front Range: *Geol. Soc. America Bull.* **83**, 195–200.

WHITE, W. A., 1961, Colloid phenomena in sedimentation of argillaceous rocks: *J. Sed. Petrology* **31**, 560–70.

1964, Origin of fissure fillings in a Pennsylvania shale in Vermillion County, Illinois: *Illinois Acad. Sci. Trans.* **57**, 208–15.

WIEGAND, GOTTFRIED, 1965, Fossile Pingos in Mitteleuropa: *Würzburger Geog. Arbeiten* **16**. (152 pp.)

WILHELMY, HERBERT, 1958, *Klimamorphologie des Massengesteine*: Braunschweig, Georg Westermann Verlag. (238 pp.)

WILLDEN, RONALD, and MABEY, D. R., 1961, Giant desiccation fissures on the Black Rock and Smoke Creek Deserts, Nevada: *Science* **133**, 1359–60.

WILLIAMS, J. R., 1965, Ground water in permafrost regions – an annotated bibliography: *U.S. Geol. Survey Water-Supply Paper* **1792**. (294 pp.)

1970, Ground water in the permafrost regions of Alaska: *U.S. Geol. Survey Prof. Paper* **696**. (83 pp.)

WILLIAMS, LLEWELYN, 1964, Regionalization of freeze–thaw activity: *Assoc. Am. Geog. Annals* **54**, 597–611.

WILLIAMS, M. Y., 1936, Frost circles: *Royal Soc. Canada Trans.* **30**(4), 129–32.

WILLIAMS, P. J., 1959*a*, The development and significance of stony earth circles: *Norske Vidensk.–Akad. Oslo, I. Mat.–naturv. Kl. 1959* **3**. (14 pp.)

1959*b*, Arctic 'vegetation arcs': *Geog. J.* **125**, 144–5.

1961, Climatic factors controlling the distribution of certain frozen ground phenomena: *Geografiska Annaler* **43**, 339–47.

1962, Quantitative investigations of soil movement in frozen ground phenomena: *Biuletyn Peryglacjalny* **11**, 353–60.

1963, Specific heats and unfrozen water content of frozen soils: 109–26 in *First Canadian Conference on Permafrost, Proc., Canada, Natl. Research Council Tech. Memo.* **76**. (231 pp.)

1966, Downslope soil movement at a sub-arctic location with regard to variations with depth: *Canadian Geotech. J.* **3**, 191–203.

1967, Properties and behaviour of freezing soils: *Norwegian Geotechnical Inst. Pub.* **72**. (119 pp.)

WILLIAMS, R. B. G., 1964, Fossil patterned ground in eastern England: *Biuletyn Peryglacjalny* **14**, 337–49.

1965, Permafrost in England during the last glacial period: *Nature* **205**, 1304–5.

1968, Some estimates of periglacial erosion in southern and eastern England: *Biuletyn Peryglacjalny* **17**, 311–35.

1969, Permafrost and temperature conditions in England during the last glacial period: 399–410 in Péwé, T. L., ed., *The periglacial environment:* Montreal, McGill–Queen's Univ. Press. (487 pp.)

WILLMAN, H. B., 1944, Resistance of Chicago area dolomites to freezing and thawing: *Illinois Geol. Survey Bull.* **68**, 249–62.

WIMAN, STEN, 1963, A preliminary study of experimental frost weathering: *Geografiska Annaler* **45**, 113–21.

WILSON, J. W., 1952, Vegetation patterns associated with soil movement on Jan Mayen Island: *J. Ecology* **40**, 249–64.

WILSON, LEE, 1968*a*, Morphogenetic classification: 717–29 in Fairbridge, R. W., ed., *The encyclopedia of geomorphology:* New York, Reinhold Book Corp. (1295 pp.)

1968*b*, Frost action: 369–81 in Fairbridge, R. W., ed., *The encyclopedia of geomorphology:* New York, Reinhold Book Corp. (1295 p.)

1969, Les relations entre les processus géomorphologiques et le climat moderne comme méthode de paleoclimatologie: *Revue de Géographie physique et de Géologie dynamique* **11**(3), 303–14.

WOLLNY, E., 1897, Untersuchungen über die Volumveränderung der Bodenarten: *Forschungen auf dem Gebiete der Agrikultur-Physik* **20** (1897–8), 1–52.

WOODCOCK, A. H., FURUMOTO, A. S., and WOOLLARD, G. P., 1970, Fossil ice in Hawaii?: *Nature* **226**, 873.

WRIGHT, H. E., JR., 1961, Late Pleistocene climate of Europe: A review: *Geol. Soc. America Bull.* **72**, 933–84.

YACHEVSKII, L. A., 1889, O vechno merzloĭ pochve v Sibiri: *Imperatorskoe Russkoe Geograficheskoe Obshchestvo Izv.* **25**(5), 341–55.

YARDLEY, D. H., 1951, Frost-thrusting in the Northwest Territories: *J. Geol.* **59**, 65–9.

YEGOROVA, G. N., 1962, Osobennosti relyefa zapadnogo Verkhovyanya v svyazi s chetvertichnym oledeniyem (basseyn reki Sobopola): *Voprosy geografii Yakutii* **2**, 83–91.

YEHLE, L. A., 1954, Soil tongues and their confusion with certain indicators of periglacial climate: *Am. J. Sci.* **252**, 532–46.

YERMOLOV, V. V., 1953, O formirovanii osnovnykh elementov relyefa okrayiny Sredne-Sibirskogo Ploskogorya mezhdu rekami Kotuy a Popigay: *Trudy Nauchno-issledovatelskogo instituta geologii Arktiki* **72**, 50–76.

YODER, E. J., 1955, Freezing-and-thawing tests on mixtures of soil and calcium chloride: 1–11 in Soil freezing, *Natl. Acad. Sci.–Natl. Research Council, Highway Research Board Bull.* **100** (35 pp.)

YONG, R. N., 1966, Soil freezing considerations in frozen soil strength: 315–19 in *Permafrost International Conference (Lafayette, Ind., 11–15 Nov. 1963) Proc., Natl. Acad. Sci.–Natl. Research Council Pub.* **1287**. (563 pp.)

YORATH, C. J., SHEARER, J., and HAVARD, C. J., 1971, Seismic and sediment studies in the Beaufort Sea: *Canada Geol. Survey Paper* **71-1** (A), 243–4.

ZAMORYEUV, V. V., 1967, Nivalnyye formy relyefa Kumylskogo golca (Yuzhnoye Zabaykalye): *Geokriologicheskiye usloviya Zabakalya i Pribaykalya*, 218–21.

ZAVARITSKIY, A. N., 1932, *Peridotitovyy massiv Rayiz v Poliarnom Urale*: Leningrad, Vsesoyuznoe geologo-razvedochnoe obyedinenie.

ZHIGAREV, L. A., 1967, *Prichiny i mekhanizm rezvitiya soliflyuktsii:* Moskva, Acad. Nauk, Izdatelstvo 'Nauka'. (158 pp.)

ZHIGAREV, L. A., and KAPLINA, T. N., 1960, Soliflukcionnye formy relyefa na Severo-Vostoke SSSR: *Trudy Inst. Merzlotovedeniya im. V. A. Obrucheva* **16**.

ZOTIKOV, I. A., 1963, Bottom melting in the central zone of the ice shield on the Antarctic continent and its influence upon the present balance of the ice mass: *Internat. Assoc. Sci. Hydrology Bull.* **8**(1), 36–44.

Index*

*Prepared by Brenda Hall, M.A., Registered Indexer of the Society of Indexers.